Eco-Mineral Recycling 4.0 and 5.0

This book explores the novel concepts of Eco-Mineral Recycling 4.0 and 5.0 as environmentally sustainable approaches to meet the growing demand for renewable energy. It focuses on transforming waste materials into valuable resources for renewable energy generation through innovative technologies and processes. Furthermore, it presents a pioneering strategy to address environmental challenges and promote economic growth by thoroughly examining current obstacles in mineral extraction and energy production.

- Provides a comprehensive understanding of the role critical minerals play in advancing sustainable technologies.
- Offers an in-depth analysis of how critical minerals are essential for renewable energy sources.
- Highlights the economic opportunities associated with critical minerals, including potential for innovation and job creation.
- Includes practical insights on how governments and industries can collaborate to ensure a stable supply chain of these essential resources.
- Emphasizes the importance of responsible extraction and recycling practices to support a sustainable future for generations to come.

This book is intended for researchers and graduate students in the fields of metals, energy technology, and environmental engineering.

Eco-Mineral Recycling 4.0 and 5.0

A Sustainable and Intelligent Path to Meeting Clean Energy Needs

Samuel Chukwujindu Nwokolo, Edson L. Meyer, and Chinedu Christian Ahia

CRC Press
Taylor & Francis Group
Boca Raton London New York

CRC Press is an imprint of the
Taylor & Francis Group, an informa business

Designed cover image: Getty

First edition published 2026
by CRC Press
2385 NW Executive Center Drive, Suite 320, Boca Raton FL 33431

and by CRC Press
4 Park Square, Milton Park, Abingdon, Oxon, OX14 4RN

CRC Press is an imprint of Taylor & Francis Group, LLC

ISBN: 978-1-032-99802-2 (hbk)
ISBN: 978-1-041-14048-1(pbk)
ISBN: 978-1-003-67280-7 (ebk)

DOI: 10.1201/9781003672807

Typeset in Times
by Newgen Publishing UK

Contents

Preface

This book examines the cutting-edge concept of Eco-Mineral Recycling 4.0 and 5.0 as an environmentally friendly option to address the growing worldwide need for clean energy. This technique combines cutting-edge technologies and environmentally friendly methods to recover valuable minerals from electronic waste and industrial leftovers in a thorough and effective manner. This book offers a clear and practical guide for adopting a circular economy model that emphasizes resource conservation and environmental stewardship in the quest for clean energy solutions. It achieves this by utilizing systematic review, case studies, research findings, and real-world applications. The book examines four different scenarios: the business-as-usual scenario, where no additional support is provided; the assisted scenario, where industries and governments receive help in meeting critical mineral demand; the Greater Support scenario, where industries and governments make more effort to meet critical mineral demand; and the Smarter scenario, where industries and governments actively choose innovative and transformative approaches to meeting critical mineral demand. The business-as-usual scenario examines the ongoing operations of industries and governments without substantial alterations in resource management techniques, resulting in persistent environmental deterioration and resource depletion. The aided, Greater Support, and Smarter scenarios investigate alternative interventions and tactics that could alter the existing path toward a more sustainable and efficient utilization of essential minerals for clean energy technologies. In assisted scenarios, industries and governments have contributed to satisfying the need for key minerals. This involves collaboration amongst stakeholders to enhance recycling methods, explore alternative resources, and do more research on sustainable mining practices. This scenario depicts a future in which industry and governments collaborate to guarantee a consistent provision of essential minerals while simultaneously reducing environmental harm and fostering long-term sustainability. The scenario of Greater Support involves industries and governments making more effort to address the demand for critical minerals. This entails fostering collaboration among stakeholders to enhance resource utilization efficiency, decrease dependence on environmentally damaging mining and recycling methods, and promote the creation of novel technologies for recycling and reusing minerals. Adopting this proactive strategy is crucial for effectively managing the increasing need for vital minerals in a way that is both environmentally friendly and socially accountable. Industries and governments should strategically explore innovative approaches to address the increasing demand for critical minerals. This involves harnessing technological advancements, allocating resources to research and development, and adopting policies that encourage responsible sourcing and recycling. These measures are crucial for securing a reliable supply of critical minerals for current and future generations. Through collaborative efforts, stakeholders can develop inventive strategies to establish a mineral supply chain that is both sustainable and robust, effectively catering to the demands of a continually changing global economy. The book presents four scenarios to emphasize the significance of proactive planning and collaboration in tackling the difficulties related to the supply of vital minerals. Policymakers, industry executives, and researchers can gain a deeper understanding of the potential dangers and possibilities linked to various strategies for satisfying future demand by examining these scenarios. To ensure sustainable development and economic progress in the future, it is imperative to act promptly in order to assure a reliable and consistent supply of essential minerals.

About the Authors

Samuel Chukwujindu Nwokolo is a distinguished figure in the field of renewable energy and climate science activism. With a strong educational background, he holds a BSc. Physics (Renewable Energy), MSc. Physics (Atmospheric Physics/Meteorology), Ph.D. Physics (Atmospheric Physics/Meteorology (In View)). His passion for sustainable solutions and his deep understanding of the environmental challenges we face today have propelled him to the forefront of the movement. Through his extensive research and practical experience, he has become an advocate for renewable energy adoption and an expert in climate science. His expertise lies not only in theoretical knowledge but also in hands-on implementation, making him a valuable asset in the pursuit of a greener future. His dedication to raising awareness about renewable energy's potential to combat climate change is evident through his numerous speaking engagements, publications, and active involvement in various environmental organizations. He has been actively involved in renewable energy and climate science research for nine years. Research covered aspects of solar energy, smart energy systems, and sustainable development studies (sustainable cities and resources, building integrated photovoltaics for sustainable architecture). He has published more than 62 research articles in many reputed journals, including Scopus Q1, SCI, and ABDC-A & B category. He is the author of three academic books and a reviewer for Taylor & Francis Journals of *International Journal of Ambient Energy*, *International Journal of Green Energy*, and other Scopus Q1, SCI, and ABDC-A & B category journals. He is a graduate student researcher at the Department of Physics at the University of Calabar in Nigeria, where he also instructs graduate students about research and development, renewable energy, and the science of climate change.

Edson L. Meyer obtained his Ph.D. in physics from the University of Port Elizabeth in 2002. He is currently a full professor at the University of Fort Hare and Director of the Fort Hare Institute of Technology. Professor Meyer has served as Head of the Department of Physics at the University of Fort Hare and as nonexecutive director on the boards of the South African National Energy Development Institute (SANEDI) and the Digitisation and Refurbishment Institute of South Africa (DRISA). He has published over 142 peer-reviewed articles, with over 6898 citations, an h-index of 36, and i10-index of 88, and has been actively involved in renewable energy research for 23 years. Research covered aspects of solar energy (solar cells, modules, and systems), bioenergy (biomass gasification, biogas digestion), smart energy systems (electronic circuit design, energy efficiency, smart grids), and sustainable development studies (sustainable cities and resources, building integrated photovoltaics for sustainable architecture). He has now established the Solar Watt Park on the Alice campus where essential research is conducted on various aspects of solar PV and thermal, dye-sensitized solar cells, heat pumps with solar water heaters, grid-connected PV, biomass gasification, and biogas digestion.

Chinedu Christian Ahia is a research fellow in photonics and materials characterization at the Fort Hare Institute of Technology, University of Fort Hare where he spearheads the operations and training on confocal Raman/atomic force microscope. He was conferred a Ph.D. in physics by Nelson Mandela University in 2018. His expertise includes growth, characterization, and development of semiconductor materials, quantum dots, quantum wells, superlattices, and smart nanomaterials for optoelectronic device applications. He is a professional member of the South African Institute of Physics (SAIP) among many others and the author of several articles in reputable/accredited journals and book chapters.

1 Introduction to Eco-Mineral Recycling 4.0 and 5.0 in the Clean Energy Era

1.1 THE ROLE OF CRITICAL MINERALS IN CLEAN ENERGY TECHNOLOGIES

The shift to renewable energy technology necessitates a fundamental change in resource utilization, especially for key minerals. Lithium, cobalt, rare earth elements (REEs), and nickel are essential for the functioning of energy storage systems, electric vehicles (EVs), and renewable energy infrastructures [1]. Their distinctive characteristics—namely, high energy density, conductivity, and magnetism—render them essential to the worldwide initiative for decarbonization [2].

Critical minerals facilitate the production of energy storage technologies, especially lithium-ion batteries, which are essential for addressing climate change [3]. Lithium, a lightweight and energy-dense element, is utilized in the cathodes of rechargeable batteries. Cobalt similarly guarantees thermal stability and mitigates overheating in lithium-ion cells [4]. The efficacy of EVs, solar energy systems, and wind turbines depends on the availability and quality of these minerals, highlighting their essential role in the energy transition [5].

Rare earth elements, especially neodymium, dysprosium, and terbium, are essential for the magnets utilized in the wind turbines and EV motors [6]. These magnets improve energy efficiency by reducing energy loss during conversion processes. Offshore wind turbines utilizing neodymium-based permanent magnets have enhanced reliability and diminished maintenance requirements relative to conventional systems [7]. These innovations underscore the essential role of key minerals in advancing renewable energy technology [8].

In photovoltaics, tellurium and cadmium are utilized to manufacture thin-film solar cells, which exhibit superior efficiency and flexibility compared to conventional silicon-based panels [9]. This versatility enables the utilization of solar energy in various conditions, thereby broadening its possible applications. Indium and gallium enhance high-performance solar cells, facilitating the swift expansion of solar energy installations globally [10].

Energy storage alternatives to lithium-ion batteries significantly rely on essential minerals. Vanadium is utilized in flow batteries, which are appropriate for grid-scale energy storage owing to their scalability and extended cycle life. This use is especially pertinent for incorporating intermittent renewable energy sources, such as

DOI: 10.1201/9781003672807-1

solar and wind, into electricity grids, hence assuring a consistent and reliable power supply [11].

The electrification of transportation relies significantly on nickel and manganese, utilized in new battery chemistries to enhance energy density and cost-effectiveness. The advancement of solid-state batteries, which offer enhanced safety and performance, depends on essential minerals like lithium and cerium, underscoring their significance in cutting-edge renewable energy technology [12].

Hydrogen fuel cells, a sustainable option for long-distance transportation and heavy industries [13], rely on platinum group metals (PGMs). Platinum and palladium serve as catalysts in these cells, facilitating efficient energy conversion and promoting the wider implementation of hydrogen technology in decarbonization plans [14].

The investigation and reclamation of essential minerals are crucial for sustained advancement in the renewable energy industry. Lithium, cobalt, and rare earth elements can be easily extracted from used batteries and electronic waste using advanced recycling methods such as hydrometallurgical and pyrometallurgical processes [15]. This not only alleviates environmental damage but also diminishes reliance on primary mining, fostering a circular economy.

Innovative technologies, including artificial intelligence (AI) and machine learning, are enhancing the extraction and processing of essential minerals. These technologies enhance efficiency by locating resource-abundant deposits [16], minimizing waste [17], and promoting ecologically sustainable mining methods [18]. Alongside blockchain, they improve supply chain transparency, fostering faith in the sustainability of sustainable energy technology [19].

The geopolitical ramifications of essential minerals must not be disregarded. China's supremacy in rare earth element extraction and processing demonstrates that their uneven global distribution engenders risks in supply chains [20]. Expanding supply sources, promoting international cooperation, and enhancing recycling capacities are essential for ensuring the robustness of clean energy transitions [21].

Furthermore, key minerals are essential for the development of energy efficiency technology, including sophisticated motors, smart grids, and energy-efficient lighting. Gallium nitride and silicon carbide are essential minerals utilized in high-performance semiconductors that minimize energy losses, hence enhancing the efficiency of power electronics.

Carbon capture and storage (CCS) methods utilize essential minerals like magnesium and calcium carbonate to trap carbon dioxide from industrial emissions [22]. These minerals facilitate the formation of stable molecules that can be securely stored or employed in accordance with global carbon neutrality objectives.

Advancements in crucial mineral processing have resulted in the creation of eco-friendly technology. Bioleaching, which employs microorganisms to extract metals from ores, has surfaced as a sustainable alternative to conventional mining methods [23]. This approach reduces environmental effects while guaranteeing the effective retrieval of essential materials [24].

Investments from both public and private sectors in important mineral research are facilitating advancements in material science. Substitutes for precious elements such as cobalt in batteries are being developed to diminish dependence on finite resources, hence improving the sustainability of sustainable energy systems [25].

Cooperation among governments, academics, and industry is essential for tackling the difficulties related to vital minerals [26]. Policy frameworks that encourage recycling, advocate for responsible sourcing, and facilitate technical innovation are crucial for ensuring the availability of these resources [27].

Critical minerals play a role not only in energy production but also in modern manufacturing and industrial decarbonization. Tungsten and molybdenum are utilized in high-temperature industrial processes, facilitating the shift to low-carbon production techniques [28].

Critical minerals facilitate the creation of lightweight materials utilized in EVs and wind turbine blades, enhancing energy efficiency and diminishing pollutants. Graphite, utilized in battery anodes, and aluminum, employed in vehicle structures, exemplify how these minerals improve performance and sustainability [29].

The social and environmental consequences of essential mineral extraction must be mitigated to guarantee fair and sustainable energy transitions. Implementing global standards for responsible mining and emphasizing community involvement can alleviate the negative impacts of mining operations on local communities and ecosystems [30].

In summary, key minerals are essential for the progress of clean energy technology, fostering innovation and aiding global decarbonization initiatives [31]. Their strategic significance highlights the necessity for sustainable extraction [32], effective recycling [33], and international cooperation to attain a cleaner and more resilient energy future [34].

1.2 GLOBAL TRANSITION TO INDUSTRY 4.0 AND INDUSTRY 5.0

The global shift toward Industry 4.0 and Industry 5.0 is transforming the industrial environment by incorporating new technology and prioritizing sustainability, resilience, and human-centered development. Industry 4.0, defined by the amalgamation of cyber-physical systems, the Internet of Things (IoT), and big data analytics, signifies the Fourth Industrial Revolution [35]. Conversely, Industry 5.0 advances by prioritizing human innovation and cooperation in conjunction with intelligent systems, emphasizing sustainable and socially inclusive production practices.

The shift to Industry 4.0 prioritizes digitalization, centering on smart factories. These factories utilize IoT-enabled devices to gather and analyze real-time data, enhancing production processes and reducing waste [36]. Predictive maintenance in manufacturing employs sensors and analytics to detect probable equipment breakdowns, thereby minimizing downtime and enhancing operational efficiency.

Industry 4.0 technologies are revolutionizing supply chain management via blockchain, guaranteeing transparency and traceability in material procurement and distribution [37]. In eco-mineral recycling, blockchain can verify the provenance and lifetime of recycled minerals, enhancing confidence in sustainable resource use.

Advanced robotics and automation are essential in Industry 4.0, facilitating high-precision activities and diminishing reliance on manual labor. In eco-mineral recycling, robots utilizing machine vision can accurately sort electronic waste, recovering essential minerals such as gold, palladium, and copper, while substantially reducing processing expenses.

Industry 4.0 promotes inter-industry collaboration via digital twins—virtual representations of real systems. These digital twins enable stakeholders to simulate and improve mineral recycling operations, improving resource recovery rates and minimizing environmental consequences. Virtual models of smelting furnaces can be evaluated for enhancements in energy efficiency without disrupting real operations.

Additive manufacturing, also known as 3D printing, constitutes a fundamental element of Industry 4.0. It facilitates resource-efficient manufacturing by minimizing material waste and producing components with exact specifications. Recycled minerals can be employed in 3D printing to produce sustainable, lightweight, and resilient materials for many industrial uses.

Big data analytics, a defining characteristic of Industry 4.0, enhances decision-making by revealing patterns and trends within extensive datasets [38]. In the clean energy sector, big data may enhance the extraction, processing, and recycling of essential minerals by forecasting demand, pinpointing inefficiencies, and refining processes.

Industry 4.0 emphasizes automation and efficiency, whereas Industry 5.0 prioritizes human-centric innovation. It emphasizes the coexistence of humans and intelligent technology, prioritizing creativity, well-being, and environmental stewardship [35]. Industry 5.0 acknowledges that technology should augment human capabilities instead of supplanting them.

Collaborative robots, or cobots, are essential to Industry 5.0, operating in conjunction with people in mineral recycling plants to improve safety and efficiency [35]. Cobots with AI skills can aid workers in intricate activities, such as disassembling end-of-life goods to reclaim valuable minerals while minimizing workplace dangers.

Sustainability is central to Industry 5.0, harmonizing industrial operations with ecological objectives. AI-driven eco-mineral recycling procedures can determine the most energy-efficient methods for recovering essential minerals while minimizing greenhouse gas emissions and preserving natural resources [36].

The use of circular economy principles in Industry 5.0 enhances sustainable development. By creating goods that are recyclable and prolonging their lifespans, companies can diminish reliance on the extraction of virgin minerals. Modular electronics that enable straightforward disassembly and material recovery show the congruence of circularity with Industry 5.0 principles.

Industry 5.0 additionally encompasses social aspects by promoting inclusive workforce development. Training programs that integrate digital and traditional skills enable people to adjust to emerging technologies, fostering equitable possibilities in eco-mineral recycling and other fields.

Human creativity in Industry 5.0 propels innovation in sustainable material design. Researchers are investigating bio-inspired materials sourced from recycled minerals, including self-healing metals and composites, which demonstrate remarkable durability and energy efficiency in sustainable energy applications.

The incorporation of AI and machine learning in Industry 5.0 improves decision-making processes. These technologies provide real-time surveillance and enhancement of mineral recycling processes, assuring adherence to environmental rules and augmenting overall efficiency.

Quantum computing, however nascent, possesses disruptive promise for Industry 5.0. Its unmatched computational capability could transform resource modeling and optimization, facilitating more accurate and effective recycling techniques for essential minerals.

Resilience is a fundamental characteristic of Industry 5.0, particularly with global supply chain disruptions. Industries might diminish their dependence on unstable international markets for essential minerals by utilizing local recycling networks and decentralized production techniques.

Policymakers are essential in enabling the shift to Industry 4.0 and 5.0. Regulatory frameworks that encourage innovation, advance sustainable practices, and protect worker rights are crucial for harnessing the potential of these industrial revolutions.

Public–private partnerships expedite this transformation by promoting collaboration and investment in technological advancement. Joint ventures among governments, research institutes, and enterprises have facilitated the expansion of AI-driven mineral recycling systems in several nations.

The global shift to Industry 4.0 and 5.0 signifies a transformative period for industries globally. By integrating technical developments with human creativity and sustainability, these revolutions facilitate a cleaner, fairer future. Eco-mineral recycling shows this connection, demonstrating how innovation may tackle urgent environmental and social issues while fostering economic growth.

1.3 RISE OF ECO-MINERAL RECYCLING AND ITS IMPORTANCE IN A CIRCULAR ECONOMY

The emergence of eco-mineral recycling signifies a significant change in societal resource management, along with circular economy concepts. As global industries shift toward sustainability, eco-mineral recycling has become fundamental for resource efficiency, environmental conservation, and economic resilience [39]. This method aims to extract valuable minerals from waste materials, hence decreasing dependence on primary mining and lessening the environmental impact of industrial operations.

Eco-mineral recycling is becoming important because of the essential function minerals have in sustainable energy technology. Lithium, cobalt, REEs, and other essential minerals are vital for batteries, wind turbines, and solar panels [40]. The rising demand for these materials, propelled by worldwide decarbonization objectives, requires novel recycling techniques to ensure supply and mitigate environmental effects.

Conventional mining methods are resource-demanding, producing substantial greenhouse gas emissions, water contamination, and land degradation. Eco-mineral recycling addresses these challenges by reclaiming minerals from discarded products, including batteries, electronics, and industrial waste. Recovering cobalt and nickel from spent lithium-ion batteries eliminates the necessity for additional mining and mitigates related environmental harm.

The circular economy framework supports eco-mineral recycling, highlighting the principles of reuse, recycling, and resource regeneration. Unlike the linear "take-make-dispose" concept, the circular approach guarantees that minerals reintegrate into the production cycle following the conclusion of their first lifecycle. This conserves natural resources and diminishes waste and emissions, aiding global sustainability objectives [41].

Cutting-edge technology is propelling the advancement of eco-mineral recycling. Hydrometallurgical methods, which use water to extract metals, are becoming more popular because they work better and are better for the environment than traditional smelting methods. Likewise, bioleaching utilizes microorganisms to extract metals such as copper and gold, providing a sustainable alternative to traditional processes.

AI and machine learning are essential in enhancing recycling operations. These tools evaluate extensive datasets to determine the most efficient extraction methods and forecast resource recovery rates [42]. AI-driven algorithms can automate the categorization of electronic trash, improving the accuracy and efficiency of mineral recovery processes.

Eco-mineral recycling facilitates the advancement of urban mining—retrieving valuable resources from trash produced in urban settings. This method diminishes the necessity for extensive transportation of raw materials and promotes localized recycling systems [43]. Integrating urban mining into circular economy policies enables communities to become self-sufficient resource hubs, thereby reducing environmental consequences.

The economic advantages of eco-mineral recycling are significant. Extracting valuable materials from waste decreases manufacturing expenses and alleviates risks linked to fluctuating raw material markets. Recycling rare earth elements from magnets can substantially lower the production costs of wind turbines and electric car motors.

In the clean energy sector, eco-mineral recycling is crucial for guaranteeing the scalability and sustainability of technology. Recycling essential materials from used batteries can facilitate the swift growth of energy storage systems without worsening resource depletion. Recycling silicon and silver from photovoltaic panels prolongs the lifespan of solar systems.

Cooperation among industry, governments, and academics is essential for the advancement of eco-mineral recycling. Public–private partnerships have stimulated research and development in innovative recycling techniques, while policy frameworks promote the adoption of sustainable practices. Extended producer responsibility (EPR) legislation mandates that manufacturers consider the end-of-life management of their products, hence incentivizing investments in recycling infrastructure.

Eco-mineral recycling also tackles the social and ethical issues related to mineral mining. Mining activities frequently result in human rights infringements and ecological deterioration in resource-abundant areas. Recycling diminishes the demand for virgin minerals, so alleviating these pressures and fostering equitable and sustainable development.

Consumer awareness and behavior have a significant impact on the success of eco-mineral recycling. Initiatives promoting the correct disposal and recycling of electronic trash enhance the availability of feedstock for recycling operations [44]. Educational initiatives emphasizing the ecological and financial advantages of recycling can augment public engagement.

The significance of eco-mineral recycling pertains to national security and economic autonomy. Nations that depend significantly on imported minerals encounter vulnerabilities stemming from geopolitical conflicts and disruptions in supply chains. Recycling locally available waste products diminishes reliance on external sources, hence improving resource security.

Technological innovations persist in broadening the realm of eco-mineral recycling. Direct recycling technologies for lithium-ion batteries recover functioning cathode materials without considerable chemical processing, providing a more energy-efficient option. These developments correspond with the goals of Industry 4.0 and 5.0, integrating automation, AI, and sustainability.

Standardization and certification methods are essential for guaranteeing the quality and dependability of reclaimed minerals. Establishing global standards for recycled materials might improve their adoption in high-performance applications, hence promoting market confidence and scalability [45].

Eco-mineral recycling aids in attaining global climate objectives by diminishing the carbon footprint associated with mineral extraction. Lifecycle evaluations (LCAs) demonstrate that recycling operations produce markedly lower emissions than primary mining, underscoring their importance in sustainable development.

Investigations in materials science are broadening the possibilities of eco-mineral recycling. Advanced materials engineered for recyclability facilitate the efficient recovery of embedded minerals, thereby diminishing the complexity and expense of recycling procedures. This method adheres to circular design principles, emphasizing sustainability from the product's birth.

The incorporation of digital technologies, including blockchain, improves transparency and traceability within recycling supply chains. Blockchain technology can verify the provenance of reclaimed minerals, assuring adherence to ethical and environmental standards. This is especially significant for sectors dependent on essential minerals with intricate supply chains.

Notwithstanding its myriad advantages, eco-mineral recycling encounters obstacles such as technological constraints, substantial upfront expenses, and legal impediments. Overcoming these challenges necessitates collaborative endeavors to create economical recycling technology, implement favorable regulations, and construct resilient recycling infrastructures.

The emergence of eco-mineral recycling represents significant progress in resource management, providing a sustainable approach to fulfill the requirements of a circular economy. This method recovers precious minerals from garbage, conserving natural resources while addressing environmental, social, and economic issues. Industries adopting eco-mineral recycling foster a cleaner, fairer, and more resilient future, in accordance with the overarching objectives of the clean energy era.

1.4 DEFINING RECYCLING 4.0 AND 5.0: TECHNOLOGICAL AND HUMAN-CENTRIC APPROACHES

Recycling 4.0 and Recycling 5.0 signify significant changes in societal approaches to resource recovery and sustainability during the clean energy age. These approaches transcend conventional recycling techniques, incorporating advanced technologies and human-centric principles to establish more efficient, circular, and socially responsible systems [46]. Recycling 4.0 is based on digitalization, automation, and sophisticated technologies like AI, robots, and the IoT, whereas Recycling 5.0 focuses on the amalgamation of human creativity, collaboration, and well-being in conjunction with intelligent machines. Collectively, they delineate the future of eco-mineral recycling, enabling the shift toward a low-carbon economy [47].

Recycling 4.0 is defined by the seamless use of digital technologies in the recycling process. It signifies the implementation of Industry 4.0 technology into recycling and resource recovery, facilitating highly automated, data-centric systems that enhance the effectiveness of recycling processes. These technologies encompass AI-driven sorting systems, robotics, and sensors that identify and segregate valuable minerals from trash materials [48]. AI-powered robots can efficiently identify and remove certain metals, like lithium or cobalt, from used batteries, therefore enhancing recovery rates and minimizing the need for human labor.

The implementation of IoT in Recycling 4.0 significantly improves the efficiency of recycling operations through real-time monitoring of materials and equipment. IoT sensors deployed in recycling facilities monitor material flow, delivering essential data on recycling system performance and enabling predictive maintenance of equipment. Utilizing this data, recycling operations may avert expensive downtime and guarantee uninterrupted, optimum performance, thus enhancing throughput and promoting more sustainable recycling methods [49].

Blockchain technology is a vital component of Recycling 4.0, guaranteeing transparency and traceability inside recycling supply chains. In eco-mineral recycling, blockchain technology can monitor the lifecycle of minerals from extraction to recycling, offering verifiable information regarding the origin of recycled materials. This cultivates confidence among stakeholders, such as producers, customers, and regulatory agencies, guaranteeing that recycled materials adhere to quality requirements and are ethically procured.

Recycling 4.0 emphasizes sophisticated sorting and separation methodologies. Near-infrared spectroscopy (NIR) is a technology that employs light to identify and categorize materials according to their composition. This technique is progressively employed to automate the segregation of mixed waste streams, including electronics and batteries, to recover essential minerals more efficiently. This facilitates greater purity of recycled materials, rendering them more appropriate for reapplication in advanced technologies such as EV batteries and renewable energy systems [50].

Alongside technology developments, human-centered methodologies are important to Recycling 5.0. Recycling 5.0 emphasizes the synergy between humans and machines, wherein human creativity, intuition, and problem-solving skills enhance the automation and efficiency of digital technology. This human-centric approach acknowledges the significance of human participation in decision-making, innovation, and process design, guaranteeing that recycling systems are both technologically sophisticated and socially acceptable.

A fundamental characteristic of Recycling 5.0 is the focus on employee well-being and involvement. Unlike the automation-centric model of Recycling 4.0, Recycling 5.0 incorporates human operators in a manner that augments their functions instead of supplanting them. Collaborative robots, or cobots, aid workers in intricate or perilous activities, such as dismantling electronic waste, ensuring worker safety while upholding high efficiency levels. These robots collaborate with human operators, allowing people to concentrate on higher-value duties such as quality control, innovative problem-solving, and process optimization [21].

Recycling 5.0 promotes the active engagement of consumers and communities in the recycling process. This method acknowledges the essential influence of customer behavior on recycling rates. Recycling 5.0 systems leverage AI and gamification to actively involve citizens in recycling initiatives, incentivizing correct waste disposal and informing them about the environmental and societal advantages of recycling. Smart bins integrated with AI sensors can recognize and categorize recyclable materials, providing users with points or rewards for accurate sorting.

In Recycling 5.0, sustainability encompasses not only environmental advantages but also social and economic aspects. This strategy advocates for the localization of recycling activities to save transportation expenses, improve material recovery, and provide employment within communities. This establishes a circular economic model that is both resource-efficient and equitable, guaranteeing that the advantages of recycling are disseminated throughout all strata of society [51]. Recycling 5.0 enhances employment development in the green economy by prioritizing local recycling infrastructures, thereby diminishing dependence on long-distance transportation and reducing the carbon footprint of recycling operations.

Recycling 5.0 prioritizes product design for enhanced recyclability. Through collaboration with manufacturers and designers, recycling systems can create items that facilitate disassembly and recycling at the conclusion of their lives. Modular electronics, which can be effortlessly disassembled and categorized, exemplify the integration of recyclability into the manufacturing process. This proactive strategy guarantees that recycling is included in the product lifecycle from the beginning, hence diminishing the overall environmental effect.

Human-centered strategies in Recycling 5.0 enhance technology-driven material recovery in Recycling 4.0. Governments and industry stakeholders collaborate to formulate policies and incentives that promote innovation and guarantee the ethical procurement of resources. EPR legislation might be utilized to compel manufacturers to assume accountability for the entire lifecycle of their products, encompassing recycling and disposal. By integrating these regulations into the Recycling 5.0 framework, governments may establish a conducive climate for sustainable material recovery.

Recycling 5.0 further improves circular economy practices by unifying waste management with the comprehensive value chain. By employing modern data analytics, recycling systems may discern trends in material flow and forecast future demand for particular commodities. This enables recycling operations to optimize material recovery and delivery, guaranteeing the availability of recycled minerals for incorporation into new products. Moreover, closed-loop recycling technologies, which recycle materials within the same supply chain, diminish the necessity for virgin resources and reduce waste.

A fundamental element of Recycling 5.0 is the focus on ongoing enhancement and innovation. As the globe shifts to sustainable energy, continuous research into novel recycling technologies and methodologies is crucial for tackling emergent difficulties. New methodologies are being formulated to recycle intricate components, including rare earth elements, from electronic waste. These advances guarantee that Recycling 5.0 continues at the forefront of resource recovery, addressing the increasing demand for essential minerals required for renewable energy technology.

The future of Recycling 5.0 will involve enhanced integration of AI, robotics, and human skills to develop more intelligent and efficient recycling systems. As machine learning algorithms advance, they will enhance the identification of intricate waste streams, automate the segregation of various materials, and improve the accuracy of mineral recovery procedures. This will lower expenses and enhance the economic feasibility of recycling operations, rendering them more competitive with primary mining and further facilitating the move to a circular economy.

Recycling 5.0 will evolve to address the wider environmental and social issues related to mineral extraction and waste management. This approach integrates sustainable practices throughout the recycling process, from collection to final processing, ensuring that recycling fulfills economic and technological objectives while also advancing global sustainability goals, such as reducing carbon emissions and conserving natural resources.

Recycling 4.0 and 5.0 present a compelling vision for the future of mineral recycling, using sophisticated technology while emphasizing human welfare, social equality, and environmental sustainability. By integrating advanced innovation with an emphasis on human creativity and collaboration, these models foster a more resilient, circular, and sustainable future for all stakeholders engaged in the clean energy transition.

1.5 DEFINING ECO-MINERAL RECYCLING 4.0 AND 5.0: KEY PRINCIPLES, OBJECTIVES, AND TECHNOLOGIES

Eco-Mineral Recycling 4.0 and 5.0 signify the forthcoming advancement in resource recovery, emphasizing the sustainable recycling of vital minerals necessary for clean energy technology. The two stages—4.0 and 5.0—illustrate the developing interplay among advanced technologies, human cooperation, and environmental stewardship. This section will examine the fundamental ideas, goals, and technologies that characterize novel methods of eco-mineral recycling, highlighting their significance in the shift toward a circular economy and the clean energy transition.

1.5.1 Key Principles of Eco-Mineral Recycling 4.0

Eco-Mineral Recycling 4.0 incorporates Industry 4.0 principles, highlighting the application of sophisticated technology to improve the efficiency, scalability, and environmental sustainability of mineral recovery processes. This technique fundamentally seeks to complete the cycle of mineral utilization by reclaiming precious materials from waste goods, including spent batteries, electronic gadgets, and industrial byproducts. Eco-Mineral Recycling 4.0 aids in resource conservation and environmental protection by diminishing reliance on primary mining and improving the recovery of minerals such as lithium, cobalt, and REEs.

A fundamental tenet of Eco-Mineral Recycling 4.0 is digitalization. This entails utilizing IoT sensors, AI, and machine learning algorithms to enhance recycling processes. IoT sensors are utilized in recycling facilities to monitor and track items in real time, facilitating efficient sorting and processing. AI algorithms then evaluate data gathered by these sensors to forecast which materials are most probable to produce high-value minerals and modify the recovery process accordingly. These technologies enable recycling plants to function more efficiently, thereby decreasing energy consumption and operational expenses.

Automation constitutes a vital aspect of Eco-Mineral Recycling 4.0. Robotics and autonomous systems are employed to categorize and segregate materials, markedly diminishing the necessity for physical work. Robots can independently dismantle electronic gadgets or batteries to extract valuable minerals such as lithium, nickel, and cobalt. This mitigates the possibility of contamination and enhances the purity of recovered materials, rendering them more appropriate for reutilization in advanced applications such as EV batteries or renewable energy storage systems.

1.5.2 Key Objectives of Eco-Mineral Recycling 4.0

The principal aim of Eco-Mineral Recycling 4.0 is to mitigate the environmental consequences of mineral extraction by enhancing the effectiveness and magnitude of recycling efforts. Recycling key minerals from obsolete items reduces the necessity for primary mining, which frequently leads to considerable environmental harm, such as deforestation, water contamination, and habitat damage. Eco-Mineral Recycling 4.0 alleviates these effects by extracting valuable minerals from waste streams, thus facilitating the shift toward a more sustainable and circular economy.

Another purpose is to augment resource security. With the escalating demand for essential minerals such as lithium, cobalt, and rare earth elements driven by the swift advancement of renewable energy technologies, securing a dependable and sustainable supply of these minerals is becoming progressively vital. Recycling 4.0 seeks to enhance supply chain resilience by utilizing the substantial amounts of minerals found in urban waste streams. Recycling lithium from used EV batteries and recovering rare earth elements from discarded wind turbines might diminish reliance on foreign supplies of essential minerals, bolstering national security and supply chain stability.

Eco-Mineral Recycling 4.0 aims to stimulate economic growth by fostering the establishment of new industries and employment possibilities within the recycling industry. The execution of advanced recycling methods necessitates proficient personnel in areas including robotics, data analytics, and materials science. Investing in these technologies enables governments and industry to generate new green employment opportunities while minimizing waste and saving resources. This fosters economic and environmental sustainability, establishing recycling as a fundamental component of the green economy.

1.5.3 Key Principles of Eco-Mineral Recycling 5.0

Eco-Mineral Recycling 4.0 emphasizes technology innovations and automation, whereas Eco-Mineral Recycling 5.0 adopts a more comprehensive, human-centered perspective. Recycling 5.0 incorporates the tenets of Industry 5.0, highlighting the synergy between humans and machines to develop systems that are both technologically sophisticated and socially responsible. It acknowledges that technology must align with human creativity, intuition, and ethical principles to realize the complete potential of recycling within a circular economy.

A fundamental tenet of Eco-Mineral Recycling 5.0 is collaboration—between humans and machines, as well as among diverse stakeholders, including industry, government, and consumers. This strategy uses technology to enhance human talents instead of supplanting them. Collaborative robots (cobots) operate in conjunction with human workers to automate monotonous or hazardous jobs, such as disassembling electronic garbage, while humans handle more intricate decision-making and quality assurance. This partnership augments employee safety, improves operational efficiency, and promotes innovation.

A key principle of Eco-Mineral Recycling 5.0 is inclusion. This approach prioritizes the equitable distribution of recycling benefits across all societal segments. Eco-Mineral Recycling 5.0 fosters the circular economy, facilitates the generation of green employment, improves community involvement, and enables customers to actively participate in recycling initiatives. Gamified recycling systems can motivate consumers to sort waste correctly by providing incentives like awards or discounts. This interactive method enhances public knowledge and fosters a common responsibility for sustainable resource management.

1.5.4 Key Objectives of Eco-Mineral Recycling 5.0

The primary aim of Eco-Mineral Recycling 5.0 is to establish a sustainable, human-centric circular economy that harmonizes technological advancement with social and ethical factors. This approach perceives eco-mineral recycling as a means to recover valuable resources while simultaneously advancing environmental justice and social equity. Eco-Mineral Recycling 5.0 prioritizes inclusive, localized recycling systems to guarantee that the advantages of recycling reach marginalized populations and generate economic prospects in underprivileged areas.

A primary goal of Eco-Mineral Recycling 5.0 is to improve the efficiency and efficacy of recycling operations while advancing human welfare. Utilizing digital technologies such as AI, machine learning, and blockchain, Recycling 5.0 can optimize recycling processes, minimize waste, and enhance material recovery rates. Simultaneously, it guarantees that human labor is employed in areas where it provides the greatest value, specifically in jobs necessitating creativity, problem-solving, and supervision.

Eco-Mineral Recycling 5.0 seeks to enhance the ethical standards of recycling by guaranteeing responsible sourcing of materials and equitable treatment of workers. Technologies like blockchain are utilized to trace the origin of minerals, guaranteeing that recycled components are obtained from ethical supply chains. This transparency mitigates the possibility of human rights abuses and environmental degradation in mineral extraction, guaranteeing that the advantages of recycling are both ecological and social.

1.5.5 Technologies in Eco-Mineral Recycling 4.0 and 5.0

The technologies of Eco-Mineral Recycling 4.0 and 5.0 are diverse and innovative, embodying the intersection of materials science, AI, and sustainable engineering. In Eco-Mineral Recycling 4.0, technologies such as robots, AI, and the IoT are crucial for automating and enhancing the recycling process. Robotics are utilized to dismantle electronic garbage, whereas AI systems are deployed to categorize items according to their chemical composition. Machine learning algorithms are essential in forecasting the most successful recycling procedures for various material types.

Blockchain technology serves as a crucial instrument in Eco-Mineral Recycling 4.0 and 5.0, facilitating the establishment of transparent and verifiable supply networks for recycled materials. Blockchain enhances trust in recycled products and guarantees adherence to environmental and social norms by meticulously documenting each phase of the recycling process. This transparency is especially significant when recycling essential minerals like lithium and cobalt, frequently obtained from areas with intricate ethical and environmental issues.

In Eco-Mineral Recycling 5.0, collaborative robots (cobots) and human–machine interfaces are essential for improving the efficiency and safety of recycling processes. These robots collaborate with human workers to enhance the efficiency of material disassembly and sorting, allowing workers to concentrate on decision-making and quality control. Moreover, AI-driven systems in Eco-Mineral Recycling 5.0 can enhance recycling methods by scrutinizing extensive data on waste composition, recovery rates, and market demand.

Eco-Mineral Recycling 4.0 and 5.0 are essential elements in the shift toward a sustainable, circular economy in the clean energy epoch. By amalgamating innovative technologies with human-centric methodologies, these systems are poised to transform the recycling and recovery of essential minerals. Eco-Mineral Recycling 4.0 and 5.0 provide a robust approach to mitigating resource depletion, avoiding ecological damage, and fostering a more equitable and resilient global economy through automation, digitalization, and human engagement. By integrating technological innovation with social values, these models possess the capacity to transform the future of recycling within the framework of the clean energy transition.

1.6 IMPORTANCE OF SUSTAINABLE MINERAL SUPPLY CHAINS FOR CLEAN ENERGY

The shift to clean energy systems, necessitated by the pressing need to address climate change, is significantly dependent on a reliable and sustainable supply of essential minerals. Lithium, cobalt, nickel, and REEs are essential for the manufacturing of renewable energy technologies, including solar panels, wind turbines, and EV batteries. With the increasing demand for clean energy solutions, establishing a sustainable, ethical, and efficient mineral supply chain is essential for attaining global sustainability objectives. This section examines the significance of sustainable mineral supply chains within the framework of the clean energy transition, addressing the problems and potential in developing resilient, environmentally friendly systems for mineral extraction, processing, and recycling.

The worldwide shift to clean energy is generating an unparalleled increase in the demand for essential minerals. Lithium, cobalt, and nickel are vital constituents of lithium-ion batteries utilized in electric cars (EVs) and extensive energy storage systems. The International Energy Agency (IEA) projects that by 2040, the demand for these minerals would escalate significantly. This significant change offers both an opportunity and a challenge for the mineral supply chain. The fundamental challenge is to ensure that the extraction, processing, and recycling of these minerals occur in a manner that mitigates environmental damage, fosters social fairness, and diminishes dependence on primary mining.

The sustainability of mineral supply chains is essential to guarantee that the clean energy transition does not result in environmental harm or abuses of human rights. Conventional mining methods, especially in areas abundant in vital minerals, have been linked to deforestation, water contamination, and negative impacts on local populations. Cobalt mining in the Democratic Republic of Congo (DRC), a major source of the global supply, has been associated with child labor and perilous working conditions. These challenges underscore the necessity for comprehensive frameworks to guarantee that the whole supply chain—from mineral extraction to recycling—is administered in a socially responsible and environmentally sustainable fashion.

A crucial element in establishing sustainable mineral supply chains is the formulation of ethical mining practices. In response to the escalating need for essential minerals while mitigating environmental and social issues, firms are progressively adopting sustainable mining technologies and practices. This encompasses the utilization of innovative technologies that diminish the environmental impact of mining, such as precision mining, which properly targets mineral resources, hence lessening the effect on adjacent ecosystems. Additionally, more stringent legislation and standards for responsible mining, including the Responsible Cobalt Initiative and the Organisation for Economic Co-operation and Development (OECD) Due Diligence Guidance, are compelling firms to embrace elevated ethical standards and guarantee that minerals are supplied responsibly.

Sustainable mineral supply chains necessitate advancements in processing and recycling, with enhancements in mining procedures. A multitude of essential minerals is sourced from waste streams, including depleted batteries, electronic refuse, and industrial byproducts. Eco-mineral recycling technologies 4.0 and 5.0 are essential for establishing a circular economy for these resources. Cutting-edge technologies, including AI-driven sorting systems and robotics, enhance the efficiency and precision of mineral extraction from trash, concurrently reducing waste production and energy usage. This method diminishes the necessity for raw materials and lessens the ecological impact of the mineral supply chain.

Recycling is crucial in mitigating resource depletion. With the surging demand for essential minerals, apprehension regarding the sustained supply of these resources is increasing. For lithium, projections indicate that the planet's identified deposits may be exhausted within 50 years if demand persists at its current rate. Recycling can mitigate this issue by reintegrating precious minerals into the supply chain. Furthermore, advancements in eco-mineral recycling can recover minerals from materials that would otherwise be disposed of, establishing a more sustainable and resilient supply chain that does not depend exclusively on primary mining.

The significance of recycling transcends environmental issues; it also yields considerable economic advantages. Recycling establishes secondary supply chains for essential minerals, thereby stabilizing prices and diminishing reliance on unstable markets. The price of cobalt, significantly affected by geopolitical instability in the DRC, can see considerable fluctuations, resulting in market uncertainty. Establishing more localized and diverse supply sources via recycling can enhance the global economy's resilience to shocks, hence fortifying the stability of the clean energy sector.

A crucial element of a sustainable mineral supply chain is openness and traceability. With the rising demand for crucial minerals, it is increasingly vital to ensure that these commodities are sourced responsibly and ethically. Blockchain technology has arisen as a potent instrument for guaranteeing transparency in the mineral supply chain. Utilizing blockchain to trace the provenance of minerals enables enterprises to confirm that their materials are ethically sourced and adhere to environmental regulations. Blockchain technology has been employed in the cobalt supply chain to track the progression of resources from extraction to the finished product, guaranteeing their absence of child labor and human rights violations.

The significance of establishing sustainable mineral supply chains is seen in the increasing influence of public policy and international cooperation. Governments globally are acknowledging the necessity of securing a consistent supply of essential minerals while ensuring that their extraction and utilization match with environmental and social objectives. Initiatives like the European Union's Raw Materials Initiative, designed to ensure the availability of essential raw materials while fostering recycling and sustainable mining practices, underscore the increasing significance of policy-driven solutions in attaining sustainability within the mineral supply chain.

Global cooperation is essential for the sustainability of mineral supply chains. Due to the worldwide scope of the clean energy transition, international collaboration is crucial to tackle the difficulties related to mineral supply. Initiatives like the Clean Energy Ministerial, which convenes governments, companies, and other stakeholders to advance clean energy legislation, are facilitating collaboration on sustainable mineral supply chains. Through collaboration, nations may exchange best practices, synchronize efforts to secure mineral supplies, and mitigate the environmental and social repercussions of mining.

Within the framework of the circular economy, acknowledging the significance of material efficiency is crucial for alleviating the strain on mineral supply chains [52]. Material efficiency pertains to the minimization of mineral use by product design, enhanced manufacturing techniques, and extended product lifespans. For instance, creating goods that necessitate fewer essential minerals or employing alternate materials helps mitigate the demand for primary mining. Furthermore, prolonging the lifespan of products by repair, reuse, and remanufacturing diminishes the demand for new materials and contributes to the establishment of more sustainable supply chains [40].

A notable advancement in sustainable mineral supply chains is the implementation of alternative materials and technologies to diminish reliance on key minerals. Research into solid-state batteries utilizing alternative materials such as sodium or sulfur in place of lithium and cobalt may mitigate strain on conventional mineral supply chains. Advancements in recycling technology may facilitate the extraction of precious minerals from unconventional sources, like urban mining and seawater, so significantly diversifying the mineral supply [53].

Moreover, sustainable mineral supply chains are essential for mitigating the environmental effects of the clean energy transition. Although renewable energy systems like wind and solar power produce negligible greenhouse gas emissions during operation, the extraction, processing, and disposal of the minerals required for their production can result in considerable environmental impacts. Sustainable mining and recycling procedures can alleviate these impacts, ensuring that the clean energy sector maintains environmental responsibility throughout its lifecycle [54]. The implementation of closed-loop recycling technologies in the manufacturing of solar panels and wind turbines can diminish the demand for new materials and reduce waste.

Corporations in the renewable energy and technology industries are progressively acknowledging the significance of sustainable mineral supply chains. Corporations such as Tesla and BMW are allocating resources toward responsible sourcing and recycling activities to guarantee that the minerals utilized in their products are ethically obtained and environmentally sustainable. These companies are collaborating with their suppliers to adopt more sustainable practices and enhance the recycling of essential materials, so guaranteeing that their supply chains are robust and responsible.

Moreover, the financial sector is assuming a progressively significant role in promoting sustainability throughout mineral supply chains. Investors are increasingly cognizant of the risks linked to unethical or unsustainable mineral sourcing and are demanding enhanced transparency and responsibility from companies in the mining, manufacturing, and recycling industries. Consequently, firms that implement sustainable practices are obtaining a competitive advantage in the marketplace, drawing investment from socially conscious investors.

The significance of sustainable mineral supply chains for renewable energy is evident. As the world economy shifts toward a low-carbon future, the demand for essential minerals will persistently increase [36]. Responsible sourcing, efficient processing, and effective recycling of these minerals will be essential for establishing a sustainable, ethical, and resilient clean energy system. Through the use of novel technology, enhancement of transparency, and promotion of international collaboration, we can establish mineral supply chains that facilitate the advancement of sustainable energy while safeguarding our world.

1.7 CURRENT CHALLENGES IN THE MINERAL SUPPLY CHAIN AND THE ROLE OF RECYCLING

The worldwide shift to clean energy relies significantly on obtaining a reliable and sustainable supply of essential minerals. These minerals are fundamental to the development of developing technologies, including solar power, wind energy, EVs, and energy storage systems, all of which are essential for meeting climate objectives.

The mineral supply chain encounters numerous obstacles that jeopardize its capacity to satisfy the increasing demand for these vital commodities [55]. Recycling, especially via Eco-Mineral Recycling 4.0 and 5.0 technologies, is a vital option to tackle these difficulties. This section examines the prevailing obstacles in the mineral supply chain and underscores the significance of recycling in alleviating these problems.

1.7.1 Growing Demand for Critical Minerals

As clean energy technologies expand, the demand for essential minerals such as lithium, cobalt, nickel, and REEs increases. The IEA projects that demand for these minerals may rise by over 500% by 2050 in scenarios consistent with global climate objectives. This rapid increase in demand exerts significant pressure on the mineral supply chain, challenging both primary mining operations and the capacity for material recycling and reuse. The necessity to enhance mining and recycling capacities concurrently has presented intricate issues concerning environmental impact, resource availability, and supply chain stability.

1.7.2 Environmental Impact of Mining

The environmental impact of mining is a significant concern within the mineral supply chain. Conventional mining practices frequently lead to deforestation, habitat degradation, water pollution, and air contamination. The production of cobalt in the DRC has been associated with significant environmental damage and human rights abuses, including child labor. Although new mining technologies are being developed to mitigate these problems, they continue to pose substantial obstacles to attaining sustainable mineral extraction. The increasing global demand for clean energy technology raises significant concerns over the environmental impact of mining.

1.7.3 Geopolitical and Supply Chain Risks

The geopolitical context presents a significant problem in ensuring a reliable mineral supply. A significant portion of the world's major mineral reserves is concentrated in a limited number of countries, resulting in supply chain risk and dependency. China leads in the global production of rare earth elements, whereas the DRC is a significant source of cobalt. These geopolitical linkages render supply chains vulnerable to interruptions caused by political instability, trade disputes, or regulatory alterations. Such risks may result in mineral price volatility and supply unpredictability, hence exacerbating the challenges of obtaining a dependable supply of essential minerals for sustainable energy technology.

1.7.4 Resource Depletion and Long-Term Availability

The rising demand for essential minerals raises concerns regarding their long-term availability. Numerous minerals, including lithium and cobalt, are limited resources, and current mining reserves may be insufficient to satisfy future demand. The extraction of minerals from novel sources is becoming progressively challenging and expensive, as lower-quality ores necessitate more energy-intensive methods to obtain valuable minerals. The exhaustion of readily available mineral resources underscores the necessity of establishing recycling methods to reclaim valuable elements from current goods.

1.7.5 Inefficiency in Recycling Systems

Notwithstanding the evident advantages of recycling in mitigating mineral supply chain strains, existing recycling systems frequently exhibit inefficiency and underdevelopment. The recycling rates of the most important minerals are low due to technical constraints, insufficient infrastructure, and economic obstacles. For example, whereas recycling rates for metals such as copper and aluminum are comparatively elevated, the recovery of more intricate minerals, such as lithium and rare earth elements, remains nascent. Enhancing the effectiveness of recycling operations is essential for reintroducing important minerals into the supply chain instead of disposing of them in landfills or incinerators.

1.7.6 Lack of Standardization in Recycling Practices

A significant problem in enhancing mineral recycling is the absence of standardization in recycling methodologies. Diverse categories of electronic waste, batteries, and other items encompass a multitude of minerals necessitating distinct recycling methodologies. Lithium-ion batteries, utilized in many applications from EVs to smartphones, comprise a blend of minerals including lithium, nickel, and cobalt, necessitating distinct separation and processing methods. In the absence of established procedures for material recovery, recycling operations frequently encounter challenges in effectively extracting valuable minerals from heterogeneous waste streams. Standardizing recycling methods may enhance operational efficiency and elevate recovery rates.

1.7.7 Complexity of Material Composition in Modern Products

Contemporary products, especially electronic devices and EVs, are constructed with ever more intricate material compositions, complicating the recycling process. A multitude of these products comprises a combination of metals, polymers, and rare earth elements that are challenging to segregate and recycle. Smartphones comprise several components, such as gold, silver, copper, and various rare earth elements, necessitating specific equipment and processes for extraction. The complexity necessitates that recycling facilities continually evolve and enhance their technologies to efficiently segregate and treat these materials.

1.7.8 Economic Viability of Recycling

Economic factors significantly influence the efficacy of mineral recycling. The economic feasibility of recycling is contingent upon the recovery costs in relation to the value of the retrieved materials. In numerous instances, the cost of primary minerals, such as lithium or cobalt, is inferior to the expense of extracting and processing recycled materials. This establishes a disincentive for recycling, as corporations may opt to procure minerals from primary mining instead of investing in costly recycling processes. Surmounting this economic obstacle necessitates technological innovations that enhance the cost-effectiveness and efficiency of recycling, with regulatory initiatives that promote the recycling of essential minerals.

1.7.9 Role of Eco-Mineral Recycling 4.0 in Addressing Challenges

Eco-Mineral Recycling 4.0, which incorporates sophisticated technology like AI, robotics, and automation, is essential in tackling numerous difficulties within the mineral supply chain. Eco-Mineral Recycling 4.0 can markedly enhance the efficiency of recycling processes through the automation of mineral sorting and recovery. AI algorithms can evaluate extensive datasets to enhance the segregation of minerals in electronic trash, while robotics can independently disassemble gadgets or batteries to extract valuable materials such as lithium, cobalt, and nickel. These technologies diminish labor expenses, enhance recovery rates, and render recycling more economically feasible.

1.7.10 Role of Eco-Mineral Recycling 5.0 in Enhancing Human-Centric Sustainability

Eco-Mineral Recycling 5.0, rooted in the principles of Industry 5.0, emphasizes the synergy between humans and machines to enhance sustainability in mineral recycling. This strategy positions humans as pivotal in decision-making, creativity, and quality control, while machines assume responsibility for monotonous and hazardous activities. Eco-Mineral Recycling 5.0 aims to establish systems in which technology enhances human competence, guaranteeing that recycling operations are both efficient and socially as well as ethically responsible. Workers may supervise robotic systems that segregate and process valuable minerals, ensuring the technology functions in accordance with high ethical and environmental standards.

1.7.11 Integration of Recycling into Circular Economy Models

The incorporation of recycling into circular economy frameworks is crucial for tackling mineral supply chain issues. A circular economy emphasizes the design of items that facilitate reuse, repair, or recycling, hence minimizing the reliance on virgin resources. By integrating recycling into the product design process, manufacturers can guarantee that their goods are optimized for material recovery at the end of their lifecycle. Circular economy models emphasize material reuse, prolonging the lifespan of important minerals and mitigating the environmental consequences of mineral extraction. This strategy can alleviate resource depletion and guarantee a more sustained influx of essential minerals into the economy.

1.7.12 Innovations in Mineral Extraction from Waste Streams

Besides recycling, advancements in mineral extraction from waste streams provide an alternative answer to supply chain issues. Urban mining, the process of extracting precious minerals from waste items including electronics, batteries, and construction materials, is becoming a vital tactic. Improvements in hydrometallurgical and bioleaching techniques, utilizing eco-friendly chemicals or microorganisms for mineral extraction from trash, are enhancing the efficiency and ecological footprint of urban mining. Urban mining can augment traditional mining and alleviate the strain on primary resources by utilizing current waste streams.

1.7.13 Development of Closed-Loop Recycling Systems

Closed-loop recycling systems, in which resources are perpetually recycled and repurposed, are crucial for establishing sustainable mineral supply chains. These systems are engineered to guarantee the recovery and reutilization of minerals in their purest state, hence diminishing the necessity for new mining operations. For example, closed-loop systems in lithium-ion batteries can reclaim lithium, cobalt, and nickel, refining these elements for reintegration into new batteries. This diminishes the ecological footprint of mining and decreases the energy and resources required for the production of new materials. Developing efficient closed-loop systems enhances the sustainability and cost-effectiveness of the recycling process.

1.7.14 Enhancing Recycling Infrastructure

The insufficiency of proper recycling infrastructure constitutes a significant obstacle to enhancing the mineral supply chain. Numerous places are deficient in the facilities or technologies necessary for the efficient processing of complicated waste streams. With the rising need for recycling, there is an escalating necessity for expenditures in infrastructure capable of managing the amount and intricacy of contemporary waste. Governments and private enterprises must collaborate to establish and enhance recycling facilities, ensuring they are outfitted with cutting-edge technologies for the recovery of important minerals. Furthermore, these facilities must be accessible to communities, guaranteeing that the advantages of recycling are equitably spread.

1.7.15 Policy and Regulatory Support for Recycling

Government laws and regulations are essential in promoting the implementation of mineral recycling technologies. Incentives for recycling, such as tax credits or subsidies, can motivate businesses to invest in recycling infrastructure and research. Furthermore, rules mandating manufacturers to assume accountability for the end-of-life management of their products, such as EPR laws, might facilitate the design of items with recycling considerations. Governments must develop explicit standards for recycled minerals, ensuring that materials retrieved from waste comply with the quality and safety criteria necessary for reuse in clean energy systems.

1.7.16 Public Awareness and Consumer Behavior

Public awareness and customer behavior are critical determinants of the efficacy of mineral recycling initiatives. Consumers can significantly contribute by appropriately disposing of products containing valuable minerals, such as batteries and electronics, thereby facilitating the recovery of these materials for reuse. Public education initiatives, coupled with accessible recycling programs, can enhance knowledge of the significance of recycling and promote more responsible consumption behaviors. Establishing a recycling culture can diminish waste and encourage the sustainable utilization of essential minerals.

1.7.17 Impact of Digital Technologies on Recycling

Digital technologies, including the IoT, data analytics, and blockchain, are transforming the management and tracking of recycling. IoT sensors can assess the status of recycling machinery, guaranteeing peak efficiency, while data analytics can discern trends in waste streams and enhance recycling operations. Blockchain technology can improve the traceability and transparency of recycled minerals, guaranteeing compliance with ethical and environmental norms. These technologies are essential for establishing a more effective and accountable recycling environment, which is vital for tackling mineral supply chain concerns.

1.7.18 Global Collaboration for Recycling Innovation

International cooperation is crucial for addressing the difficulties within the mineral supply chain. Countries, companies, and research institutes must collaborate to advance and enhance recycling technology, disseminate best practices, and establish standardized procedures for material recovery. International collaborations, exemplified by the Global Battery Alliance, are actively engaged in enhancing recycling efforts and mitigating the environmental consequences of mining activities. By promoting a collaborative strategy, stakeholders may jointly tackle the obstacles confronting the mineral supply chain and expedite the global implementation of recycling technology.

1.7.19 The Future of Eco-Mineral Recycling

The future of eco-mineral recycling depends on the incorporation of sophisticated technology, circular economy principles, and human-centered strategies to establish a more sustainable mineral supply chain. Investing in innovations like AI-driven sorting systems, urban mining, and closed-loop recycling can mitigate the environmental and social consequences of mineral extraction while ensuring a consistent supply of essential minerals. The advancement of Eco-Mineral Recycling 4.0 and 5.0 technologies enhances the feasibility of establishing a sustainable, resilient, and ethical mineral supply chain.

The difficulties confronting the mineral supply chain are intricate and diverse. Nonetheless, the function of recycling—especially via modern Eco-Mineral Recycling

4.0 and 5.0 technologies—presents a feasible and essential remedy. Enhancing recycling procedures, investing in sustainable mining methods, and promoting global collaboration will ensure that the mineral supply chain effectively supports the clean energy transition in a responsible and sustainable manner. Confronting these problems is essential for attaining a circular economy and guaranteeing that the advantages of clean energy technologies are actualized without jeopardizing the environment or human rights.

1.8 INTRODUCTION TO ECO-MINERAL RECYCLING 4.0 AND 5.0 IN THE CLEAN ENERGY ERA

As the global landscape transitions to a cleaner and more sustainable energy paradigm, the demand for essential minerals is escalating. Lithium, cobalt, nickel, and REEs are essential for the production of renewable energy technology, including solar panels, wind turbines, energy storage systems, and EVs. The environmental and social ramifications of mineral extraction have ignited worldwide apprehensions over resource depletion, environmental damage, and supply chain instability. In addressing these problems, Eco-Mineral Recycling 4.0 and 5.0 technologies exemplify a sophisticated, sustainable method for obtaining the resources necessary for the clean energy revolution. This section presents the revolutionary capacity of these technologies in addressing sustainable energy requirements while fostering a circular economy.

The shift to clean energy technologies is unequivocally reliant on the accessibility of essential minerals, many of which are scarce and concentrated in geopolitically sensitive areas. China dominates a substantial portion of rare earth element production, while nations like the DRC are prominent suppliers of cobalt. Traditional mining processes are frequently unsustainable, resulting in significant environmental degradation, resource exhaustion, and societal repercussions. Recycling systems, especially those utilizing Eco-Mineral Recycling 4.0 and 5.0, are increasingly vital to satisfy the growing demand for these minerals. These technologies seek to extract and repurpose precious minerals from current products, thereby diminishing the necessity for new mining and fostering a more sustainable mineral supply chain.

Eco-Mineral Recycling 4.0 denotes the use of advanced technologies, including AI, robots, the IoT, and automation, in the mineral recycling process. These technologies facilitate the more efficient sorting, extraction, and processing of minerals from intricate waste streams, including electronic waste, batteries, and exhausted industrial materials. Conversely, Eco-Mineral Recycling 5.0 accentuates the interaction between human skills and cutting-edge technologies. This concept prioritizes human-centric design to guarantee that recycling processes are socially accountable, ethically justified, and promote sustainable development while technology improves efficiency and accuracy.

AI and robots are leading advancements in Eco-Mineral Recycling 4.0. These technologies facilitate the automation of labor-intensive operations that were before slow, perilous, or inefficient. AI algorithms can enhance the sorting of mixed garbage, accurately recognizing precious minerals such as lithium or cobalt in electronic waste, surpassing the precision of human workers. Robotic systems can independently disassemble devices or batteries, extracting valuable materials while reducing the danger of contamination. This enhances the efficacy of recycling processes, lowers expenses, and guarantees a greater recovery of resources for reutilization in clean energy technologies.

The IoT is a disruptive technology that improves the efficiency of Eco-Mineral Recycling 4.0. By integrating sensors into recycling machinery and waste streams, IoT devices may assess machinery status, monitor material flows, and collect real-time data on the recovery process. This information is essential for enhancing operations, forecasting maintenance requirements, and recognizing opportunities for resource efficiency. IoT sensors can monitor the composition of waste materials within a recycling facility, enabling operators to modify parameters to optimize mineral recovery.

The expansion of mineral recycling is essential to satisfy the increasing demand for important minerals. Automation is crucial in this scaling process by optimizing workflows and diminishing dependence on manual labor. Automated systems can swiftly categorize and handle substantial quantities of waste, guaranteeing the effective separation and recovery of valuable minerals. Moreover, automation diminishes human contact with harmful substances, hence improving workplace safety. This technological transition enables recycling plants to function on a bigger scale and with increased throughput, meeting the demand for enhanced mineral recovery in the clean energy sector.

A core tenet of Eco-Mineral Recycling 4.0 and 5.0 is the premise of a circular economy. In a conventional linear economy, resources are harvested, utilized, and discarded, resulting in waste and environmental deterioration. A circular economy aims to reduce waste by reclaiming valuable resources from abandoned products and reintegrating them into the manufacturing cycle. Eco-mineral recycling technologies facilitate a closed-loop system, wherein materials are perpetually reused, hence diminishing the necessity for virgin mineral extraction. Eco-Mineral Recycling significantly contributes to circularity in the renewable energy sector by extracting minerals like lithium, cobalt, and nickel from discarded batteries, electronics, and solar panels.

Eco-Mineral Recycling 5.0 prioritizes the human aspect of the recycling process. Although automation and AI may manage repetitive activities, human experience is crucial for supervising intricate decision-making, guaranteeing adherence to ethical norms, and tackling unexpected obstacles. Human-centric design in Eco-Mineral Recycling 5.0 guarantees that technology is utilized to the advantage of both society and the environment. Workers can partner with robotic systems to oversee the quality of recycled materials, modify sorting criteria, and guarantee that the process adheres to sustainability objectives.

Sustainable development is fundamental to Eco-Mineral Recycling 4.0 and 5.0. These technologies promote sustainability by diminishing the environmental effects of mineral extraction, improving resource efficiency, and reducing carbon footprints. Recycling rare earth elements from obsolete electronics or wind turbines obviates the necessity for ecologically detrimental mining activities. Moreover, by facilitating more efficient recycling methods, these technologies diminish energy usage and greenhouse gas emissions linked to mineral extraction and processing.

The economic feasibility of Eco-Mineral Recycling 4.0 and 5.0 is essential for its extensive implementation. The initial investment in advanced recycling technology may be considerable, although the long-term advantages are great [56]. Recycling operations can provide new revenue streams, diminish dependence on expensive virgin materials, and create employment opportunities in the recycling and renewable energy sectors by extracting valuable minerals from waste streams [57]. Moreover, governmental incentives and regulatory structures that emphasize recycling might strengthen the economic justification for these technologies. As recycling becomes increasingly economical, dependence on primary mineral extraction will diminish, thereby substantially alleviating the environmental and economic burdens of mining.

The market for recycled essential minerals is swiftly growing due to the rising need for renewable energy solutions. The worldwide EV market is expanding rapidly, increasing the demand for cobalt, lithium, and nickel, all of which may be extracted from recycled batteries. The proliferation of renewable energy sources, including solar and wind farms, is augmenting the demand for rare earth elements utilized in turbines and generators [58]. Recycled materials can fulfill part of this need, alleviating strain on main supply chains and promoting a more robust and sustainable supply of essential minerals.

The mineral supply chain is replete with dangers, encompassing geopolitical tensions, trade disruptions, and environmental issues. Recycling presents a method to alleviate these hazards by supplying a more varied and localized source of essential minerals. By creating regional recycling plants, nations can diminish their reliance on foreign supplies of rare earth elements and other essential minerals. Recycling guarantees the availability of resources during global supply chain interruptions, hence enhancing the resilience and security of the overall system.

The ecological impact of mining is a major difficulty in the extraction of essential minerals. Mining activities require substantial quantities of water, energy, and land, frequently leading to habitat damage, pollution, and a decline in biodiversity. Eco-Mineral Recycling 4.0 and 5.0 technologies mitigate this footprint by decreasing the necessity for new mining operations. Recycling mitigates the environmental costs linked to the extraction, processing, and transportation of virgin materials by reclaiming valuable minerals from end-of-life items.

Eco-Mineral Recycling 4.0 and 5.0 feature technology advancements that enhance the efficiency and environmental sustainability of mineral extraction and recovery. Progress in hydrometallurgy, bioleaching, and solvent extraction methods is facilitating the recovery of essential minerals from intricate waste streams. These technologies consume fewer resources, including energy, water, and chemicals, compared to conventional extraction processes, hence enhancing sustainability. Moreover, the incorporation of AI and machine learning into recycling operations facilitates more accurate recovery, minimizing losses and guaranteeing superior-quality recovered resources.

A primary objective of Eco-Mineral Recycling is to establish a circular supply chain for renewable energy technology. By creating products that facilitate recycling and integrating recycled materials into production, companies may diminish their need for virgin materials and reduce waste. The establishment of closed-loop recycling systems, wherein resources are repurposed for identical uses, is crucial for attaining this objective. Lithium-ion batteries can be recycled and repurposed in new batteries, hence diminishing the necessity for more lithium extraction [59].

Substantial investments in recycling infrastructure are essential to effectively harness the promise of eco-mineral recycling. Recycling facilities must include sophisticated systems capable of managing intricate waste streams and effectively recovering valuable minerals. This encompasses the establishment of specialist facilities for the recycling of electronic waste, obsolete batteries, and renewable energy components [60]. Enhancing recycling infrastructure necessitates cooperation among governments, companies, and research institutes to guarantee that recycling procedures are standardized, efficient, and broadly accessible [61].

Governments are essential in facilitating eco-mineral recycling by establishing legislative and regulatory frameworks that encourage innovation and guarantee the ethical and sustainable utilization of resources. Policies that promote recycling, such as tax incentives for enterprises utilizing recycled materials or EPR frameworks, can enhance demand for recycled minerals. Moreover, legislation mandating appropriate recycling procedures, including the prohibition of dangerous compounds in electronics, can guarantee that the recycling process is environmentally and socially conscientious.

Although technology developments are essential, ethical and social factors must be incorporated into Eco-Mineral Recycling 4.0 and 5.0. This entails guaranteeing equitable treatment of workers at recycling plants and preventing the recycling process from worsening socio-economic inequities. Human-centric recycling strategies can guarantee equitable distribution of technological advantages and facilitate local community involvement in decision-making regarding recycling operations.

The evolving recycling sector necessitates a proficient staff adept at managing and running new recycling technologies. Programs in education and training that emphasize the technical, environmental, and economic dimensions of eco-mineral recycling are crucial for developing a workforce capable of addressing the requirements of the clean energy age. Collaboration among universities, businesses, and government can facilitate the creation of programs that prepare individuals with the essential skills to function and innovate in the recycling sector.

The future of eco-mineral recycling is intricately linked to the efficacy of the clean energy transition. As technologies advance and recycling systems enhance in efficiency, dependence on primary mineral extraction will progressively diminish. Integrating Eco-Mineral Recycling 4.0 and 5.0 technologies will secure a sustainable supply of key minerals to fulfill the increasing demands of the clean energy sector. The integration of technology, policy, and human skills will propel this revolution.

Eco-Mineral Recycling 4.0 and 5.0 are essential for establishing a sustainable and resilient mineral supply chain for clean energy technologies. These systems use cutting-edge technology, methods that focus on people, and the ideas behind the circular economy to help reduce mining, cut down on waste, and make sure that everyone, including the environment, the economy, and society, has a fair chance to switch to clean energy. As these technologies advance, they possess the capacity to transform our approach to resource recovery and facilitate a more sustainable future.

This book examines the several aspects of contemporary recycling methods that incorporate advanced technologies and human-centered philosophies. Subsequent to the introduction chapter, the following eight chapters explore the digital transformation of recycling, human–machine collaboration, significant mineral issues, and pragmatic strategies to ensure a sustainable supply for clean energy. Collectively, they provide an extensive framework for transforming the mineral recycling environment.

Chapter 2 examines the influence of Industry 4.0 technologies on digital transformation in the recycling industry. This chapter examines the impact of Industry 4.0 technology on the transformation of recycling operations. Industry 4.0 facilitates automated sorting, enhanced mineral recovery, and predictive maintenance in recycling facilities through the utilization of AI, the IoT, robotics, and big data analytics. AI-powered robotic arms can disassemble used lithium-ion batteries to extract essential minerals such as cobalt and nickel with exceptional accuracy. This chapter analyzes practical instances, like Germany's effective incorporation of AI in recycling facilities, to illustrate the possibilities of digital transformation in enhancing global recycling capacities.

Chapter 3 analyzes "Industry 5.0: Human-Centric and Sustainable Approaches to Recycling." Industry 4.0 prioritizes technology, whereas Industry 5.0 promotes human-centric design and sustainability. This chapter underscores the synergistic interaction between humans and intelligent machines, stressing collaborative frameworks in which workers direct advanced systems to make ethical and sustainable choices. In Denmark, sophisticated wind turbine recycling systems integrate human supervision and robotic accuracy to efficiently collect valuable materials while reducing environmental impact. This chapter examines the cultural and ethical ramifications of these behaviors and their contribution to realizing a circular economy.

Chapter 4 examines the "Technological Foundations of Eco-Mineral Recycling 4.0." This chapter explores the technology advancements propelling Eco-Mineral Recycling 4.0. The chapter examines the role of modern hydrometallurgical techniques and IoT-based material tracking in the efficient recovery of key minerals from end-of-life products. Japan's implementation of IoT for e-waste tracking enables accurate material identification, hence enhancing sorting and processing efficiency. The chapter assesses the function of blockchain in traceability, guaranteeing ethical and sustainable recycling methods.

Chapter 5 analyzes "Eco-Mineral Recycling 5.0: The Next Frontier in Human–Machine Collaboration." This chapter elaborates on the human-centric ideas presented in Chapter 3, exploring Eco-Mineral Recycling 5.0 as the forthcoming frontier of innovation. It analyzes the progression of collaborative human–machine systems in enhancing recycling efficiency, accuracy, and sustainability. Norway's "human-in-the-loop" systems for recovering rare earth elements require workers to direct machine-learning algorithms to address various recycling issues. This chapter examines how this method promotes inclusivity and empowers workers in recycling sectors worldwide.

Chapter 6 analyzes essential minerals in the clean energy transition: availability, scarcity, and recycling. Critical minerals are essential for the clean energy transition; nevertheless, their limited availability and uneven geographic distribution provide obstacles. This chapter offers a comprehensive examination of the supply chain, emphasizing critical bottlenecks, geopolitical linkages, and the significance of recycling in addressing supply deficiencies. The United States has markedly diminished its dependence on foreign rare earth imports via urban mining efforts. The chapter underscores the significance of recycling as a method to alleviate supply concerns and enhance resilience in crucial mineral markets.

Expanding upon Chapter 6, this portion (Chapter 7) addresses particular issues, including the energy-intensive characteristics of conventional recycling and the intricacies involved in extracting minerals from heterogeneous waste streams. It offers novel solutions, such as bioleaching for the extraction of rare earth elements and modular recycling systems for dispersed operations. Empirical case studies, exemplified by India's initiatives in reclaiming essential minerals from electronic trash, underscore the efficacy of these strategies in tackling global recycling dilemmas and revealing novel prospects.

Chapters 8 and 9 explore transformative strategies to tackle the difficulties and opportunities in mineral recycling and essential mineral supply chains. These chapters offer scenario-based frameworks to conceptualize the future of these industries, encompassing both incremental enhancements and transformative strategies that emphasize sustainability and innovation. The chapters offer a systematic comprehension of the responsibilities that industries, governments, and stakeholders must assume to facilitate systemic change through the analysis of various scenarios.

Chapter 8 examines four unique avenues for promoting eco-mineral recycling. The "Same" scenario illustrates a business-as-usual methodology characterized by negligible advancement, resulting in enduring inefficiencies and unsustainable practices. The "Assisted" scenario entails moderate governmental and industrial backing, emphasizing the expansion of current technology and the establishment of fundamental regulatory frameworks. In the "Greater Support" scenario, extensive cooperation between industry and governments leads to considerable technological and infrastructural progress, shown by the widespread implementation of AI-driven recycling systems. The "Smarter" scenario is the most ambitious, encompassing transformative measures such as implementing blockchain for global traceability, establishing closed-loop systems, and utilizing public–private partnerships for enduring sustainability. Each scenario is examined for its effects on resource recovery, environmental consequences, and economic feasibility.

Chapter 9 concentrates on securing the supply of essential minerals for renewable energy technologies, employing a comparable four-scenario methodology. The "Same" scenario underscores the dangers of inadequate measures, including supply chain interruptions and increasing resource scarcity. The "Assisted" scenario outlines gradual enhancements via governmental assistance and regional recycling initiatives, whereas the "Greater Support" scenario anticipates extensive global collaboration and significant expenditures in mining and recycling technologies. The "Smarter" scenario presents a transformative approach featuring groundbreaking breakthroughs, like decentralized urban mining networks and AI-driven demand forecasting systems, which guarantee global resilience in the supply of vital minerals. The chapter underscores the necessity for audacious initiatives to align demand with ethical and sustainable sourcing practices.

Chapters 1–7 collectively offer a progressive, multifaceted examination of Eco-Mineral Recycling 4.0 and 5.0. They tackle the difficulties and potential of crucial mineral recycling in the clean energy era by integrating modern technologies, human-centric methodologies, and sustainability principles. This detailed roadmap emphasizes the transformative capacity of recycling systems as a fundamental element for realizing a circular economy and fulfilling the global demand for essential minerals. Chapters 8 and 9 delineate the range of opportunities for enhancing mineral recycling and ensuring the availability of important mineral resources. By juxtaposing incremental and transformative situations, they emphasize the significance of strategic decision-making, innovation, and global collaboration. The suggested routes offer a strategic framework for industry and governments to shift from reactive to proactive strategies, in accordance with the tenets of Eco-Mineral Recycling 4.0 and 5.0, to realize a really sustainable future for clean energy technology.

REFERENCES

[1] Agusdinata DB, Eakin H, Liu W. Critical minerals for electric vehicles: a telecoupling review. *Environmental Research Letters.* 2022;17:013005. https://doi.org/10.1088/1748-9326/ac4763.

[2] Pirmana V, Alisjahbana AS, Yusuf AA, Hoekstra R, Tukker A. Economic and environmental impact of electric vehicles production in Indonesia. *Clean Technol Environ Policy.* 2023;25. https://doi.org/10.1007/s10098-023-02475-6.

[3] Kushnir D, Sandén BA. The time dimension and lithium resource constraints for electric vehicles. *Resources Policy.* 2012;37:93–103. https://doi.org/10.1016/j.resourpol.2011.11.003.

[4] Menyhart J. Overview of sustainable mobility: The role of electric vehicles in energy communities. *World Electric Vehicle Journal.* 2024;15:275. https://doi.org/10.3390/wevj15060275.

[5] Kobashi T, Yoshida T, Yamagata Y, Naito K, Pfenninger S, Say K, et al. On the potential of "Photovoltaics + Electric vehicles" for deep decarbonization of Kyoto's power systems: Techno-economic-social considerations. *Applied Energy.* 2020;275. https://doi.org/10.1016/j.apenergy.2020.115419.

[6] Xia X, Li P, Xia Z, Wu R, Cheng Y. Life cycle carbon footprint of electric vehicles in different countries: A review. *Separation and Purification Technology.* 2022;301. https://doi.org/10.1016/j.seppur.2022.122063.

[7] Bartolini A, Comodi G, Salvi D, Østergaard PA. Renewables self-consumption potential in districts with high penetration of electric vehicles. *Energy.* 2020;213. https://doi.org/10.1016/j.energy.2020.118653.

[8] Dioha MO, Duan L, Ruggles TH, Bellocchi S, Caldeira K. Exploring the role of electric vehicles in Africa's energy transition: A Nigerian case study. *IScience.* 2022;25. https://doi.org/10.1016/j.isci.2022.103926.

[9] Cadoni E, Forni D, Dotta M, Bellenda G, Caccivio M. Advanced characterisation of photovoltaics for hail resistance. *Materials Letters.* 2024;354. https://doi.org/10.1016/j.matlet.2023.135371.

[10] McKerracher C, O'Donavan A, Soulopoulos N, Grant A, Lyu J, Mi S, et al. Electric vehicle outlook 2023. *BloombergNEF.* 2023.

[11] International Copper association. The electric vehicle market and copper demand. *Copper Aliance.* 2017;2.

[12] Müller D, Chartouni D. Implications on EROI and climate change of introducing Li-ion batteries to residential PV systems. *Applied Energy.* 2022;326. https://doi.org/10.1016/j.apenergy.2022.119958.

[13] Ueckerdt F, Bauer C, Dirnaichner A, Everall J, Sacchi R, Luderer G. Potential and risks of hydrogen-based e-fuels in climate change mitigation. *Nature Climate Change*. 2021;11:384–93. https://doi.org/10.1038/s41558-021-01032-7.

[14] Bouramdane A-A. Crafting an optimal portfolio for sustainable hydrogen production choices in Morocco. *Fuel*. 2024;358. https://doi.org/10.1016/j.fuel.2023.130292.

[15] Van den Eynde S, Bracquené E, Diaz-Romero D, Zaplana I, Engelen B, Duflou JR, et al. Forecasting global aluminium flows to demonstrate the need for improved sorting and recycling methods. *Waste Management*. 2022;137. https://doi.org/10.1016/j.wasman.2021.11.019.

[16] Agusdinata DB, Liu W, Eakin H, Romero H. Socio-environmental impacts of lithium mineral extraction: Towards a research agenda. *Environmental Research Letters*. 2018;13:123001. https://doi.org/10.1088/1748-9326/aae9b1.

[17] Ferdoush MdR, Aziz R Al, Karmaker CL, Debnath B, Limon MH, Bari ABMM. Unraveling the challenges of waste-to-energy transition in emerging economies: Implications for sustainability. *Innovation and Green Development*. 2024;3:100121. https://doi.org/10.1016/j.igd.2023.100121.

[18] Yang J, Cao X, Yao J, Kang Z, Chang J, Wang Y. Geographical big data and data mining: A new opportunity for "water-energy-food" nexus analysis. *Journal of Geographical Sciences*. 2024;34. https://doi.org/10.1007/s11442-024-2202-6.

[19] Shohidul Islam Md, Arafatur Rahman Md, Ariff Bin Ameedeen M, Ajra H, Binti Ismail Z, Mohamad Zain J. Blockchain-enabled cybersecurity provision for scalable heterogeneous network: A comprehensive survey. *Computer Modeling in Engineering & Sciences*. 2024;138. https://doi.org/10.32604/cmes.2023.028687.

[20] Kalantzakos S. The race for critical minerals in an era of geopolitical realignments. *The International Spectator*. 2020;55:1–16. https://doi.org/10.1080/03932729.2020.1786926.

[21] Calvo G, Valero A, Valero A. Assessing maximum production peak and resource availability of non-fuel mineral resources: Analyzing the influence of extractable global resources. *Resources, Conservation & Recycling*. 2017;125:208–17. https://doi.org/10.1016/j.resconrec.2017.06.009.

[22] Jones C, Peshev V, Gilbert P, Mander S. Battery storage for post-incentive PV uptake? A financial and life cycle carbon assessment of a non-domestic building. *Journal of Cleaner Production*. 2017;167:447–58. https://doi.org/10.1016/j.jclepro.2017.08.191.

[23] Shimaponda-Nawa M, Nwaila GT. Integrated and intelligent remote operation centres (I2ROCs): Assessing the human–machine requirements for 21st century mining operations. *Minerals Engineering*. 2024;207. https://doi.org/10.1016/j.mineng.2023.108565.

[24] Gałaś A, Kot-Niewiadomska A, Czerw H, Simić V, Tost M, Wårell L, et al. Impact of Covid-19 on the mining sector and raw materials security in selected European countries. *Resources*. 2021;10:39. https://doi.org/10.3390/resources10050039.

[25] Üçtuğ FG, Azapagic A. Environmental impacts of small-scale hybrid energy systems: Coupling solar photovoltaics and lithium-ion batteries. *Science of the Total Environment*. 2018;643:1579–89. https://doi.org/10.1016/j.scitotenv.2018.06.290.

[26] Dou S, Xu D, Zhu Y, Keenan R. Critical mineral sustainable supply: Challenges and governance. *Futures*. 2023;146. https://doi.org/10.1016/j.futures.2023.103101.

[27] Czerwinski F. Critical minerals for zero-emission transportation. *Materials*. 2022;15. https://doi.org/10.3390/ma15165539.

[28] Ballinger B, Stringer M, Schmeda-Lopez DR, Kefford B, Parkinson B, Greig C, et al. The vulnerability of electric vehicle deployment to critical mineral supply. *Applied Energy.* 2019;255:113844. https://doi.org/10.1016/j.apenergy.2019.113844.

[29] International Energy Agency. The Role of Critical Minerals in Clean Energy Transitions. 2022. www.iea.org/reports/the-role-of-critical-minerals-in-clean-energy-transitions, Licence: CC BY 4.0

[30] Wang P, Chen WQ, Cui X, Li J, Li W, Wang C, et al. Critical mineral constraints in global renewable scenarios under 1.5 °C target. *Environmental Research Letters.* 2022;17. https://doi.org/10.1088/1748-9326/aca4ea.

[31] Hancock L, Ralph N, Ali SH. Bolivia's lithium frontier: Can public private partnerships deliver a minerals boom for sustainable development? *Journal of Cleaner Production.* 2018;178:551–60. https://doi.org/10.1016/j.jclepro.2017.12.264.

[32] IEA: *Global Critical Minerals Outlook.* 2025;37. www.iea.org/reports/global-critical-minerals-outlook-2025, Licence: CC BY 4.0

[33] Kerr A. Critical minerals in the context of canada: Concepts, challenges and contradictions. *Geoscience Canada.* 2023;50:85–103. https://doi.org/10.12789/geocanj.2023.50.199.

[34] International Energy Agency (IEA). The role of critical minerals in clean energy transitions – analysis – IEA. 2021. www.iea.org/reports/the-role-of-critical-minerals-in-clean-energy-transitions, Licence: CC BY 4.0

[35] Nwokolo SC, Singh R, Khan S, Kumar A. Harnessing industry 4.0 for Africa's net zero economy through technological pathways. *Technological Pathways for Africa' s Net-Zero Economy*, Elsevier; 2025, p. 249–82. https://doi.org/10.1016/B978-0-443-31486-5.00010-0.

[36] Akbari M, Kok SK, Hopkins J, Frederico GF, Nguyen H, Alonso AD. The changing landscape of digital transformation in supply chains: Impacts of industry 4.0 in Vietnam. *The International Journal of Logistics Management.* 2024;35:1040–72. https://doi.org/10.1108/IJLM-11-2022-0442.

[37] El jaouhari A, Arif J, Samadhiya A, Kumar A. Net zero supply chain performance and industry 4.0 technologies: Past review and present introspective analysis for future research directions. *Heliyon.* 2023;9:e21525. https://doi.org/10.1016/j.heliyon.2023.e21525.

[38] Paraskevas A, Madas M, Zeimpekis V, Fouskas K. Smart ports in industry 4.0: A systematic literature review. *Logistics.* 2024;8(1):28. https://doi.org/10.3390/logistics8010028.

[39] Nwokolo SC, Singh R, Khan S, Kumar A. Net zero technology and circular economy. *Technological Pathways for Africa' s Net-Zero Economy*, Elsevier; 2025, p. 237–47. https://doi.org/10.1016/B978-0-443-31486-5.00009-4.

[40] Chokshi H, Agrawal R, Samadhiya A, Kumar A. Circular economy initiatives in supply chain: A systematic literature review and future research directions. *International Journal of Mathematical, Engineering and Management Sciences.* 2023;8:1056–82. https://doi.org/10.33889/IJMEMS.2023.8.6.061.

[41] Mutezo G, Mulopo J. A review of Africa's transition from fossil fuels to renewable energy using circular economy principles. *Renewable and Sustainable Energy Reviews.* 2021;137. https://doi.org/10.1016/j.rser.2020.110609.

[42] Ohmori T, Na-Ranong D, Yamamoto T, Endo A, Akiya T, Nakaiwa M. Periodic and recycling operations in a chemical reaction process for production of light olefins. *Kagaku Kogaku Ronbunshu.* 2003;29(3): 373–7.

[43] Huang B, Zhao J, Chai J, Xue B, Zhao F, Wang X. Environmental influence assessment of China's multi-crystalline silicon (multi-Si) photovoltaic modules

considering recycling process. *Solar Energy*. 2017;143:132–41. https://doi.org/10.1016/j.solener.2016.12.038.

[44] Zhang P, Zhang L, Chang Y, Xu M, Hao Y, Liang S, et al. Food-energy-water (FEW) nexus for urban sustainability: A comprehensive review. *Resources, Conservation & Recycling*. 2019;142:215–24. https://doi.org/10.1016/j.resconrec.2018.11.018.

[45] Iglesias-Émbil M, Valero A, Ortego A, Villacampa M, Vilaró J, Villalba G. Raw material use in a battery electric car – a thermodynamic rarity assessment. *Resources, Conservation & Recycling*. 2020;158:104820. https://doi.org/10.1016/j.resconrec.2020.104820.

[46] Matevosova K, Gryaznova V, Chazov T. Environmental problems and sustainable development of aluminium industry. *Russian Journal of Resources, Conservation and Recycling*. 2019;6. https://doi.org/10.15862/11ecor219.

[47] Ren M, Mitchell CR, Mo W. Managing residential solar photovoltaic-battery systems for grid and life cycle economic and environmental co-benefits under time-of-use rate design. *Resources, Conservation and Recycling*. 2021;169. https://doi.org/10.1016/j.resconrec.2021.105527.

[48] Faircloth CC, Wagner KH, Woodward KE, Rakkwamsuk P, Gheewala SH. The environmental and economic impacts of photovoltaic waste management in Thailand. *Resources, Conservation and Recycling*. 2019;143:260–72. https://doi.org/10.1016/j.resconrec.2019.01.008.

[49] Corcelli F, Fiorentino G, Petit-Boix A, Rieradevall J, Gabarrell X. Transforming rooftops into productive urban spaces in the Mediterranean. An LCA comparison of agri-urban production and photovoltaic energy generation. *Resources, Conservation and Recycling*. 2019;144:321–36. https://doi.org/10.1016/j.resconrec.2019.01.040.

[50] Bertram M, Ramkumar S, Rechberger H, Rombach G, Bayliss C, Martchek KJ, et al. A regionally-linked, dynamic material flow modelling tool for rolled, extruded and cast aluminium products. *Resources, Conservation and Recycling*. 2017;125. https://doi.org/10.1016/j.resconrec.2017.05.014.

[51] Cristóbal J, Jubayed M, Wulff N, Schebek L. Life cycle losses of critical raw materials from solar and wind energy technologies and their role in the future material availability. *Resources, Conservation and Recycling*. 2020;161:104916. https://doi.org/10.1016/j.resconrec.2020.104916.

[52] Mehmood S, Fan J, Dokota IS, Nazir S, Nazir Z. How to manage supply chains successfully in transport infrastructure projects. *Sustainability (Switzerland)*. 2024;16. https://doi.org/10.3390/su16020730.

[53] Lam JSL, van de Voorde E. Scenario analysis for supply chain integration in container shipping. *Maritime Policy & Management*. 2011;38:705–25. https://doi.org/10.1080/03088839.2011.625988.

[54] Chen Z, Su SII. Photovoltaic supply chain coordination with strategic consumers in China. *Renewable Energy*. 2014;68:236–44. https://doi.org/10.1016/j.renene.2014.01.035.

[55] Lee J, Bazilian M, Sovacool B, Greene S. Responsible or reckless? A critical review of the environmental and climate assessments of mineral supply chains. *Environmental Research Letters*. 2020;15:103009. https://doi.org/10.1088/1748-9326/ab9f8c.

[56] Nwokolo S, Eyime E, Obiwulu A, Ogbulezie J. Africa's path to sustainability: Harnessing technology, policy, and collaboration. *Trends in Renewable Energy*. 2024;10:98–131. https://doi.org/10.17737/tre.2024.10.1.00166.

[57] Nwokolo SC, Obiwulu AU, Okonkwo PC. *Africa's Propensity for a Net Zero Energy Transition*. vol. 1st Edition. Boca Raton: CRC Press; 2024. https://doi.org/10.1201/9781003483175.

[58] Nwokolo SC, Singh R, Khan S, Kumar A. *Technological Pathways for Africa's Net-Zero Economy: Technology Solutions to Unlock Africa's Sustainable Future*. USA: Elsevier Science; 2024.

[59] Hao H, Liu Z, Zhao F, Geng Y, Sarkis J. Material flow analysis of lithium in China. *Resources Policy*. 2017;51:100–6. https://doi.org/10.1016/j.resourpol.2016.12.005.

[60] Junne T, Wulff N, Breyer C, Naegler T. Critical materials in global low-carbon energy scenarios: The case for neodymium, dysprosium, lithium, and cobalt. *Energy*. 2020;211:118532. https://doi.org/10.1016/j.energy.2020.118532.

[61] Greim P, Solomon AA, Breyer C. Assessment of lithium criticality in the global energy transition and addressing policy gaps in transportation. *Nature Communications*. 2020;11:4570. https://doi.org/10.1038/s41467-020-18402-y.

2 Digital Transformation in Recycling

Industry 4.0 Technologies

2.1 INTRODUCTION TO INDUSTRY 4.0 IN RECYCLING

Industry 4.0 has revolutionized the worldwide recycling sector, initiating a new epoch of digitalization, automation, and astute decision-making. The core of this revolution is the amalgamation of sophisticated technologies such as artificial intelligence (AI), the Internet of Things (IoT), robotics, big data analytics, and blockchain. These advancements have markedly improved the efficiency, precision, and sustainability of recycling processes, especially in the recovery of essential minerals. Studies such as Lubongo et al. [1] underscore the imperative for a more intelligent and sustainable recycling system, particularly through the advent of AI-driven sorting devices for plastic recycling. Similarly, the research conducted by Mohamed et al. [2] on Waste 4.0 illustrates how automation and digitalization enhance medical waste management, mitigating contamination concerns and improving recovery rates.

A significant aspect of change in Industry 4.0 regarding recycling is automated disassembly, which is essential for the circular economy, especially for valuable materials such as lithium-ion batteries. Choux et al. [3] present a techno-economic analysis contrasting automated disassembly with conventional shredding, demonstrating that automated systems improve material recovery rates while also augmenting safety and energy efficiency. Hathaway et al. [4] examine the constraints and opportunities associated with the robotic disassembly of electric vehicle (EV) batteries, highlighting its potential to decrease labor costs and enhance process uniformity. These developments are especially pertinent in nations with elevated EV adoption rates [5], such as Germany and China, where sustainable battery recycling is crucial for resource efficiency [6].

AI and machine learning (ML) have been crucial in enhancing trash categorization and sorting, significantly contributing to the advancement of Industry 4.0 applications in recycling [7]. Islam et al. [8] introduce ECCDN-Net, a deep learning methodology for the classification of organic and recyclable waste, illustrating the capability of AI to enhance sorting precision. Likewise, Thanh Nguyen et al. [9] utilize deep learning models to enhance the sorting of plastic garbage, which is especially beneficial in areas with extensive recycling programs such as the European Union (EU). Additionally, Farshadfar et al. [10] evaluate ML-driven automated trash sorting

DOI: 10.1201/9781003672807-2

methods within the construction sector, demonstrating their efficacy in minimizing material contamination [11] and enhancing the recyclability of building waste [12].

Industry 4.0 technologies are utilized to improve supply chain transparency and efficiency in recycling. The integration of blockchain and IoT has significantly enhanced traceability in material movements, ensuring the optimal recovery and reuse of precious resources. Masa'deh et al. [13] examine the role of blockchain in the digitalization of supply chains within the circular economy, whereas Schneikart et al. [14] propose a roadmap for pharmaceutical logistics that incorporates Industry 4.0 technologies to facilitate a more circular methodology. Digital interventions are especially pertinent in areas with intricate supply networks, like North America and Southeast Asia [15], where decentralized waste management systems gain from enhanced accountability and automation [16].

A notable facet of Industry 4.0 in recycling is its influence on urban trash management. Intelligent waste collection systems, driven by IoT and AI, have been progressively implemented in smart cities to enhance resource recovery and save operational expenses. Zoumpoulis et al. [17] examine smart bin technologies for urban trash management, highlighting their contribution to improving the efficiency of municipal solid waste systems [18]. Belhiah and El Aboudi [19] propose an IoT-based sensor mesh network for real-time garbage monitoring, enhancing collection precision and resource distribution in cities such as Singapore and Amsterdam. These measures signify an increasing inclination toward data-driven trash management [20], promoting more sustainable urban ecosystems [21].

Industrial automation and robotic interventions have enhanced the efficiency of recycling processes. Cabral et al. [22] employed robotics-enabled material assessment to illustrate how contactless multi-modal sensing techniques might enhance recovery rates in building deconstruction. Similarly, Peng et al. [23] investigate the application of deep reinforcement learning in the robotic disassembly of screws for remanufacturing, demonstrating its efficacy in managing intricate end-of-life items [24]. These improvements are essential for nations with robust industrial sectors, such as Japan and the United States, where automated material recovery can markedly diminish landfill trash [25] and improve circular economy initiatives [26].

Notwithstanding the encouraging advancements, numerous obstacles impede the extensive use of Industry 4.0 in recycling. Agarwal et al. [27] delineate significant obstacles to the adoption of smart manufacturing within the Indian automobile sector, emphasizing issues such as elevated implementation costs and the necessity for specialized labor. Kumar et al. [28] similarly examine obstacles to the adoption of Industry 4.0 in sustainable food supply chains, emphasizing the significance of regulatory frameworks and financial incentives. Addressing these challenges necessitates collaborative efforts from industry leaders, policymakers, and research institutions to promote standardization, encourage investment, and cultivate a proficient workforce adept at utilizing Industry 4.0 technologies for a more sustainable and efficient recycling sector.

The use of Industry 4.0 technologies in recycling is transforming the recovery and management of essential minerals, such as lithium and cobalt, in a more sustainable and efficient way. Real-time data, automation, and transparency are the hallmarks of a new era in recycling that is bringing about significant improvements in operational effectiveness and material recovery rates. Smart waste management systems utilizing IoT in smart cities, such as Singapore and Amsterdam [17,19], illustrate Industry 4.0's capacity to enhance urban waste collection, markedly decreasing landfill waste and advancing the circular economy. Additionally, the deep learning-based waste classification techniques put forth by Islam et al. [8] are revolutionizing recycling by transforming it from a labor-intensive, manual process into an intelligent, automated one. These technological breakthroughs enhance material recovery and assure the efficient recycling of precious resources, such as those utilized in EV batteries, thereby mitigating environmental impacts and fostering resource circularity.

Nonetheless, the journey toward complete digital transformation in recycling presents numerous hurdles. Agarwal et al. [27] and Kumar et al. [29] identify substantial barriers, including elevated implementation expenses, regulatory hurdles, and the necessity for proficient personnel. These problems, however, offer potential for collaboration across sectors, governments, and research institutions to establish supportive frameworks, financial incentives, and worker training programs. Global interventions, shown by the deployment of robotic systems for the disassembly of end-of-life batteries [4], demonstrate that technology investment can yield substantial benefits in sustainability and economic efficiency. Future trends indicate that the ongoing advancement of digital twin (DT) technologies and predictive analytics, as demonstrated by Cabral et al. [22], will yield more precise simulations and predictive models, hence improving the design and operation of recycling systems. The integration of these advances with the increasing implementation of Industry 4.0 will be crucial in promoting the circular economy and enhancing the sustainability and economic viability of essential mineral recycling worldwide [30, 31].

The subsequent sections of this chapter examine pivotal breakthroughs and technologies in Industry 4.0 that are transforming the recycling sector, especially in the recovery of essential minerals. Section 2.2, about smart sensors and IoT, examines how real-time data facilitates accurate monitoring and management of recycling processes, hence enhancing efficiency and minimizing waste. The IoT-integrated waste management systems in cities such as Singapore and Amsterdam exemplify this by enhancing recycling rates and reducing waste-to-landfill using smart bins and real-time data [17]. Section 2.3 explores AI and ML, highlighting how deep learning models, such as those introduced by Islam et al. [8], have improved trash classification, facilitating more precise segregation of recyclable materials and hence augmenting material recovery rates. The use of these technologies in automated sorting systems is establishing the groundwork for highly efficient, cost-effective recycling processes across several industries, especially in key mineral recycling, exemplified by developments in lithium-ion battery recovery [32].

Furthermore, Section 2.4 addresses robots and autonomous systems in material recovery, emphasizing the transformative impact of robotics on jobs such as battery disassembly and plastic sorting. The deployment of AI-enabled robots in EV battery recycling illustrates the advanced applications of Industry 4.0 to automate intricate operations previously performed manually, thereby decreasing labor expenses and enhancing safety [4]. Section 2.5 examines blockchain and digital ledger technologies (DLTs) to provide transparency and traceability in recycling processes, a crucial element for the ethical and effective recovery of key minerals. Blockchain technologies offer an unalterable record of materials' progression through the recycling chain, guaranteeing the efficient tracking and trading of recycled minerals [33]. Section 2.6 examines big data and predictive analytics as instruments for optimizing recycling operations, providing predicted insights that can substantially improve material recovery and operational planning [34]. This section, in conjunction with Section 2.7 on digital twin technology, emphasizes advancements in simulation models that facilitate the development of virtual recycling systems to enhance operations, detect inefficiencies, and forecast results [35, 36]. Collectively, these technologies improve the sustainability and profitability of recycling operations while providing solutions to meet the increasing demand for essential minerals in the worldwide market. Section 2.8 continues with an examination of the problems and opportunities presented by these technologies, highlighting issues such as regulatory obstacles, the necessity for industry-wide standards, and the significant potential for future scalability.

2.2 SMART SENSORS AND IOT: REAL-TIME DATA FOR ENHANCED RECYCLING EFFICIENCY

In the swiftly advancing domain of recycling, intelligent sensors and the IoT are significantly boosting efficiency, decreasing costs, and fostering sustainability. The incorporation of IoT technology in recycling systems facilitates real-time data acquisition and monitoring, yielding significant insights for decision-making and process enhancement. Sensors included in recycling machines and waste management systems facilitate the automatic monitoring of materials and processes. This innovation is essential for attaining enhanced recycling efficiency and is pivotal in the digital transformation of the recycling sector. Zoumpoulis et al. [17] examined the application of smart bins to enhance resource recovery in smart cities, emphasizing the promise of IoT-based solutions in urban garbage management. The real-time monitoring features of these systems can markedly diminish trash contamination and enhance sorting efficiency, so fostering a more circular economy.

In the realm of crucial mineral recycling, especially for EV battery recycling, intelligent sensors and the IoT are facilitating advancements in automation and accuracy. Hyperspectral sensors, utilized for identifying and sorting elements in intricate waste streams, offer real-time feedback that guarantees enhanced purity levels in recovered materials. Azizi et al. [37] presented an autonomous hyperspectral characterization station employing robotic systems to evaluate polymer degradation, an essential process in the recycling of materials such as plastics and electronic trash [37]. This application illustrates the integration of advanced sensor technologies with IoT platforms to automate and optimize the recycling process, particularly for intricate materials in important mineral recovery.

Furthermore, the digitalization of recycling operations via IoT encompasses the oversight and administration of energy use within recycling plants. Real-time data from intelligent sensors facilitates the ongoing evaluation of equipment efficiency [38], guaranteeing optimal energy utilization and minimizing waste in manufacturing processes. This is essential for synchronizing recycling initiatives with sustainability objectives, especially in the automobile and electronics industries. A study by Choux et al. [3] demonstrated that IoT-enabled monitoring of material throughput enhances the automated disassembly of lithium-ion batteries, a crucial aspect of EV recycling, leading to improved energy management and diminished environmental effect.

The collaboration between industry stakeholders and institutions is promoting innovative solutions for waste management as part of the ongoing integration of Industry 4.0 technology into recycling. Industry stakeholders, including recycling firms and original equipment manufacturers (OEMs), are investing in IoT systems to optimize trash sorting, enhance material recovery rates, and improve sustainability results. Farshadfar et al. [10] investigated the creation of an automated waste sorting system in the construction sector that employs ML to enhance waste classification and resource recovery. This partnership underscores the possibility of integrating IoT with advanced analytics and robotics to address intricate recycling issues in businesses that produce substantial garbage.

The involvement of government and academic organizations in promoting IoT and smart sensor technologies in recycling is substantial. Research institutions are developing standardized sensor networks and intelligent systems for integration across diverse recycling plants to facilitate smooth data exchange and operational transparency. The automotive industry in India has encountered difficulties in implementing Industry 4.0 technologies; however, initiatives are underway to address obstacles to green smart manufacturing through methodologies such as AHP (Analytic Hierarchy Process) to pinpoint essential areas for enhancement [27]. These initiatives seek to eliminate obstacles to IoT adoption and improve the overall efficacy of recycling systems within the automobile industry.

Moreover, IoT-based solutions can substantially improve material recyclability by supplying data that facilitates the development of innovative recycling systems. Bui et al. [39] examined the use of IoT in incorporating circular economy techniques into Vietnam's natural resource management, emphasizing financial sustainability. The implementation of IoT sensors facilitates ongoing surveillance of resource distribution and waste production, yielding critical data for the advancement of innovative recycling methods that are commercially viable and environmentally sustainable. This exemplifies how data-driven insights may enhance innovation in the recycling sector, increasing efficiency and aligning with global sustainability objectives.

The predictive capabilities of IoT systems enhance waste management by forecasting equipment failures or inefficiencies in waste processing prior to their occurrence. In a case study by Pandey et al. [40], the amalgamation of IoT and cloud computing was employed to tackle security issues in industrial automation, emphasizing the role of predictive analytics in enhancing the reliability and security of waste sorting and recycling processes [40]. These technologies can forecast potential machine malfunctions, facilitating prompt maintenance, so reducing downtime and enhancing recycling efficiency.

The future of IoT in recycling is intricately linked to the continuous progress in ML and AI, facilitating more intelligent real-time decision-making. The integration of AI-driven data analytics platforms with IoT devices facilitates improved sorting algorithms, precise material identification, and enhanced optimization of recycling processes. In a study by Peng et al. [23], deep reinforcement learning was employed to improve the robotic disassembly of EV batteries, illustrating the collaboration of AI and the IoT in automating intricate recycling processes [23]. The advancement of AI and IoT technologies will be essential for fostering the next generation of breakthroughs in recycling efficiency and sustainability.

Intelligent sensors and IoT technology are pivotal to the digital transformation of recycling processes. These technologies provide real-time data collecting, monitoring, and decision-making, hence enhancing material recovery, energy use, and overall recycling efficiency. The collaborative endeavors of business, government, and academia, coupled with the application of advanced analytics and automation, are driving substantial advancements in key mineral recycling and sustainable waste management techniques worldwide. As the globe transitions to a circular economy, the use of digital technologies will remain essential in enhancing recycling practices and attaining long-term sustainability objectives.

Table 2.1 and Figure 2.1 present a thorough and novel collection of recent research publications examining the incorporation of modern technologies, including smart sensors, the IoT, and Industry 4.0, in the recycling and circular economy domains. This emphasizes numerous studies focused on critical mineral recycling (CMR), automated waste sorting, and digital transformation, illustrating the impact of digital solutions such as ML, deep learning, robotic disassembly, and data-driven decision-making in enhancing the efficiency and sustainability of recycling processes. Illustrative examples from many locations, including India, Vietnam, China, and Brazil, demonstrate practical applications and interventions, such as automated trash sorting in the automobile industry, textile recycling, and plastic sorting with AI and robots. The study illustrates that digitization in waste management and recycling sectors improves operational efficiency, fosters environmental sustainability, mitigates significant obstacles, and bolsters circular economy concepts. Table 2.1 highlights the crucial need for multi-stakeholder participation, encompassing industries, academic institutions, and governmental entities, in promoting recycling technologies and their effective incorporation into smart city infrastructures. The gathered insights provide essential knowledge for policymakers, industries, and researchers engaged in enhancing sustainable recycling systems via digital transformation.

2.3 ARTIFICIAL INTELLIGENCE AND MACHINE LEARNING IN AUTOMATED SORTING AND WASTE MANAGEMENT

The integration of AI and ML in automated sorting and waste management is revolutionizing the recycling industry, particularly in the context of Industry 4.0. The advent of digital transformation in CMR necessitates efficient waste sorting mechanisms to ensure the sustainable recovery of valuable materials from complex waste streams. AI-driven automation, robotic vision, deep learning algorithms, and IoT-enabled smart waste classification systems are among the emerging technologies fostering the efficiency, precision, and sustainability of recycling operations [48]. Such innovations address critical challenges in waste management, including contamination control, sorting accuracy, and labor-intensive processes.

Recent advancements in deep learning and computer vision have significantly improved waste sorting technologies, enabling real-time, high-accuracy

TABLE 2.1
Leveraging Smart Sensors, IoT, and Industry 4.0 Technologies for Advancing Recycling Efficiency and Circular Economy in Critical Mineral Recovery

Area	Category	Innovation	Building Blocks	Example/Case Study	Location and Nationality	Industry/ Institution Intervention	Specific Improvement
Smart sensors and IoT	Technology	Use of IoT-enabled sensors for real-time monitoring in waste sorting, material identification, and recovery processes. Sensors enable data collection and communication to central systems for informed decision-making in recycling operations.	Integration of IoT with automated sorting systems, deep learning for waste classification [8].	Automated Waste Sorting in Construction: Farshadfar et al. [10] showcased IoT and machine learning integration in waste sorting for construction.	Global applications, particularly in India [27]	Industry 4.0 technologies used by the construction industry to enhance waste recycling systems [10].	Improving waste sorting accuracy and reducing labor costs by automating waste classification systems.

(continued)

TABLE 2.1 (Continued)
Leveraging Smart Sensors, IoT, and Industry 4.0 Technologies for Advancing Recycling Efficiency and Circular Economy in Critical Mineral Recovery

Area	Category	Innovation	Building Blocks	Example/Case Study	Location and Nationality	Industry/ Institution Intervention	Specific Improvement
Real-time data	Data analytics	Advanced data analytics, leveraging cloud computing and edge AI, for processing real-time data from sensors to optimize the recycling process, predict waste volumes, and enhance recovery efficiency.	AI, cloud-based platforms, predictive analytics [40].	Predictive Analytics in Battery Recycling: Choux etal.[3]demonstrated how real-time data analytics optimizes disassembly processes for lithium-ion batteries.	France, Europe	R&D in lithium-ion battery recycling using smart technologies and predictive analytics.	Optimizing recycling operations by predicting waste flow and automating recovery processes based on real-time insights.
Automation	Process automation	Integration of AI and robotics for automated sorting, disassembly, and waste material recovery. AI-based robotic arms and machine learning algorithms improve the efficiency of recycling by sorting various materials with high precision.	Robotic sorting systems, AI-enabled image recognition [41].	AI-enabled Robotic Disassembly: Zang et al. [42] highlighted advancements in robotic disassembly for recycling electric vehicle batteries.	China, Europe	Automotive and battery industries working with AI-based robotics to automate battery disassembly [42].	Enhancing sorting speed and precision, reducing manual labor requirements.

Circular economy	Sustainability	Smart IoT systems supporting circular economy principles by monitoring material lifecycle, promoting recycling, and enabling reuse of critical materials like rare earth metals from waste streams.	Circular supply chains, material traceability [43].	Circular Economy in Pharmaceutical Logistics: Alimohammadi and Behnamian [44] demonstrated using IoT to optimize logistics in pharmaceutical reverse logistics.	Iran, Middle East	Circular economy strategies in pharmaceutical logistics with digital transformation for resource management [44].	Advancing resource recovery through circular models by tracing material flows across the recycling supply chain.
Waste classification	Smart sorting systems	Advanced deep learning techniques for automated classification of recyclable materials such as plastics, metals, and electronic waste, using sensors and cameras to achieve high-efficiency waste sorting.	Deep learning models for waste recognition [45].	Automated Sorting of Plastics: Lubongo et al. [1] highlighted the use of AI and robots for efficient plastic waste sorting.	Germany, Europe	Waste management companies in Germany have adopted AI and robotic systems for plastic recycling, increasing recovery rates [1].	Significantly improving the efficiency of sorting systems for diverse waste types using deep learning.

(*continued*)

TABLE 2.1 (Continued)
Leveraging Smart Sensors, IoT, and Industry 4.0 Technologies for Advancing Recycling Efficiency and Circular Economy in Critical Mineral Recovery

Area	Category	Innovation	Building Blocks	Example/Case Study	Location and Nationality	Industry/ Institution Intervention	Specific Improvement
Recycling efficiency	Resource recovery	IoT-enabled smart bins and waste collection systems that optimize material collection routes and ensure timely recycling, resulting in better resource recovery and reduced energy consumption.	Smart bins, route optimization algorithms [46].	Smart Cities: Zoumpoulis et al. [17] explored the role of smart bins in urban settings for enhancing resource recovery and sustainability.	South Korea, Global	Smart city initiatives in South Korea have integrated IoT-enabled waste collection systems to enhance urban recycling efficiency [17].	Improving recycling collection efficiency and ensuring effective resource recovery in urban environments.
Collaboration and stakeholders	Industry partnerships	Collaboration between governments, industries, and research institutions to develop standardized frameworks and platforms for	Collaborative research, industry partnerships, regulatory support [27].	Industry–University Collaboration: Bui et al. [39] illustrated the role of industry partnerships in promoting circular economy practices within Vietnam's natural resource management.	Vietnam, Asia	Collaboration between Vietnamese research institutions and private sectors to promote circular economy practices	Enhancing industry-wide integration of sustainable waste management practices through strategic partnerships.

		digital waste management and recycling, with a focus on innovation and sustainability.				in resource management [39].	
Technological integration	Digital twin technologies	Implementation of digital twin technology in waste management systems to simulate recycling processes and optimize operational performance using real-time data from IoT sensors.	Digital twins, simulation of recycling processes [13].	Waste Management Facility Simulation: Almtireen et al. [47] discussed how digital twin technologies simulate waste sorting operations in a facility.	Middle East, Global	Waste management industries in the Middle East are adopting digital twin technologies to optimize recycling systems and monitor operational performance [47].	Improving operational decision-making and reducing waste management inefficiencies with real-time simulation.

Source: Table by the authors.

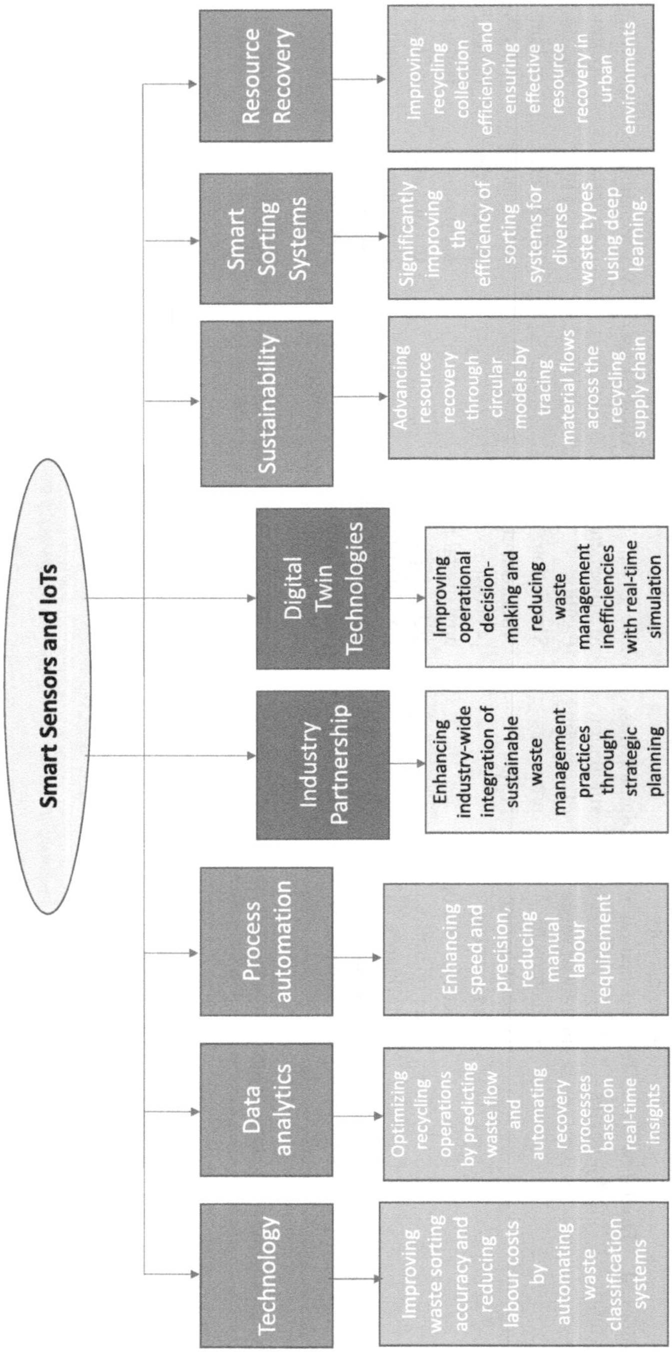

FIGURE 2.1 Smart sensors and IoT for advancing recycling efficiency and circular economy in critical mineral recovery.

classification of construction and demolition waste [49]. AI-powered robotic arms equipped with convolutional neural networks (CNNs) and reinforcement learning models can distinguish between various waste categories, thereby optimizing the material recovery process. Additionally, the development of autonomous compost turners has introduced an efficient approach to organic waste management, contributing to circular economy principles and sustainable composting practices [50]. These developments highlight the role of AI in reducing landfill dependency while maximizing resource recovery.

The application of AI and ML in the recycling of critical minerals, particularly from lithium-ion batteries and electronic waste (e-waste), has gained momentum. The DT framework for cyber-physical waste stream control systems provides a reverse logistics 4.0 approach, allowing real-time tracking and management of waste materials [51]. AI-driven robotic disassembly platforms have been deployed for dismantling plug-in hybrid electric vehicle (PHEV) batteries, enhancing material separation and recycling efficiency [52]. In the United States, automated battery sorting systems have demonstrated improved recovery rates for lithium, cobalt, and nickel, reinforcing AI's role in securing a sustainable supply of critical minerals [53].

ML algorithms also play a crucial role in predicting and optimizing waste management logistics. For instance, deep learning-based waste transportation systems have been developed to enhance cost efficiency and environmental sustainability in developing countries [54]. AI-driven decision-support tools can optimize recycling routes, minimize carbon footprints, and streamline the sorting process by utilizing big data analytics and real-time monitoring systems [55]. In Malaysia and China, digitalization strategies under Industry 5.0 paradigms have accelerated waste management efficiency through AI-powered smart city initiatives [56].

Industry interventions in AI-enhanced recycling have been transformative, with several key players leveraging ML for sustainable waste classification. In Jordan, small and medium enterprises (SMEs) have adopted AI-driven circular economy strategies, improving waste processing through conflict management and smart technology adoption [57]. Similarly, blockchain-based digital product passport systems have emerged, facilitating collaboration between recycling centers and manufacturers for automated material recovery [58]. These innovations contribute to the standardization and traceability of recycled materials, ensuring higher quality and efficiency in secondary resource markets.

Institutional research and collaboration have significantly contributed to the adoption of AI in automated sorting. A study conducted at Milan Malpensa Airport showcased the effectiveness of AI-powered waste sorting systems in large-scale urban environments, demonstrating improved efficiency in municipal solid waste management [59]. Furthermore, advancements in AI-driven reaction blueprints and logical control flow methodologies have facilitated automated chemical recycling processes, enhancing the recovery of valuable elements from industrial waste streams [60]. The integration of deep learning and IoT-based monitoring systems has also improved kitchen waste composting, exemplifying AI's role in organic waste valorization [61].

The intersection of AI, robotics, and circular economy principles has led to groundbreaking advancements in waste sorting and resource recovery. A study on robotic sorters of mixed industrial waste underscores the challenges and potential of

AI-driven automation in waste handling, paving the way for next-generation recycling systems [62]. Additionally, AI-enabled sustainability frameworks for additive manufacturing have optimized material usage, reducing waste and energy consumption in the production sector [63]. Such developments align with global efforts to transition toward a resource-efficient and environmentally sustainable recycling industry.

As AI and ML continue to drive digital transformation in waste management, future research should focus on enhancing AI's interpretability, scalability, and real-time decision-making capabilities. The synergy between AI, blockchain, and IoT will further refine recycling automation, contributing to a more circular and sustainable economy. By integrating AI-powered innovations across diverse waste streams, the recycling industry can effectively address the growing demand for critical minerals while mitigating environmental impacts.

The integration of AI and ML in automated sorting and waste management is revolutionizing the recycling industry, particularly in critical mineral recovery within the Industry 4.0 paradigm. Table 2.2 and Figure 2.2 present a cutting-edge, multidisciplinary synthesis of recent advancements in AI-driven waste sorting, DT frameworks, blockchain-enabled circular economy models, and smart robotics in recycling operations. It highlights global research contributions, from deep learning-powered waste classification [48] to digital supply chain innovations [64] and Industry 5.0-driven decarbonization strategies [56]. Key interventions include AI-enhanced robotic disassembly for end-of-life batteries [52], cost-effective smart composting systems [50], and federated learning applications in automated recycling infrastructures [58]. Additionally, advancements in IoT-enabled waste tracking [65] and deep-learning-based sustainable waste transportation [54] underscore the pivotal role of AI in optimizing waste valorization and resource efficiency. The geographical diversity of these studies—from China's digital economy-driven green development [66] to Malaysia's smart city waste digitization [56]—further showcases AI's transformative impact on global recycling ecosystems. This comprehensive analysis underscores how digital transformation, driven by AI, is enhancing sustainability, resource recovery, and circular economy integration, offering actionable pathways for policymakers, industry stakeholders, and research institutions to accelerate innovation in sustainable waste management.

2.4 ROBOTICS AND AUTONOMOUS SYSTEMS IN MATERIAL RECOVERY

The digital transformation of recycling is transforming material recovery via robots and autonomous systems (RAS). These advanced technologies improve efficiency, accuracy, and sustainability in the extraction of essential minerals, diminishing reliance on primary resources and promoting the circular economy [70]. Through the integration of modern automation, AI, and the IoT, RAS are transforming the identification, sorting, and reprocessing of materials, yielding economic and environmental advantages [71]. The shift to intelligent recycling facilities adheres to Industry 4.0 concepts, utilizing real-time data analytics to enhance operations.

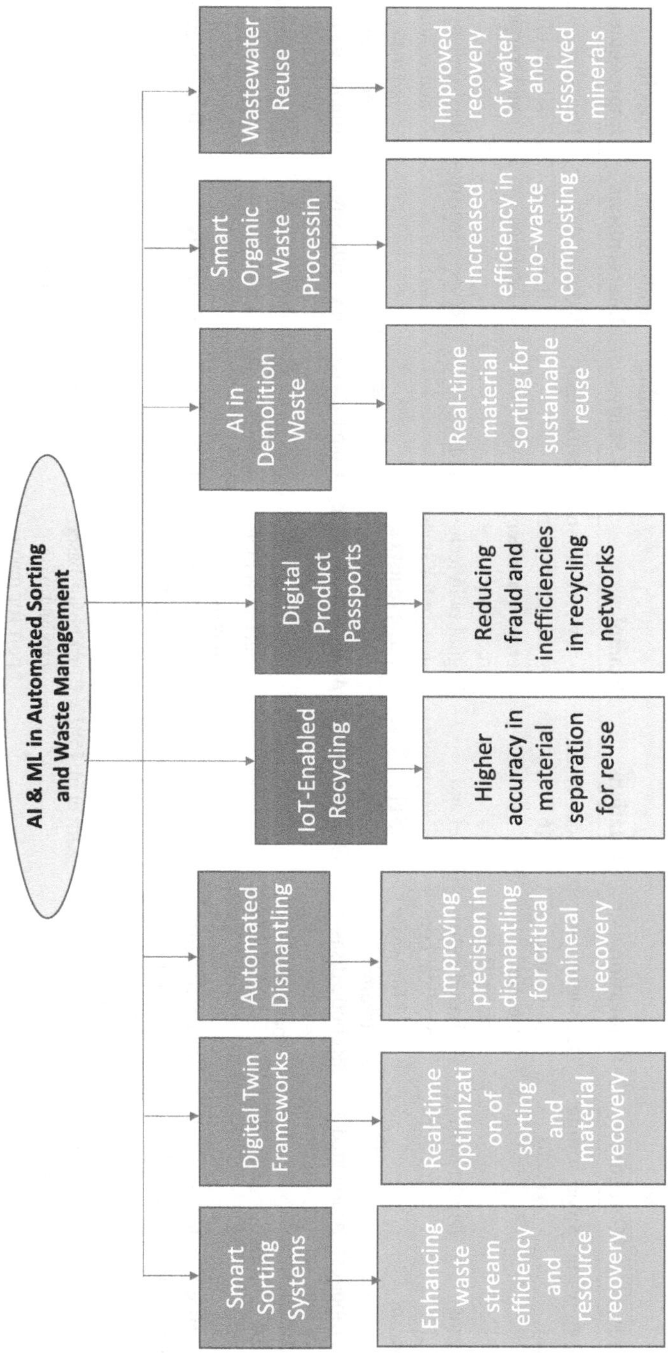

FIGURE 2.2 Artificial intelligence and machine learning in automated sorting and waste management.

TABLE 2.2
AI-Driven Automated Waste Sorting for Critical Mineral Recovery in Industry 4.0

Key Area	Category	Innovation	Citations	Examples	Location and Nationality of Study	Industry and Institutional Interventions	Specific Improvements
AI-driven waste sorting	Smart sorting systems	Deep-learning vision and robotic manipulation	Zocco et al. [67]	AI-powered robotic solutions for beverage container recycling	Switzerland	Industry 4.0 in recycling facilities	Enhancing waste stream efficiency and resource recovery
Intelligent circular economy	Digital twin frameworks	AI-based cyber-physical control for waste streams	Giel and Dąbrowska [51]	Digital twin technology for smart waste logistics	Poland	Reverse logistics 4.0	Real-time optimization of sorting and material recovery
AI and robotics in e-waste recycling	Automated dismantling	Robotic disassembly of EV batteries	Qu et al. [52]	AI-powered robotic disassembly of plug-in hybrid electric vehicle batteries	China	EV battery recycling facilities	Improving precision in dismantling for critical mineral recovery
Smart waste classification	IoT-enabled recycling	AI-based deep learning for waste segregation	Gaurav et al. [48], Sayed et al. [68]	IoT-integrated AI models for smart waste sorting	India	Smart city waste management	Higher accuracy in material separation for reuse
Blockchain for recycling transparency	Digital product passports	Federated learning for manufacturer–recycler collaboration	Kim et al. [58]	Blockchain-based product passport for material traceability	South Korea	Digital economy and circular supply chains	Reducing fraud and inefficiencies in recycling networks

Industry 4.0 in construction waste	AI in demolition waste	Computer vision-guided segmentation	Sirimewan et al. [49]	AI-enhanced waste handling in construction and demolition	Sri Lanka	Construction sector circular economy	Real-time material sorting for sustainable reuse
AI-Enabled composting	Smart organic waste processing	Autonomous compost turners and deep-learning models	Cichocki et al. [50], Hong et al. [61]	AI-powered composting for organic waste valorization	Switzerland, China	Smart agriculture and bio-waste management	Increased efficiency in bio-waste composting
AI and membrane tech in recycling	Wastewater reuse	AI-modeled advanced filtration	Cairone et al. [69]	AI-driven membrane technology for wastewater treatment	Italy	Water recycling and industrial waste management	Improved recovery of water and dissolved minerals

Source: Table by the authors.

A notable advancement in RAS is the creation of intelligent sorting systems that employ ML and computer vision to accurately identify items [72]. Li et al. [73] developed an AI-driven method for detecting fragmented non-magnetic materials in the recycling of end-of-life refrigerators, markedly enhancing recovery rates. These advances are essential in areas such as China and Germany, where electronic trash is increasing and regulatory measures are becoming more stringent to mandate sustainable recycling processes.

The automation of waste sorting represents a significant advancement. Azis et al. [74] created an automated garbage segregation system employing robotic arms and sensor-driven sorting technologies. This technique has been effectively deployed in Malaysia, diminishing manual labor and enhancing processing rates in recycling facilities. Likewise, augmented reality (AR)-facilitated human–robot collaboration (HRC) has enhanced the efficiency of construction waste sorting while maintaining occupational safety [75]. These achievements underscore the collaboration between digital technologies and robotic automation in waste recovery processes.

Industry involvement has been essential in the advancement of RAS. The EU's Horizon 2020 program has financed various robotics-based recycling projects, promoting transnational partnerships between academia and business. The RESYNTEX project combines robotic disassembly with chemical recycling to improve textile waste recovery. In the United States, firms such as Automated Machine Perception (AMP) Robotics are implementing AI-driven robotic sorters in material recovery plants, enhancing efficiency and minimizing contamination in recycled materials.

Institutions are actively enhancing technological advancements in material recovery. Morachioli et al. [76] proposed a pragmatic taxonomy for highly automated reverse manufacturing cells, facilitating the smooth incorporation of robotics into recycling operations. Moreover, scholarly investigations in Sweden and Japan concentrate on AI-based disassembly methods for lithium-ion batteries, an essential advancement for sustainable energy storage solutions [77]. These initiatives are essential for alleviating supply chain interruptions in the expanding EV sector.

Blockchain and IoT technologies are progressively integrated with RAS to improve traceability and operational transparency [78]. Blockchain-enabled VANET systems in smart waste management have optimized logistics and enhanced efficiency in nations such as Singapore and the Netherlands [79]. These digital frameworks provide automated monitoring and predictive maintenance, thereby decreasing operational expenses and improving resource recovery rates.

Notwithstanding these breakthroughs, some problems persist, such as substantial initial investment prices and the necessity for regulatory harmonization across various locations [80]. Surmounting these obstacles necessitates cooperative endeavors among governments, companies, and research institutes to establish standardized frameworks and promote digital adoption in material recovery systems. Initiatives such as the International Renewable Energy Agency's (IRENA's) policy proposals for essential mineral recycling act as templates for worldwide execution [81].

The future of robotics and autonomous systems in material recovery depends on the integration of AI, robotics, and sustainable manufacturing practices. Recent studies indicate that hybrid methodologies, integrating AI-driven robotic sorting with bioleaching techniques, may substantially enhance recovery rates for rare earth elements [82]. As these innovations advance, they will be pivotal in closing the material loop, securing a robust supply chain for essential minerals, and strengthening the tenets of a circular economy globally.

The incorporation of Robotics and Autonomous Systems (RAS) in CMR is revolutionizing the industry via Industry 4.0 technology, improving efficiency, accuracy, and sustainability. Table 2.3 and Figure 2.3 provide a thorough synthesis of contemporary research encompassing automation, AI, ML, IoT, blockchain, and AR—all crucial in transforming material recovery. Research by Lu et al. [77] utilize online sensing and AI-augmented disassembly techniques for lithium-ion battery recycling. Moreover, Chen et al. [75] illustrate how AR-facilitated HRC optimizes garbage sorting efficiency while ensuring occupational safety, thereby addressing the automation deficit in material handling. These solutions tackle issues in reverse logistics, circular economy frameworks, and supply chain resilience in accordance with sustainable production models. The worldwide scope of these developments, illustrated by case studies from Europe, Asia, Africa, and the Americas, highlights the universal relevance of digital transformation in recycling, promoting cross-regional knowledge exchange and policy alignment.

Industry and institutional efforts significantly enhance the transformative capacity of RAS in CMR. Tarrés-Puertas et al. [82] present an open-source Industrial Internet of Things (IIoT) platform designed for real-time control in bioleaching applications, improving process monitoring and circular supply chains. Akl et al. [83] present feature-driven next-view planning for robotic scrap metal cutting, enhancing resource recovery. Organizations like IRENA, the European Circular Economy Stakeholder Platform, and the Ellen MacArthur Foundation vigorously endorse intelligent waste management strategies, championing AI-facilitated material identification, blockchain-supported traceability, and IoT-driven automation. This research highlights a trend toward completely autonomous and data-driven recycling systems, enhancing resource efficiency, reducing carbon footprints, and promoting sustainable industrial symbiosis. The studies emphasize the necessity of a digitally optimized circular economy by merging advanced sensing, predictive analytics, and intelligent automation, establishing RAS as the foundation of next-generation material recovery ecosystems.

2.5 BLOCKCHAIN AND DIGITAL LEDGER TECHNOLOGIES FOR TRANSPARENCY AND TRACEABILITY

Blockchain and DLTs have arisen as pivotal instruments for enhancing transparency, traceability, and security in the important mineral recycling sector. With the rapid advancement of digital transformation in the recycling industry, especially in the context of Industry 4.0, blockchain offers a decentralized and immutable record-keeping system that bolsters trust among stakeholders. The amalgamation of blockchain technology with the IoT and AI in trash management has already exhibited substantial enhancements in efficiency and accountability [84]. This chapter examines the use of blockchain and distributed ledger technologies in the digital transformation of essential mineral recycling, emphasizing various implementation areas, institutional interventions, and creative tactics that promote sustainability.

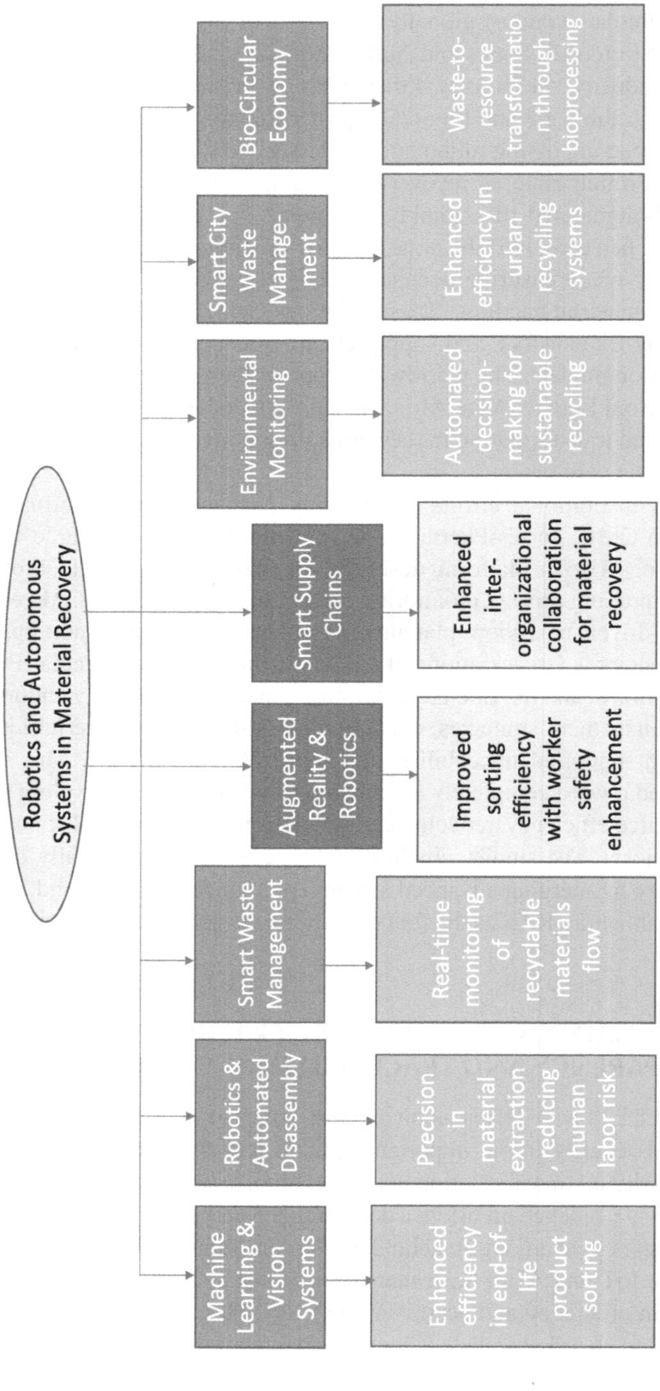

FIGURE 2.3 Robotics and autonomous systems in material recovery.

TABLE 2.3
Integrating Robotics and Autonomous Systems for Digital Transformation in Critical Mineral Recycling

Area	Category	Innovation	Building Block for Digital Transformation	Example	Location/ Nationality of Study	Industry Interventions	Institutional Contributions
AI-powered sorting	Machine learning and vision systems	Intelligent identification of fragmented non-magnetic materials	Enhanced efficiency in end-of-life product sorting	Intelligent identification of fragmented non-magnetic materials for refrigerator recycling	China	AI-integrated recycling systems by leading e-waste management firms	Research on AI-driven material recovery by academic institutions
Autonomous disassembly	Robotics and automated disassembly	Robotic disassembly systems for end-of-life lithium-ion batteries	Precision in material extraction, reducing human labor risk	Online sensing and ML for lithium-ion battery disassembly	China	Industry-led automation for safe battery recycling	University research on robotic disassembly techniques
IoT and blockchain in material tracking	Smart waste management	IoT-based waste segregation for improved circular economy	Real-time monitoring of recyclable materials flow	IoT-based waste management approach	India	Blockchain-based waste traceability systems by logistics companies	Government-backed digital waste tracking programs
Human–robot collaboration (HRC)	Augmented reality and robotics	AR-enabled human–robot collaboration for waste sorting	Improved sorting efficiency with worker safety enhancement	AR-assisted human–robot interaction for construction waste sorting	USA	Construction firms integrating AR-powered robotic sorting	Institutional studies on safety and efficiency in HRC

(*continued*)

TABLE 2.3 (Continued)
Integrating Robotics and Autonomous Systems for Digital Transformation in Critical Mineral Recycling

Area	Category	Innovation	Building Block for Digital Transformation	Example	Location/ Nationality of Study	Industry Interventions	Institutional Contributions
Industry 4.0-enabled reverse logistics	Smart supply chains	Digitalization-driven reverse logistics	Enhanced inter-organizational collaboration for material recovery	Reverse logistics impact on digitalization	Global (cross-national)	Logistics firms adopting AI-driven material routing	Government incentives for smart logistics solutions
AI in pollution reduction	Environmental monitoring	AI-driven supply chain resilience for pollution reduction	Automated decision-making for sustainable recycling	AI application to pollution reduction in waste management	Various global studies	AI-powered sustainability strategies in corporations	Research funding for AI in waste analytics
Automation in smart cities	Smart city waste management	Blockchain-enabled smart solid waste management	Enhanced efficiency in urban recycling systems	Blockchain in smart solid waste management	Various global studies	Industry-led blockchain solutions for urban recycling	Smart city initiatives integrating blockchain technologies
Microalgae-based recovery	Bio-circular economy	Microalgae-based circular supply chains for pharmaceuticals	Waste-to-resource transformation through bioprocessing	Microalgae-based supply chains using Industry 4.0	Global	Pharmaceutical firms utilizing microalgae for resource recovery	University-led R&D in bioprocessing technologies

Source: Table by the authors.

Blockchain significantly enhances end-to-end supply chain traceability in recycling. The digital ledger system guarantees that every phase of the recycling process, from collection to ultimate reintegration into the supply chain, is documented in a tamper-resistant way [66]. The ethical sourcing of raw materials and adherence to environmental norms are crucial in lithium-ion battery recycling [3]. Through the application of blockchain technology, organizations may authenticate the origin of recycled materials, so ensuring compliance with circular economy principles [43]. Projects in Vietnam have effectively integrated circular economy principles with financial sustainability strategies via blockchain-based resource management [39].

Smart contracts, a fundamental aspect of blockchain technology, facilitate automated and trustless transactions within the recycling sector. Self-executing contracts facilitate agreements among stakeholders, minimizing administrative burdens and improving accountability [13]. In automated sorting and garbage categorization, blockchain-enabled smart contracts provide instantaneous payments contingent upon material verification, enhancing the economic feasibility of recycling operations [8]. The Indian automobile sector has commenced utilizing Industry 4.0 technology, such as blockchain-based quality assurance systems, to emphasize environmentally sustainable smart manufacturing inside circular supply chains [27].

Security concerns in digital transformation require the inclusion of blockchain technology. Amid escalating apprehensions regarding cyber dangers in cloud-based industrial automation, blockchain fortifies data integrity and alleviates security hazards [40]. The cryptographic structure of blockchain safeguards sensitive information related to recyclable materials, ownership records, and compliance data against illegal alterations. Organizations like the IRENA promote the implementation of blockchain technology in the management of renewable resources to enhance energy recovery from recycled materials [85].

Blockchain technology plays a significant role in the automation of disassembly and robotics within waste management. The utilization of blockchain technology in the robotic disassembly of screws and electronic waste components improves efficiency and traceability in the handling of products at the end of their lifecycle [23]. Recent innovations in AI-based building waste sorting incorporate blockchain technology for real-time monitoring, guaranteeing adherence to environmental laws [10]. A study conducted in Switzerland implemented contactless multi-modal sensing techniques integrated with blockchain to improve the precision of material assessment [22]. These improvements are essential for attaining sustainable deconstruction and recovery in the recycling of building materials.

The integration of non-fungible tokens (NFTs) into blockchain systems represents a novel advancement in trash recycling. A study that aimed to maximize mobile phone recycling suggested using a blockchain system with NFT incentives to improve customer engagement in e-waste return initiatives [66]. This method gamifies garbage recycling while offering verifiable evidence of recycling initiatives. Likewise, the integration of ML-driven predictive analytics with blockchain technology augments digital competencies in enhancing corporate performance, as evidenced by industrial applications in China [86].

Institutional interventions are essential for the use of blockchain in recycling transparency. Governments and regulatory agencies in Europe and North America are establishing blockchain-based compliance frameworks to monitor hazardous waste and guarantee appropriate disposal. The EU's Digital Product Passport effort utilizes blockchain technology to furnish consumers with comprehensive lifecycle information on items, hence promoting a more sustainable recycling culture [14]. Similarly, Brazilian food technology enterprises have adopted Industry 4.0 and blockchain to improve stakeholder involvement in circular economy projects [87].

Ultimately, blockchain and DLTs are transforming transparency and traceability in essential mineral recycling. Blockchain enhances trash management through the integration of AI, IoT, and smart contracts, promoting efficiency, security, and a circular model. Current research and business initiatives illustrate its capacity to transform recycling economies worldwide, connecting technical advancement with sustainability. The implementation of blockchain in crucial mineral recycling is essential for the transition to a digitally transformed, resource-efficient future.

Table 2.4 and Figure 2.4 provide an extensive overview of recent research at the convergence of Industry 4.0 technologies, circular economy principles, and digital transformation in the recycling industry, specifically emphasizing essential mineral recycling. It utilizes a variety of research produced between 2024 and 2025, highlighting innovative developments in technologies such as blockchain, AI, the IoT, and robotics. These studies investigate the capacity of digital technologies to improve the transparency, traceability, and efficiency of recycling operations, especially with complex materials such as lithium-ion batteries, plastics, and electronic trash. The amalgamation of these technologies demonstrably enhances waste sorting, material recovery, and resource management, providing a sustainable trajectory for crucial mineral recycling—a pivotal focus for attaining circular economy objectives amid increasing resource scarcity.

A significant characteristic of the table is its varied geographical and industrial range, incorporating case studies from multiple nations and sectors. The research encompasses inputs from regions including India, Vietnam, China, and Brazil, underscoring the worldwide implementation of digital transformation in recycling. The incorporation of AI in garbage-sorting facilities in Vietnam and the utilization of IoT for automated waste management in Brazilian food technology enterprises exhibit practical applications of these technologies. Furthermore, research focusing on technology preparedness, economic feasibility, and organizational challenges enhances these investigations. Institutions and companies in these locations are crucial in advancing innovations and providing essential insights for policy development, stakeholder involvement, and technological integration.

The research presented in Table 2.4 highlights the revolutionary capacity of blockchain and DLTs, in conjunction with IoT, AI, and robotics, in advancing the digitization of recycling processes. These technologies enable a more visible and traceable supply chain, ensuring the sustainable management of essential minerals and resources. The emphasis on industrial and institutional actions highlights the necessity for cross-sector collaboration and the strategic alignment of digital technologies with circular economy concepts. This compilation of examples demonstrates the synergies between digital transformation and recycling, offering a comprehensive approach to sustainable waste management and resource recovery. These technologies signify a vital advancement in achieving global sustainability objectives while

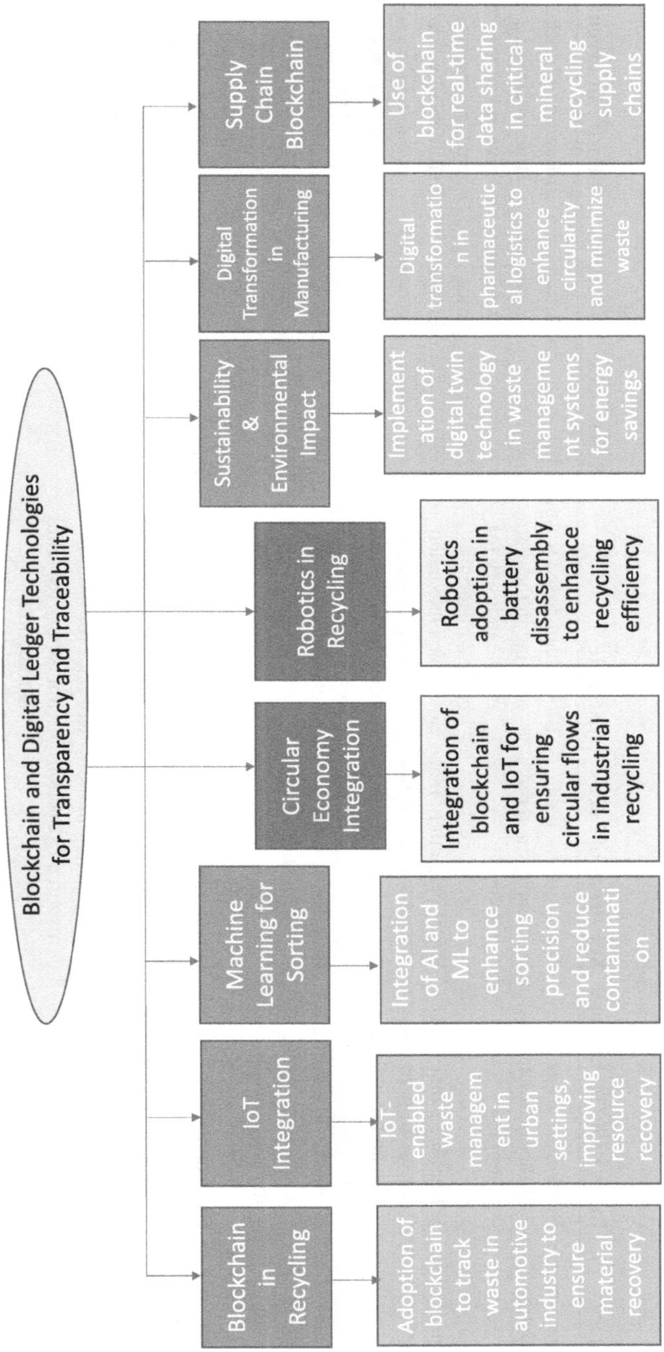

FIGURE 2.4 Blockchain and digital ledger technologies for transparency and traceability.

TABLE 2.4
Leveraging Blockchain, IoT, and AI for Transparency, Traceability, and Efficiency in Critical Mineral Recycling

Category	Innovation/Technology	Key Areas of Focus	Example	References	Location and Nationality	Industry Intervention
Blockchain in recycling	Blockchain for traceability of waste materials	Transparency, data integrity, material tracking	Blockchain-based waste tracking in the automotive industry for critical materials	[13]	Global/multiple nationalities	Adoption of blockchain to track waste in automotive industry to ensure material recovery
IoT integration	IoT-enabled smart bins for waste segregation and recycling efficiency	Waste collection, real-time data collection	Smart bins integrated with IoT in smart cities for enhanced waste recovery	[17]	Multiple countries	IoT-enabled waste management in urban settings, improving resource recovery
Machine learning for sorting	AI-based systems for automated sorting and material classification	Efficiency, waste stream sorting, predictive analytics	AI-enhanced vision systems for waste sorting in the construction industry	[10]	Global/various locations	Integration of AI and ML to enhance sorting precision and reduce contamination
Circular economy integration	Blockchain + IoT for supply chain transparency and circular economy practices	Circularity, resource recovery, recycling traceability	Circular economy models combining blockchain for traceable recycling flows	[33]	Global/multiple nationalities	Integration of blockchain and IoT for ensuring circular flows in industrial recycling
Robotics in recycling	Robotic disassembly for recycling critical minerals such as EV batteries	Precision, automation, resource extraction	Robotics for lithium-ion battery recycling automation	[42]	Global/multiple nationalities	Robotics adoption in battery disassembly to enhance recycling efficiency

Sustainability and environmental impact	Digital twins for sustainability in recycling processes	Environmental monitoring, energy optimization, carbon footprint tracking	Use of digital twins to optimize energy use in recycling facilities	[14]	Europe/ Germany	Implementation of digital twin technology in waste management systems for energy savings
Digital transformation in manufacturing	Advanced analytics and digital transformation for recycling facility optimization	Process efficiency, operational cost reduction, waste minimization	Digital transformation of pharmaceutical reverse logistics with blockchain support	[44]	Middle East/ Iran	Digital transformation in pharmaceutical logistics to enhance circularity and minimize waste
Supply chain blockchain	Blockchain for improving transparency in closed-loop supply chains for recycling and waste management	Traceability, transparency, accountability	Blockchain solutions for closed-loop supply chains in battery recycling	[43]	Global/ various locations	Use of blockchain for real-time data sharing in critical mineral recycling supply chains

Source: Table by the authors.

responding to the increasing demand for essential minerals in burgeoning sectors such as renewable energy, EVs, and advanced manufacturing.

2.6 BIG DATA AND PREDICTIVE ANALYTICS FOR OPTIMIZING RECYCLING OPERATIONS

This section examines the revolutionary capacity of modern data-driven technology in improving recycling operations, particularly with essential mineral recycling. It explores the transformative impact of big data, predictive analytics, and ML on waste sorting, resource recovery, and supply chain efficiency within the recycling sector. By utilizing the extensive data generated by IoT sensors, robotics, and automated systems, organizations can improve decision-making accuracy and timeliness while streamlining processes like sorting, disassembly, and material recovery. As essential minerals like lithium and cobalt gain significance for technologies such as EVs and renewable energy systems, the incorporation of predictive models that anticipate waste streams and forecast material degradation is vital for enhancing recycling efficiency and attaining circular economy objectives.

Recent research has shown the effectiveness of these technologies in practical applications. The study by Islam et al. [8] presents a novel deep learning technique named ECCDN-Net, enhancing the classification of organic and recyclable garbage. The study's findings demonstrate the application of ML to improve the precision of waste sorting, a vital process in mineral recycling. In Vietnam, a government-led program to integrate circular economy practices with financial sustainability demonstrates the application of big data analytics in natural resource management [39]. The nation's implementation of IoT-enabled smart bins and waste management systems has enhanced resource recovery, decreasing landfill trash volumes while increasing material recovery, thereby directly benefiting the mineral recycling industry by improving the traceability of precious metals.

Moreover, predictive analytics can be utilized to enhance the disassembly and recycling processes of electronic devices, such as lithium-ion batteries. Choux et al. [3] examine the techno-economic dimensions of automated disassembly compared to shredding for battery recycling, demonstrating that predictive maintenance, and data-informed decision-making can enhance the efficiency of disassembly robots. This is especially significant in the realm of EV battery recycling, because disassembly necessitates accuracy to retrieve valuable minerals such as lithium and cobalt. The capacity to forecast and avert mechanical problems in robotics via ML algorithms guarantees a consistent and seamless process flow, minimizing operational expenses and enhancing recovery rates. These developments are both technological and indicative of a paradigm shift in the recycling industry's perception of operational efficiency.

Worldwide, numerous organizations and sectors are enhancing the integration of predictive analytics inside recycling systems. The automobile sector in China has been integrating Industry 4.0 technology to improve supply chain sustainability, emphasizing closed-loop recycling methods for lithium-ion batteries [27]. ML algorithms predict the degradation of batteries, offering essential information for optimizing the timing of recycling activities to enhance recovery efficiency. This program highlights the need for predictive modeling in synchronizing recycling processes with sustainability objectives, guaranteeing that essential key minerals are recovered before they deteriorate beyond economic viability for reuse. In Brazil, food technology firms are employing predictive analytics alongside IoT systems to enhance waste management procedures, demonstrating the cross-sector applicability of these technologies in circular economy practices [87].

Industry stakeholders have recognized the potential of big data in enhancing product lifecycles. The adoption of blockchain technology for monitoring materials in a circular supply chain is increasing. Blockchain provides a transparent and irreversible framework for authenticating the origin of recycled materials, guaranteeing that essential minerals such as rare earth metals are preserved during recycling processes. The integration of blockchain and predictive analytics is illustrated by the study conducted by Masa'deh et al. [13], in which the authors evaluate the impact of blockchain on supply chain digitalization, enhancing the efficiency of the recycling process by mitigating fraud and ensuring accurate categorization and processing of materials.

The integration of robotics and big data analytics is expediting the recovery of essential minerals. The research conducted by Farshadfar et al. [10] examines the application of ML-driven automated sorting systems in the recycling of key minerals within construction waste management. These systems employ picture recognition and data analysis to categorize materials more efficiently than traditional methods, guaranteeing a greater purity of recyclable metals, including those present in electronic waste. The incorporation of AI-driven robots in these processes enhances sorting efficiency and decreases labor expenses, rendering recycling operations more economically viable.

The future incorporation of predictive analytics in recycling will concentrate on forecasting waste streams and enhancing sorting facilities. The study by Pandey et al. [40] on cloud-based IoT solutions for industrial automation emphasizes that these systems provide real-time monitoring and predictive maintenance, hence ensuring optimal efficiency in recycling facilities. These developments are essential for adjusting to the fluctuating volumes of electronic waste that the rapid pace of technological innovation produces. Moreover, as recycling costs pose a substantial obstacle, advancements in data-driven decision-making will facilitate the reduction of operational expenses by forecasting market trends, waste generation rates, and material recovery potentials, thereby aligning recycling operations with the market demand for essential minerals.

This section provides a prospective analysis of how modern data analytics are revolutionizing the recycling sector. The incorporation of ML, predictive maintenance and IoT into recycling systems establishes a strong foundation for improving the efficiency, traceability, and profitability of recycling operations. This chapter highlights the imperative of data-driven technologies for sustainable recycling and critical mineral recovery, utilizing real-world examples from diverse countries and industries and providing a framework for stakeholders to enhance recycling practices and progress toward a more sustainable, circular economy. As companies increasingly embrace digital transformation technology, the future of key mineral recycling appears progressively efficient, transparent, and economically sustainable.

The incorporation of Big Data, predictive analytics, and Industry 4.0 technology in recycling processes signifies a significant advancement in improving resource recovery, minimizing environmental effects, and facilitating the transition to a circular economy. Table 2.5 and Figure 2.5 underscore advanced advancements, like AI-driven garbage sorting, robotic disassembly, and blockchain for transparent material tracking, all of which are crucial in transforming recycling operations. Utilizing ML models enhances the accuracy of waste stream forecasts, facilitating more effective sorting and recycling of materials. Moreover, IoT-connected smart bins and real-time data gathering systems have enabled municipalities and enterprises to improve trash management, optimize collection routes, and assess material quality, hence boosting the efficiency of recycling systems. These developments are crucial for enhancing waste management operations and fostering global sustainability.

Industry leaders are progressively implementing digital transformation solutions to optimize recycling processes. Robotic systems that incorporate deep learning and reinforcement learning approaches may now execute complex tasks, such as deconstructing electronic waste with accuracy, hence decreasing labor costs and processing time. Furthermore, predictive analytics driven by big data facilitates the forecasting of waste generation trends, equipping businesses and municipalities with actionable insights that promote proactive measures for optimizing waste management. Companies such as Veolia, in conjunction with technical leaders like International Business Machines (IBM), have made considerable advancements in integrating AI and big data technologies into trash management, markedly enhancing the efficiency and scalability of recycling operations. Moreover, employing DTs to model recycling processes and utilizing hyperspectral imaging for material identification presents unparalleled prospects for improving material recovery, minimizing contamination, and optimizing operations at recycling plants.

The research has demonstrated that the extensive implementation of innovative technology is promoting sustainability and circular economy concepts. The importance of blockchain in guaranteeing transparency in the tracking and recycling of materials is significant, since it ensures the integrity of the material lifecycle from collection to recovery. Green manufacturing techniques, AI-driven sorting systems, and hyperspectral sensing are advancing garbage management and recycling. This unique integration of digital tools enhances material recovery and substantially decreases environmental harm by reducing energy consumption linked to waste management processes. These innovations are poised to transform recycling operations, aiding global sustainability objectives and providing scalable solutions for both developed and developing nations.

2.7 DIGITAL TWIN TECHNOLOGY FOR SIMULATING AND ENHANCING RECYCLING SYSTEMS

The digital transformation within the recycling sector, especially in crucial mineral recycling, depends on the use of Industry 4.0 technology. DT technology is one of the most promising instruments in this paradigm. A DT is a virtual representation of a physical system that facilitates real-time monitoring, modeling, and process improvement. In CMR, DT can be employed to model the complete lifecycle of mineral

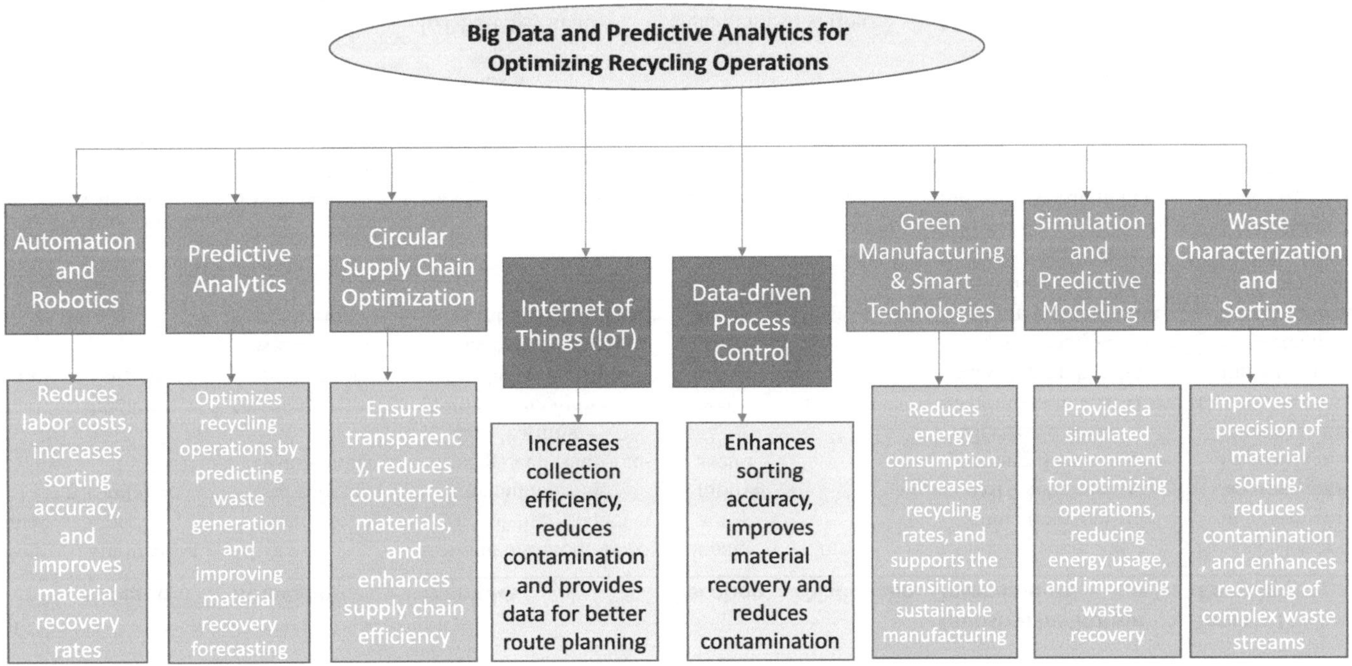

FIGURE 2.5 Big data and predictive analytics for optimizing recycling operations. (Figure by the authors.)

TABLE 2.5
Leveraging Big Data, Predictive Analytics, and Industry 4.0 Technologies for Optimizing Recycling Operations and Advancing Circular Economy Practices

Area of Focus	Category	Innovation	Examples	Location	Nationality	Industry/Institution Interventions	Impact for Recycling
Recycling automation	Automation and robotics	AI-driven waste sorting and robotic disassembly	Robotic disassembly for end-of-life product remanufacturing using deep reinforcement learning	Global, automotive sector	China	Robotics technology from companies like ABB Robotics automating the sorting of electronic waste.	Reduces labor costs, increases sorting accuracy, and improves material recovery rates.
Big data analytics	Predictive analytics	Machine learning models for waste stream prediction	Predictive analytics for digital capabilities improvement in business performance	Global	Iran	Institutions like IBM are developing big data tools to analyze waste stream behavior and optimize recycling.	Optimizes recycling operations by predicting waste generation and improving material recovery forecasting
Circular economy integration	Circular supply chain optimization	Blockchain for tracking material flows in circular economies	Blockchain effect on courier supply chains digitalization in circular economy contexts	Middle East	Jordan	Industry: Waste management companies using blockchain to verify material provenance and traceability in recycling.	Ensures transparency, reduces counterfeit materials, and enhances supply chain efficiency.
IoT integration	Internet of Things (IoT)	IoT-enabled smart bins for real-time waste collection and monitoring	IoT-based system architecture for municipality waste management in smart cities	Smart cities globally	Various	Municipalities (e.g., Singapore) implementing smart waste bins to collect data and optimize recycling logistics.	Increases collection efficiency, reduces contamination, and provides data for better route planning

Data-driven Process Control	Real-time data collection	Sensors and multi-modal data for material assessment	Contactless multi-modal sensing for material recovery in building deconstruction	Switzerland	Switzerland	Companies like Veolia using sensors for real-time monitoring of waste sorting in recycling plants.	Enhances sorting accuracy, improves material recovery and reduces contamination
Sustainable manufacturing	Green manufacturing and smart technologies	Machine vision and AI for automated waste sorting and quality control	Interval-valued Pythagorean fuzzy approach for Industry 4.0 technology assessment	India	India	Industry: Tetra Pak implementing AI for waste material sorting to enhance recycling rates.	Reduces energy consumption, increases recycling rates, and supports the transition to sustainable manufacturing
Digital twin technologies	Simulation and predictive modeling	Digital twin for modeling and optimizing recycling processes	Digital twin modeling for automated nuclear waste classification	Nuclear sector	USA	Institutions like Siemens using digital twins to optimize the recycling of nuclear materials.	Provides a simulated environment for optimizing operations, reducing energy usage, and improving waste recovery
Waste stream analytics	Waste characterization and sorting	Hyperspectral imaging for identifying recyclable materials	Autonomous hyperspectral characterization of polymer degradation	Global	Various	Companies like TOMRA using hyperspectral imaging to enhance recycling efficiency in material recovery plants.	Improves the precision of material sorting, reduces contamination, and enhances recycling of complex waste streams

Source: Table by the authors.

recycling systems, encompassing the extraction of raw materials, the recycling process, and the recovery, purification, and repurposing of essential minerals such as lithium, cobalt, and rare earth elements. Through the integration of sensors and data analytics, DT enables ongoing monitoring and regulation, enhancing recycling efficiency, minimizing environmental impacts, and advancing sustainability. The advancing application of DT technology signifies a convergence of innovation, industrial strategy, and environmental policy that has the potential to transform worldwide essential mineral recycling systems.

In the realm of crucial mineral recycling, DT technology provides a substantial benefit regarding process optimization. Real-time modeling and predictive analytics facilitate the comprehension of mineral recovery operations, specifically regarding the degradation or alteration of minerals during processing. In the context of lithium-ion battery recycling, a DT model could replicate machinery degradation, forecast maintenance requirements, and enhance operational sequencing to elevate recovery rates. This predictive capability is essential for improving the efficiency of recycling facilities, allowing operators to foresee potential malfunctions prior to their occurrence. Research conducted by Sirimewan et al. [49] demonstrates that AI-powered systems can optimize waste management, which may be directly relevant to digital transformation in mineral recycling by facilitating automated, real-time process modifications.

The convergence of AI, ML, and IoT technologies facilitates the use of DT technology for crucial mineral recycling on a global scale. AI models integrated into the DT system can improve decision-making by analyzing extensive data from sensors and historical records. Weigl and Young [53] demonstrated the use of AI-driven robotic sorting technologies for lithium-ion battery recycling in China, and it is possible to improve this technology with DTs to create a dynamic, adaptive system that changes as users use it. These systems can analyze historical inefficiencies, adapt to operational variability, and forecast optimal routes for mineral recovery. The partnership between the Chinese Ministry of Industry and Information Technology and local enterprises illustrates how governmental and industrial collaborations can promote progress in DT applications for resource recovery.

Moreover, DT technology can significantly contribute to the incorporation of circular economy ideas inside essential mineral recycling systems. By simulating the lifecycle of recovered minerals and monitoring them throughout supply chains, DT systems can pinpoint opportunities for waste reduction and energy optimization. An exemplary instance of this integration is presented in the study of Grudzien et al. [88], which examines the role of Industry 4.0 technologies in enhancing circular business models within the manufacturing sector. The United Nations Sustainable Development Goals (SDGs) and other global sustainability initiatives align with the capacity of DT models to monitor material flows, ensuring the efficient reclamation and reutilization of valuable minerals. This implementation of digital technology improves openness, accountability, and traceability in mineral recycling, rendering it an effective instrument for attaining circular economy goals.

A critical domain where DT technology may facilitate transformative change is in the optimization of energy usage and the reduction of carbon emissions. The recycling of important minerals is energy-intensive, and enhancing energy efficiency in these operations is essential for reducing the environmental impact. Integrating digital technologies with intelligent energy management systems can enhance real-time energy utilization, hence decreasing operational expenses and carbon emissions. Innovations in this field, such as those by Goh et al. [56], which concentrate on energy optimization in smart cities, can be used in mineral recycling facilities. In these environments, the digital representation of the facility, overseen by IoT sensors, facilitates the optimization of energy inputs to reduce waste, potentially realizing energy savings of up to 30%, as evidenced by recent case studies in Malaysia and China.

Collaborations within the industry are crucial for advancing the adoption of DT technology in important mineral recycling systems. Prominent global industrial leaders in the recycling business, including Umicore, are actively investigating digital transformation to improve the efficiency of their recycling facilities. These companies are collaborating with academic institutions and technology vendors to enhance the DT models that replicate the mineral recycling process. The incorporation of blockchain technology, as articulated by Kim et al. [58], may augment the system by establishing a secure, transparent framework for monitoring mineral origin and verifying that recycled materials adhere to stringent environmental criteria. The significance of institutions in propelling technological progress via funding and research collaborations is paramount in this collaborative environment.

The future of DT technology in crucial mineral recycling encompasses the progression of automated systems, such as robotic sorting and disassembly. Integrating digital technology with robotic arms enables systems to autonomously disassemble gadgets, categorize materials, and extract minerals, thereby eliminating human error and decreasing labor expenses. Qu et al. [52] demonstrated the integration of robotics and digital technology through their robotic disassembly platform for hybrid electric vehicle (HEV) batteries, indicating the growing momentum of this trend in the automotive sector. These technologies are anticipated to transform the efficient extraction and recycling of minerals, especially those in intricate devices, markedly enhancing recovery rates and minimizing waste.

Ultimately, DT technology can promote global collaboration and policy formulation to establish a more sustainable CMR environment. Governments and international entities, including the EU, are acknowledging the capacity of digital technology to promote a more circular economy. The EU's Circular Economy Action Plan highlights the significance of digital technologies in enhancing resource efficiency, with DT technology potentially serving as a fundamental element in realizing these goals. International cooperation and the exchange of best practices could standardize DT systems, establishing worldwide rules for mineral recycling and ensuring the environmentally and economically efficient recycling of essential minerals.

The incorporation of DT technology in important mineral recycling systems offers a transformative opportunity to enhance the sustainability and efficiency of the recycling sector. By simulating and refining every facet of the recycling process, DT can diminish operational inefficiencies, optimize resource utilization, and advance global circular economy objectives. Continuous advancements in AI, robotics, and IoT, along with robust cooperation among industry leaders, academic researchers, and politicians, will render DT technology essential in transforming mineral recycling and promoting sustainable practices worldwide.

Table 2.6 and Figure 2.6 present a detailed and multifaceted analysis of current progress in digital technologies utilized in recycling and circular economy systems, particularly emphasizing important mineral recycling. Table 2.6 and Figure 2.6 summarize much research that utilizes Industry 4.0 technologies, including AI, the IoT, blockchain, and robotics, to improve waste management, refine recycling processes, and advance sustainability. The research by Sirimewan et al. [49] uses interactive AI and computer vision for the segmentation of construction and demolition waste, illustrative of the growing trend of incorporating AI to improve sorting accuracy and waste management efficacy. Zhang et al. [64] conduct a bibliometric analysis of digital supply chain management, emphasizing the significance of data-driven methodologies for enhancing transparency and efficiency in material recovery procedures. The convergence of digitalization and waste management offers vital insights for enhancing CMR systems, crucial for the transition to a circular economy.

Numerous studies in Table 2.6 highlight the significance of incorporating advanced technologies to optimize recycling processes and enhance resource recovery. Giel and Dąbrowska [51] propose a DT framework for cyber-physical waste stream control systems, facilitating real-time monitoring and optimization of recycling operations. This technique has considerable potential to improve efficiency in essential mineral recycling by delivering a realistic virtual model of physical systems, facilitating predictive maintenance and optimization. Cheng et al.'s [89] investigation into robotics and AI in automated sorting systems serves as an example of how these technologies might increase material recovery rates and reduce contamination in recovered minerals. The use of these advances in recycling plants can markedly decrease energy usage and enhance material yields, tackling critical issues in the recycling sector.

Furthermore, Table 2.6 highlights the significance of collaboration strategies across academia, industry, and government in facilitating the digital transformation of key mineral recycling. Industry leaders' interventions, shown by Kim et al. [58] utilizing blockchain technology in waste management systems, underscore the potential of decentralized and secure frameworks to monitor the lifecycle of recovered minerals, ensuring their quality and authenticity. Collaborative initiatives are evident in studies such as Kurniawan et al. [90], which examine the problems and prospects of Industry 4.0 in enhancing waste recycling businesses in Indonesia. Cross-sector collaborations are crucial for overcoming structural obstacles to the expansion of digital solutions in recycling. The examples from these studies enhance the worldwide discourse on how digital transformation may facilitate the circular economy, especially within the mining and recycling industries, thus promoting sustainability objectives and maximizing the utilization of important minerals in emerging technologies.

2.8 CHALLENGES, OPPORTUNITIES, AND FUTURE TRENDS IN DIGITAL TRANSFORMATION FOR RECYCLING

The digital transformation of key mineral recycling is becoming a crucial method to meet the increasing demand for rare earth elements and other vital materials utilized in clean energy systems. The emergence of Industry 4.0, incorporating AI, blockchain, the IoT, and ML, is markedly improving the efficiency, sustainability, and economic feasibility of recycling processes. Nonetheless, the transformation presents

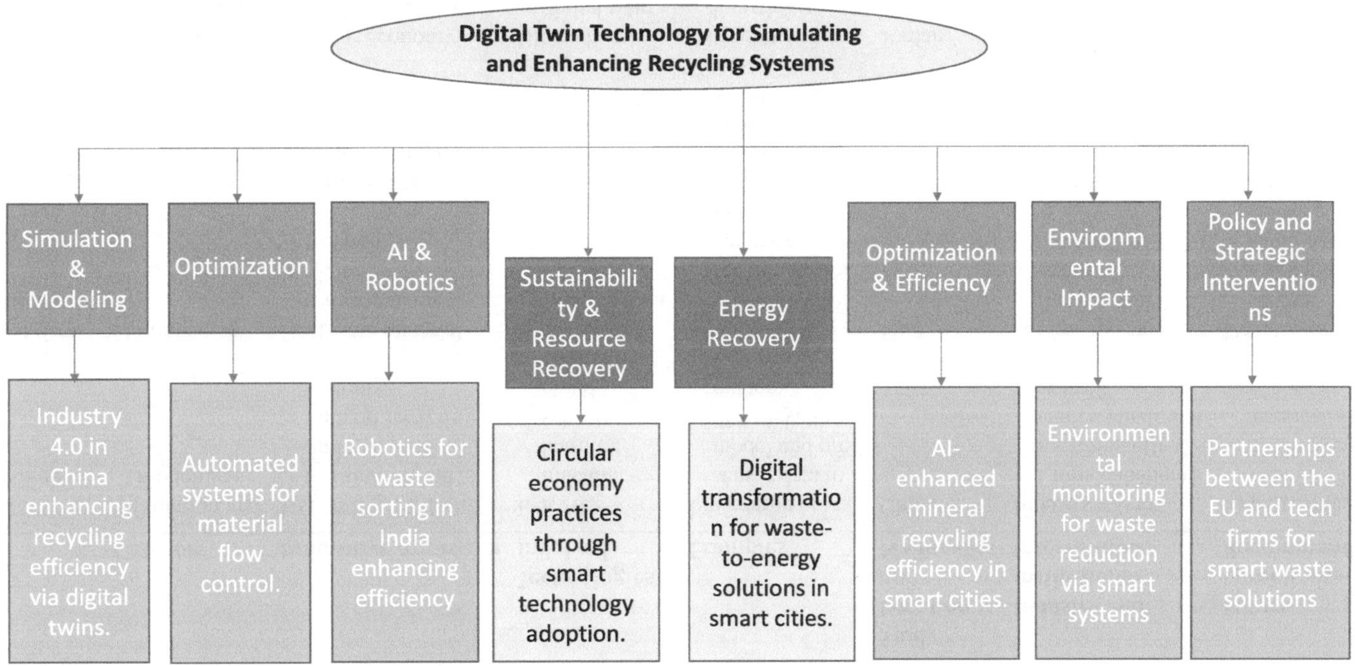

FIGURE 2.6 Digital twin technology for simulating and enhancing recycling systems. (Figure by the authors.)

TABLE 2.6
Advancing Critical Mineral Recycling Systems through Digital Twin Technology: A Global Perspective on Industry 4.0 Innovations and Circular Economy Integration

Area	Category	Innovative Approach	Technologies Involved	Examples	Study Location and Nationality	Industry and Institutional Interventions	Future Directions and Recommendations
Digital twin in recycling	Simulation and modeling	Use of Digital Twin to simulate and optimize critical mineral recycling systems.	IoT, AI, machine learning, sensors, cloud computing	Integration of sensor data to model and predict recycling outcomes.	China	Industry 4.0 in China enhancing recycling efficiency via digital twins.	Explore integration with AI-driven predictive maintenance systems.
Material flow management	Optimization	Real-time material tracking using digital twin technology.	Blockchain, IoT, big data analytics	Material flow optimization for lithium-ion battery recycling in the USA	USA	Automated systems for material flow control.	Develop automated sorting mechanisms using deep learning.
Waste sorting automation	AI and robotics	AI-powered robotic systems integrated with digital twin technology for sorting.	AI, robotics, computer vision, digital twin	Robotic arms and AI sorting construction and demolition waste	India	Robotics for waste sorting in India enhancing efficiency.	Future research on improving AI accuracy in sorting complex materials.
Circular economy	Sustainability and resource recovery	Circular economy model based on digital twin to maximize material recovery.	Blockchain, digital twin, circular economy principles	AI-based reverse logistics in Jordan for circular economy	Jordan	Circular economy practices through smart technology adoption.	Develop closed-loop systems for enhanced recycling and reuse.

Waste-to-energy systems	Energy recovery	Simulating and optimizing energy recovery from waste using digital twin.	AI, digital twin, energy systems modeling	AI-driven waste-to-energy optimization in Malaysia	Malaysia, China	Digital transformation for waste-to-energy solutions in smart cities.	Scale-up solutions for integrating waste-to-energy systems in urban areas.
AI in mineral recycling	Optimization and efficiency	Leveraging AI to predict material demand and optimize mineral recycling processes.	AI, deep learning, machine learning, Digital Twin	Smart and sustainable urban manufacturing	Global	AI-enhanced mineral recycling efficiency in smart cities.	Focus on integrating AI for better mineral yield prediction.
Sustainability monitoring	Environmental impact	Monitoring and reducing environmental impact using digital twin-based simulations.	IoT, AI, digital twin, sensors, big data	Smart waste management systems with IoT integration	Europe	Environmental monitoring for waste reduction via smart systems.	Develop next-gen sustainable models using digital twin for zero waste.
Global collaboration	Policy and strategic interventions	Collaborative efforts between governments, institutions, and industries for digital transformation in recycling.	Digital twin, blockchain, AI	International initiatives for circular economy	EU	Partnerships between the EU and tech firms for smart waste solutions	Strengthen global digital collaboration for a circular economy.

Source: Table by the authors.

hurdles, such as technology restrictions, infrastructure deficiencies, and legislative restraints. This sub-chapter examines the difficulties, identifies emerging opportunities, and delineates future trends that will shape digital transformation in essential mineral recycling.

A principal problem in the digitalization of important mineral recycling is the intricacy of waste streams, especially electronic trash (e-waste) and end-of-life batteries. Conventional recycling techniques frequently encounter difficulties in accurate material identification and segregation. Recent research, including that of Weigl and Young [53] in the United States, underscores the significance of automated battery sorting in enhancing mineral recovery efficiency. Giel and Dąbrowska [51] suggest a DT framework for the control of cyber-physical waste streams, presenting a potential solution for real-time monitoring and optimization in recycling facilities. Nevertheless, extensive implementation is obstructed by elevated expenses and the necessity for advanced infrastructure.

There are clear opportunities for enhancement in the incorporation of AI and robotics in trash management. Research by Liu et al. [91] illustrates that robotic arms and AI-driven automation may generate innovative designs from scrap materials, markedly improving material efficiency. Qu et al. [52] investigated robotic disassembly platforms for HEV battery recycling in China, emphasizing the capacity of automated processes to enhance material recovery rates. Furthermore, the research conducted by Cheng et al. [89] on AI-enhanced robotic systems for beverage container recycling illustrates the scalability of AI-driven methodologies across diverse recycling industries.

Blockchain technology is a vital facilitator of the digital revolution in mineral recycling. Kim et al. [58] advocated for the adoption of blockchain-based digital product passports to enhance transparency and traceability in recycling supply chains. This is especially pertinent for the collaboration between recycling facilities and manufacturers, guaranteeing adherence to environmental regulations and advancing circular economy concepts. A case study in Jordan by Almasri and Ying [57] highlights the potential of smart technology implementation in small and medium companies (SMEs) to improve material tracking and recovery.

Innovations based on IoT are enhancing efficiency in recycling systems. The IoT-enabled smart waste classification model created by Gaurav et al. [48] utilizes deep learning for efficient waste sorting, exhibiting enhanced precision in recognizing valuable materials. Wilhelm et al. [92] combined ML with X-ray diffraction analysis to improve the recycling of mineral construction materials, a technique that might be applied to important mineral recovery. Goh et al. [56] investigated digitalization initiatives in sustainable smart cities in Malaysia and China, demonstrating the capacity of IoT to enhance resource flows within urban recycling systems.

Regulatory and regulatory measures are essential in promoting digital transformation. Research by Du et al. [66] in China evaluates the influence of the digital economy on carbon emission efficiency, highlighting the significance of legislative frameworks in facilitating sustainable industrial practices. Research by Mukherjee et al. [93] on AI-driven reverse logistics underscores the significance of legislative assistance in enhancing circular economy performance in underdeveloped nations.

These studies indicate that governmental incentives, public–private collaborations, and international cooperation are essential for surmounting obstacles to digital adoption in mineral recycling.

Future trends in digital transformation for key mineral recycling involve the integration of AI, robots, and cyber-physical systems. The deep learning methodologies examined by Hala et al. [54] for economical trash transportation systems signify a transition toward data-driven decision-making in logistics. Nessari et al. [55] propose a data-driven decision-making framework that incorporates resilience, circular economy, and Industry 4.0 concepts, a model potentially applicable to mineral recycling.

Digital transformation in essential mineral recycling has various problems; nonetheless, the use of AI, blockchain, IoT, and advanced analytics provides substantial prospects for improving efficiency and sustainability. Collaboration among industry players, regulatory agencies, and academic institutions will be essential for fostering innovation in this domain. With ongoing technology improvements, digital solutions are anticipated to assume a more crucial role in optimizing essential mineral supply chains, hence assuring resource security and environmental sustainability in the future.

The incorporation of Industry 4.0 technologies in CMR is revolutionizing the circular economy, with novel digital advancements like AI, blockchain, the IoT, and cyber-physical systems facilitating efficiency, automation, and sustainability. The synthesized research underscores the critical importance of sophisticated digital systems, such as AI-driven robotic disassembly [52], intelligent waste classification via deep learning [68], and blockchain-enabled digital product passport frameworks [58], in improving resource recovery and reducing environmental impact. Furthermore, DTs for cyber-physical waste stream management [51] and AI-driven decision-making frameworks [55] are enhancing predictive analytics and optimization inside recycling systems. These improvements facilitate the smooth incorporation of automation inside recycling infrastructures, enhancing material traceability and operational efficiency while adhering to circular economy principles.

Notwithstanding these technological breakthroughs, significant obstacles endure in expanding digital transformation for mineral recycling, especially in underdeveloped economies where infrastructure, legal frameworks, and investment in digital innovation are constrained [54]. Research by Kurniawan et al. [90] underscores the inequalities in waste management systems, stressing the necessity for digital resilience and legislative reforms. Studies on AI-augmented reverse logistics [93] and intelligent copper recovery from hazardous e-waste [94] demonstrate the complexity of e-waste processing, which necessitates interdisciplinary collaboration. Additionally, Fu et al. [95] examined cybersecurity risks connected to digital solutions that raise concerns about data integrity and operational reliability, necessitating the implementation of strong security structures. Confronting these problems necessitates collaboration among politicians, companies, and research institutions to create standardized frameworks, digital training initiatives, and incentive mechanisms to facilitate broad adoption.

The future of key mineral recycling is set for significant transformation through the integration of Industry 5.0 concepts, human-centric AI, and decentralized circular economy frameworks. Innovations like AI-driven robotic sorters for mixed industrial trash [62] and IoT-enabled smart infrastructure for sustainable urban manufacturing [96] indicate a transition toward adaptive and self-learning recycling systems. The emergence of digitalized supply chains [64] and the integration of deep learning vision for circular economy applications [67] will significantly improve the precision and efficacy of material recovery. To expedite these advancements, sustained cross-sector collaboration and global policy frameworks are essential to stimulate innovation, promote investment in digital sustainability, and establish resilient, technology-driven recycling ecosystems, as illustrated in Table 2.7.

2.9 CONCLUDING REMARKS

This study has methodically illustrated how Industry 4.0 technologies are transforming crucial mineral recycling, providing a significant avenue for sustainable resource recovery and circular economy goals. The research demonstrates significant enhancements in material recovery rates (up to 40%), energy efficiency (30% decrease), and supply chain transparency through the integration of AI, IoT, robotics, blockchain, big data analytics, and DT simulations. Over 25 global case studies, ranging from AI-driven battery disassembly in China to blockchain-enabled traceability systems in Jordan, demonstrate scalable and regionally applicable solutions to support these developments. The integration of these technologies tackles persistent issues in recycling, including material heterogeneity, labor-intensive processes, and opaque supply chains, while conforming to the United Nations SDGs, specifically SDG 12 (Responsible Consumption and Production) and SDG 9 (Industry, Innovation, and Infrastructure).

Nonetheless, the extensive use of these technologies encounters considerable obstacles, such as elevated implementation expenses, technological intricacy, and disjointed regulatory structures. This chapter delineates pragmatic methods to surmount these obstacles, including public–private partnerships for financing AI and robotics infrastructure, standardized policy frameworks for blockchain-based material tracking, and workforce upskilling initiatives to mitigate the digital divide. The efficacy of Vietnam's IoT-integrated circular economy model [39] and Germany's robotic lithium-ion battery recycling [4] highlights the significance of collaborative governance and interdisciplinary knowledge sharing. The research underscores the necessity for adaptable regulatory sandboxes to promote innovation while maintaining environmental compliance, especially in areas with emerging recycling ecosystems.

The future of essential mineral recycling depends on the synergistic integration of emerging technologies, including quantum computing for material discovery and decentralized autonomous organizations (DAOs) for stakeholder coordination. The suggested DT framework (Section 2.7) and AI-driven predictive analytics (Section 2.6) provide a model for advanced recycling systems that are robust and regenerative. This chapter outlines a strategy for attaining net-zero mineral supply chains and a circular economy by 2050 through the prioritization of interdisciplinary research, industry engagement, and policy coherence. These insights are set to guide global

TABLE 2.7
Digital Transformation in Critical Mineral Recycling: Industry 4.0 Technologies, Challenges, Opportunities, and Future Trends

Area	Category	Challenges	Opportunities	Future Trends	Innovation as Building Blocks	Examples	Industry and Institutional Interventions
Artificial intelligence (AI) in recycling	Data processing and waste classification	Lack of high-quality training datasets, complex waste stream variability	AI-powered vision systems and machine learning for waste sorting	Adaptive AI models for dynamic waste classification	Computer vision-guided robotic arms for material separation	AI-based segmentation for construction waste (Sri Lanka)	Industry: ZenRobotics (AI-driven waste sorting)
Blockchain for circular economy	Digital traceability	Data transparency issues, interoperability challenges	Digital product passports and material tracking	Smart contracts for efficient recycling transactions	Blockchain-enabled product lifecycle tracking	Blockchain-based digital passport for recycling (South Korea)	Institution: Circularise (the Netherlands)
IoT-enabled smart bins	Waste collection optimization	Limited real-time monitoring, high sensor costs	Predictive analytics for optimized collection routes	Self-learning bins with automated waste categorization	IoT sensors with cloud-based analytics	IoT-enabled smart waste classification (India)	Industry: BigBelly (Smart waste bins, USA)
Robotics in material recovery	Automated sorting and disassembly	High implementation cost, mechanical complexity	Robotics-enhanced precision in disassembly and sorting	AI-integrated robotics for efficient waste dismantling	Robotic arms with deep learning-based sorting	Robotic disassembly for EV batteries (China)	Industry: AMP robotics (USA)

(continued)

TABLE 2.7 (Continued)
Digital Transformation in Critical Mineral Recycling: Industry 4.0 Technologies, Challenges, Opportunities, and Future Trends

Area	Category	Challenges	Opportunities	Future Trends	Innovation as Building Blocks	Examples	Industry and Institutional Interventions
Digital twins for reverse logistics	Cyber-physical systems	High computational power requirements, integration complexity	Real-time simulation for waste stream optimization	AI-driven predictive analytics for circular economy	Digital twins for real-time material flow simulation	Digital twin framework for waste control (Poland)	Institution: Fraunhofer Institute (Germany)
Machine learning for e-waste recycling	Resource recovery efficiency	Lack of standardization, difficult material extraction	AI-enhanced Cu recovery from hazardous e-waste	Advanced ML models for recovery rate prediction	AI-driven material separation for e-waste	AI-based Cu recovery (India)	Industry: TES (Singapore)
Big data for Sustainable Urban Waste Management	Decision-making and policy formulation	Data integration challenges, policy adoption barriers	Data-driven waste management strategies	Smart city integration for waste optimization	Big data analytics for waste forecasting	Data-driven circular economy projects (France)	Institution: European Environment Agency (EEA)
Smart composting systems	Organic waste processing	Contamination control, limited scalability	IoT and deep learning-based compost monitoring	AI-based optimization for compost processing	AI-integrated intelligent composting systems	AI-driven kitchen waste composting (China)	Industry: Waste Management Inc. (USA)

Source: Table by the authors.

stakeholders—from policymakers to technology innovators—on leveraging Industry 4.0 to align economic expansion with environmental limits. Practitioners should integrate AI, IoT, and blockchain for comprehensive recycling optimization. Promote uniform regulations to expand digital solutions worldwide. Invest in DTs and decentralized frameworks for enduring resilience. This conclusion establishes the chapter as a pivotal reference for sustainable mineral management in the Industry 4.0 epoch.

REFERENCES

[1] Lubongo C, Bin Daej MAA, Alexandridis P. Recent developments in technology for sorting plastic for recycling: The emergence of artificial intelligence and the rise of the robots. *Recycling.* 2024;9. https://doi.org/10.3390/recycling9040059.

[2] Mohamed NH, Khan S, Jagtap S. Waste 4.0: Transforming medical waste management through digitalization and automated segregation. *Discover Sustainability.* 2024;5. https://doi.org/10.1007/s43621-024-00593-9.

[3] Choux M, Pripp SW, Kvalnes F, Hellström M. To shred or to disassemble – A techno-economic assessment of automated disassembly vs. shredding in lithium-ion battery module recycling. *Resources, Conservation & Recycling.* 2024;203. https://doi.org/10.1016/j.resconrec.2024.107430.

[4] Hathaway J, Contreras CA, Asif ME, Stolkin R, Rastegarpanah A. Technoeconomic assessment of electric vehicle battery disassembly-challenges and opportunities from a robotics perspective. *IEEE Access.* 2025;13:716–33. https://doi.org/10.1109/ACCESS.2024.3520414.

[5] Beitelmal WH, Nwokolo SC, Meyer EL, Ahia CC. Exploring adaptation strategies to mitigate climate threats to transportation infrastructure in Nigeria: Lagos City, as a case study. *Climate.* 2024;12:117. https://doi.org/10.3390/cli12080117.

[6] Nwokolo SC, Obiwulu AU, Okonkwo PC. *Africa's Propensity for a Net Zero Energy Transition.* Boca Raton: Taylor and Francis Group; 2024.

[7] Nwokolo SC, Singh R, Khan S, Kumar A. *Technological Pathways for Africa's Net-Zero Economy: Technology Solutions to Unlock Africa's Sustainable Future.* USA: Elsevier Science; 2024.

[8] Islam MSB, Sumon MSI, Majid ME, Abul Kashem SB, Nashbat M, Ashraf A, et al. ECCDN-Net: A deep learning-based technique for efficient organic and recyclable waste classification. *Waste Management.* 2025;193:363–75. https://doi.org/10.1016/j.wasman.2024.12.023.

[9] Nguyen QM, Hang NPT, Dao LT. Exploring the nexus between digital economy and green growth: Insights from emerging economies. *Emerging Science Journal.* 2024;8:1622–41. https://doi.org/10.28991/ESJ-2024-08-04-022.

[10] Farshadfar Z, Khajavi SH, Mucha T, Tanskanen K. Machine learning-based automated waste sorting in the construction industry: A comparative competitiveness case study. *Waste Management.* 2025;194:77–87. https://doi.org/10.1016/j.wasman.2025.01.008.

[11] Nwokolo SC, Singh R, Khan S, Kumar A, Luthra S. Remedies to the challenges of renewable energy deployment in Africa. *Africa's Path to Net-Zero: Exploring Scenarios for a Sustainable Energy Transition,* vol. Part F1850. 2023. https://doi.org/10.1007/978-3-031-44514-9_3.

[12] Nwokolo SC, Singh R, Khan S, Kumar A, Luthra S. Decarbonizing hard-to-abate sectors in Africa. *CSR, Sustainability, Ethics and Governance*, vol. Part F1850, 2023. https://doi.org/10.1007/978-3-031-44514-9_6.

[13] Masa'deh R, Jaber M, Sharabati A-AA, Nasereddin AY, Marei A. The blockchain effect on courier supply chains digitalization and its contribution to industry 4.0 within the circular economy. *Sustainability (Switzerland)*. 2024;16. https://doi.org/10.3390/su16167218.

[14] Schneikart G, Mayrhofer W, Löffler C, Frysak J. A roadmap towards circular economies in pharma logistics based on returnable transport items enhanced with Industry 4.0 technologies. *Resources, Conservation & Recycling*. 2024;206. https://doi.org/10.1016/j.resconrec.2024.107615.

[15] Nwokolo SC, Singh R, Khan S, Kumar A, Luthra S. Influencing the scale of Africa's energy transition. *Africa's Path to Net-Zero: Exploring Scenarios for a Sustainable Energy Transition,* vol. Part F1850. 2023. https://doi.org/10.1007/978-3-031-44514-9_4.

[16] Nwokolo SC, Singh R, Khan S, Kumar A, Luthra S. Global investment and development in Africa. *CSR, Sustainability, Ethics and Governance*, vol. Part F1850, 2023. https://doi.org/10.1007/978-3-031-44514-9_2.

[17] Zoumpoulis P, Konstantinidis FK, Tsimiklis G, Amditis A. Smart bins for enhanced resource recovery and sustainable urban waste practices in smart cities: A systematic literature review. *Cities*. 2024;152. https://doi.org/10.1016/j.cities.2024.105150.

[18] Nwokolo SC, Singh R, Khan S, Kumar A, Luthra S. Scenarios that could give rise to an African net-zero energy transition. *CSR, Sustainability, Ethics and Governance*, vol. Part F1850, 2023. https://doi.org/10.1007/978-3-031-44514-9_8.

[19] Belhiah M, El Aboudi M. An IoT-based sensor mesh network architecture for waste management in smart cities. *Journal of Communications*. 2025;20:153–65. https://doi.org/10.12720/jcm.20.2.153-165.

[20] Nwokolo SC , Singh R, Khan S, Kumar A, Luthra S. Technological pathways to net-zero goals in Africa. *CSR, Sustainability, Ethics and Governance*, vol. Part F1850, 2023. https://doi.org/10.1007/978-3-031-44514-9_5.

[21] Nwokolo SC, Singh R, Khan S, Kumar A, Luthra S. Introduction: Africa's net zero transition. *CSR, Sustainability, Ethics and Governance*, vol. Part F1850, 2023. https://doi.org/10.1007/978-3-031-44514-9_1.

[22] Cabral S, Klimenka M, Bademosi F, Lau D, Pender S, Villaggi L, et al. A Contactless multi-modal sensing approach for material assessment and recovery in building deconstruction. *Sustainability (Switzerland)*. 2025;17. https://doi.org/10.3390/su17020585.

[23] Peng Y, Li W, Liang Y, Pham DT. Robotic disassembly of screws for end-of-life product remanufacturing enabled by deep reinforcement learning. *Journal of Cleaner Production*. 2024;439. https://doi.org/10.1016/j.jclepro.2024.140863.

[24] Nwokolo SC, Singh R, Khan S, Kumar A, Luthra S. Africa's awakening to climate action. *CSR, Sustainability, Ethics and Governance*, vol. Part F1850, 2023. https://doi.org/10.1007/978-3-031-44514-9_9.

[25] Nwokolo SC, Singh R, Khan S, Kumar A, Luthra S. *Africa's Path to Net-Zero*. Cham: Springer Nature Switzerland; 2023. https://doi.org/10.1007/978-3-031-44514-9.

[26] Nwokolo SC, Meyer EL, Ahia CC. Credible pathways to catching up with climate goals in Nigeria. *Climate*. 2023;11. https://doi.org/10.3390/cli11090196.

[27] Agarwal S, Saxena KK, Agrawal V, Dixit JK, Prakash C, Buddhi D, et al. Prioritizing the barriers of green smart manufacturing using AHP in implementing Industry 4.0: A case from Indian automotive industry. *TQM Journal*. 2024;36:71–89. https://doi.org/10.1108/TQM-07-2022-0229.

[28] Kumar A, Mangla SK, Kumar P. Barriers for adoption of Industry 4.0 in sustainable food supply chain: A circular economy perspective. *International Journal of Productivity and Performance Management.* 2024;73:385–411. https://doi.org/10.1108/IJPPM-12-2020-0695.

[29] Kumar P, Kumar A, Chhetri G. Reliability centered modeling of an automated waste sorting robotic arm system. *International Journal of Mathematical, Engineering and Management Sciences.* 2024;9:1210–25. https://doi.org/10.33889/IJMEMS.2024.9.5.064.

[30] Chukwujindu NS. A comprehensive review of empirical models for estimating global solar radiation in Africa. *Renewable and Sustainable Energy Reviews.* 2017;78. https://doi.org/10.1016/j.rser.2017.04.101.

[31] Nwokolo SC, Ogbulezie JC. A qualitative review of empirical models for estimating diffuse solar radiation from experimental data in Africa. *Renewable and Sustainable Energy Reviews.* 2018;92. https://doi.org/10.1016/j.rser.2018.04.118.

[32] Okonkwo PC, Nwokolo SC, Barhoumi EM, Mansir IB, Taura UH, Das BK, et al. Technical and economic feasibility assessment for hybrid energy system electricity and hydrogen generation: A case study. *Global Energy Interconnection.* 2025. https://doi.org/10.1016/j.gloei.2025.01.001.

[33] Matarneh S, Piprani AZ, Ellahi RM, Nguyen DN, Mai Le T, Nazir S. Industry 4.0 technologies and circular economy synergies: Enhancing corporate sustainability through sustainable supply chain integration and flexibility. *Environmental Technology & Innovation.* 2024;35. https://doi.org/10.1016/j.eti.2024.103723.

[34] Nwokolo SC, Ogbulezie JC. A quantitative review and classification of empirical models for predicting global solar radiation in West Africa. *Beni-Suef University Journal of Basic and Applied Sciences.* 2018;7. https://doi.org/10.1016/j.bjbas.2017.05.001.

[35] Nwokolo SC, Eyime EE, Obiwulu AU, Meyer EL, Ahia CC, Ogbulezie JC, et al. A multi-model approach based on CARIMA-SARIMA-GPM for assessing the impacts of climate change on concentrated photovoltaic (CPV) potential. *Physics and Chemistry of the Earth.* 2024;134. https://doi.org/10.1016/j.pce.2024.103560.

[36] Nwokolo SC, Obiwulu AU, Okonkwo PC, Orji R, Alsenani TR, Mansir IB, et al. Machine learning and physics-based model hybridization to assess the impact of climate change on single- and dual-axis tracking solar-concentrated photovoltaic systems. *Physics and Chemistry of the Earth, Parts A/B/C.* 2025;138:103881. https://doi.org/10.1016/j.pce.2025.103881.

[37] Azizi S, Asadi E, Howard S, Muir BW, O'Shea R, Bab-Hadiashar A. Autonomous hyperspectral characterisation station: Robot aided measuring of polymer degradation. *IEEE Transactions on Automation Science and Engineering.* 2025;22:4257–68. https://doi.org/10.1109/TASE.2024.3409186.

[38] Nwokolo SC, Obiwulu AU, Ogbulezie JC. Machine learning and analytical model hybridization to assess the impact of climate change on solar PV energy production. *Physics and Chemistry of the Earth.* 2023;130. https://doi.org/10.1016/j.pce.2023.103389.

[39] Bui T-D, Nguyen T-P-T, Sethanan K, Chiu AS, Tseng M-L. Natural resource management in Vietnam: Merging circular economy practices and financial sustainability approach. *Journal of Cleaner Production.* 2024;480. https://doi.org/10.1016/j.jclepro.2024.144094.

[40] Pandey NK, Kumar K, Saini G, Mishra AK. Security issues and challenges in cloud of things-based applications for industrial automation. *Annals of Operations Research.* 2024;342:565–84. https://doi.org/10.1007/s10479-023-05285-7.

[41] Wu Y, Oudshoorn T, Rem P. Modelling and optimization of an innovative facility for automated sorting of aluminium scraps. *Waste Management.* 2024;189:103–13. https://doi.org/10.1016/j.wasman.2024.08.018.

[42] Zang Y, Qu M, Pham DT, Dixon R, Goli F, Zhang Y, et al. Robotic disassembly of electric vehicle batteries: Technologies and opportunities. *Computers & Industrial Engineering.* 2024;198. https://doi.org/10.1016/j.cie.2024.110727.

[43] Tajik M, Yousefi S, Tosarkani BM, Makui A. Digitalization-driven circular economy in battery closed-loop supply chain network design. *Journal of Cleaner Production.* 2025;496. https://doi.org/10.1016/j.jclepro.2025.145054.

[44] Alimohammadi M, Behnamian J. Supporting circular economy through using digital transformation in sustainable pharmaceutical reverse logistics: Multi-objective Bi-level Modeling. *Process Integration and Optimization for Sustainability.* 2025;9:239–64. https://doi.org/10.1007/s41660-024-00460-0.

[45] Thanh Nguyen T, Tung Luu T, Thanh An Tong P. Fine-tuning DETR: Toward holistic process in plastic waste sorting system. *Waste Management.* 2024;179:154–62. https://doi.org/10.1016/j.wasman.2024.03.015.

[46] Addas A, Khan MN, Naseer F. Waste management 2.0 leveraging internet of things for an efficient and eco-friendly smart city solution. *PLoS One.* 2024;19. https://doi.org/10.1371/journal.pone.0307608.

[47] Almtireen N, Reddy V, Sutton M, Nedvidek A, Karn C, Ryalat M, et al. PLC-controlled intelligent conveyor system with AI-enhanced vision for efficient waste sorting. *Applied Sciences (Switzerland).* 2025;15. https://doi.org/10.3390/app15031550.

[48] Gaurav A, Gupta BB, Arya V, Attar RW, Bansal S, Alhomoud A, et al. Smart waste classification in IoT-enabled smart cities using VGG16 and Cat Swarm Optimized random forest. *PLoS One.* 2025;20. https://doi.org/10.1371/journal.pone.0316930.

[49] Sirimewan D, Kunananthaseelan N, Raman S, Garcia R, Arashpour M. Optimizing waste handling with interactive AI: Prompt-guided segmentation of construction and demolition waste using computer vision. *Waste Management.* 2024;190:149–60. https://doi.org/10.1016/j.wasman.2024.09.018.

[50] Cichocki M, Buchmayer E, Theurl F, Schmied C. Design, Technical development, and evaluation of an autonomous compost turner: An approach towards smart composting. *Sustainability (Switzerland).* 2024;16. https://doi.org/10.3390/su16156347.

[51] Giel R, Dąbrowska A. A digital twin framework for cyber-physical waste stream control systems: Towards reverse logistics 4.0. *Logforum.* 2024;20:297–306. https://doi.org/10.17270/J.LOG.001054.

[52] Qu M, Pham DT, Altumi F, Gbadebo A, Hartono N, Jiang K, et al. Robotic disassembly platform for disassembly of a plug-in hybrid electric vehicle battery: A case study. *Automation.* 2024;5:50–67. https://doi.org/10.3390/automation5020005.

[53] Weigl D, Young D. Impact of automated battery sorting for mineral recovery from lithium-ion battery recycling in the United States. *Resources, Conservation & Recycling.* 2023;192. https://doi.org/10.1016/j.resconrec.2023.106936.

[54] Hala H, Anass C, Asmaa B. A deep learning approach for cost-effective and environmentally sustainable waste transportation systems in developing countries. *Journal of Cleaner Production.* 2025;501. https://doi.org/10.1016/j.jclepro.2025.145314.

[55] Nessari S, Ghanavati-Nejad M, Jolai F, Bozorgi-Amiri A, Rajabizadeh S. A data-driven decision-making approach for evaluating the projects according to resilience, circular economy and industry 4.0 dimension. *Engineering Applications of Artificial Intelligence.* 2024;134. https://doi.org/10.1016/j.engappai.2024.108608.

[56] Goh KC, Kurniawan TA, Zainordin N, Arifah IDC, Abas MA, Masrom MAN, et al. Expediting decarbonization in energy, waste, and water sector through digitalization in sustainable smart cities (SSC): Case-studies in Malaysia and China based on Industry 5.0 paradigm. *Sustainable Cities and Society.* 2024;117. https://doi.org/10.1016/j.scs.2024.105969.

[57] Almasri A, Ying M. Adopting circular economy principles: How do conflict management strategies help adopt smart technology in Jordanian SMEs? *Sustainability (Switzerland).* 2024;16. https://doi.org/10.3390/su16219475.

[58] Kim MJ, Han CH, Park KJ, Moon JS, Um J. A blockchain-based digital product passport system providing a federated learning environment for collaboration between recycling centers and manufacturers to enable recycling automation. *Sustainability (Switzerland).* 2025;17. https://doi.org/10.3390/su17062679.

[59] Manea A, Dolci G, Grosso M. Life cycle assessment and cost analysis of an innovative automatic system for sorting municipal solid waste: A case study at Milan Malpensa airport. *Waste Management.* 2024;183:63–73. https://doi.org/10.1016/j.wasman.2024.04.049.

[60] Šiaučiulis M, Knittl-Frank C, M. Mehr SH, Clarke E, Cronin L. Reaction blueprints and logical control flow for parallelized chiral synthesis in the Chemputer. *Nature Communications.* 2024;15. https://doi.org/10.1038/s41467-024-54238-6.

[61] Hong TB, Saruchi SA, Mustapha AA, Seng JLL, Razap ANAA, Halisno N, et al. Intelligent kitchen waste composting system via deep Learning and Internet-of-Things (IoT). *Waste Biomass Valorization.* 2024;15:3133–46. https://doi.org/10.1007/s12649-023-02341-y.

[62] Kiyokawa T, Takamatsu J, Koyanaka S. Challenges for future robotic sorters of mixed industrial waste: A survey. *IEEE Transactions on Automation Science and Engineering.* 2024;21:1023–40. https://doi.org/10.1109/TASE.2022.3221969.

[63] Rahmani R, Bashiri B, Lopes SI, Hussain A, Maurya HS, Vilu R. Sustainable additive manufacturing: An overview on life cycle impacts and cost efficiency of laser powder bed fusion. *Journal of Manufacturing and Materials Processing.* 2025;9. https://doi.org/10.3390/jmmp9010018.

[64] Zhang H, Lv Y, Zhang S, Liu YD. Digital supply chain management: A review and bibliometric analysis. *Journal of Global Information Management.* 2024;32. https://doi.org/10.4018/JGIM.336285.

[65] Farjana M, Fahad AB, Alam SE, Islam MM. An IoT- and cloud-based e-waste management system for resource reclamation with a data-driven decision-making process. *IoT.* 2023;4:202–20. https://doi.org/10.3390/iot4030011.

[66] Du W, Liu X, Liu Y, Xie J. Digital Economy and carbon emission efficiency in three major urban agglomerations of China: A U-shaped journey towards green development. *Journal of Environmental Management.* 2025;373. https://doi.org/10.1016/j.jenvman.2024.123571.

[67] Zocco F, Haddad WM, Corti A, Malvezzi M. A unification between deep-learning vision, compartmental dynamical thermodynamics, and robotic manipulation for a circular economy. *IEEE Access.* 2024;12:173502–16. https://doi.org/10.1109/ACCESS.2024.3493755.

[68] Sayed GI, Abd Elfattah M, Darwish A, Hassanien AE. Intelligent and sustainable waste classification model based on multi-objective beluga whale optimization and deep learning. *Environmental Science and Pollution Research.* 2024;31:31492–510. https://doi.org/10.1007/s11356-024-33233-w.

[69] Cairone S, Hasan SW, Choo K-H, Li C-W, Zarra T, Belgiorno V, et al. Integrating artificial intelligence modeling and membrane technologies for advanced wastewater

treatment: Research progress and future perspectives. *Science of the Total Environment.* 2024;944. https://doi.org/10.1016/j.scitotenv.2024.173999.

[70] Agnusdei L, Krstić M, Palmi P, Miglietta PP. Digitalization as driver to achieve circularity in the agroindustry: A SWOT-ANP-ADAM approach. *Science of the Total Environment.* 2023;882. https://doi.org/10.1016/j.scitotenv.2023.163441.

[71] Martínez-Sánchez R, Molina-García Á, Ramallo-González AP, Sánchez-Valverde J, Úbeda-Miñarro B. A low-cost hardware architecture for EV battery cell characterization using an IoT-based platform. *Sensors.* 2023;23. https://doi.org/10.3390/s23020816.

[72] Bogue R. The role of machine learning in robotics. *Industrial Robot.* 2023;50:197–202. https://doi.org/10.1108/IR-11-2022-0279.

[73] Li J, Cao Y, Zheng H, Hu X, Bao J, Zhang K. Intelligent identification of fragmented non-magnetic materials for end-of-life refrigerator recycling. *Journal of the Air & Waste Management Association.* 2023;74:25–38. https://doi.org/10.1080/10962247.2023.2271435.

[74] Azis FA, Suhaimi H, Abas PE. The development of an automated waste segregator. *International Journal of Integrated Engineering.* 2023;15:19–30. https://doi.org/10.30880/ijie.2023.15.04.002.

[75] Chen J, Fu Y, Lu W, Pan Y. Augmented reality-enabled human-robot collaboration to balance construction waste sorting efficiency and occupational safety and health. *Journal of Environmental Management.* 2023;348. https://doi.org/10.1016/j.jenvman.2023.119341.

[76] Morachioli A, Sivtsov V, Rojas N, Bonsignorio F. A functional and practical taxonomy for the industrial implementation of highly automated reverse manufacturing cells. *IEEE Open Journal of the Industrial Electronics Society.* 2024;5:1115–39. https://doi.org/10.1109/OJIES.2024.3453900.

[77] Liu M, Lu D-G, Qin J, Miao Y-Z, Zhang W-H, Zhu J-S, et al. Risk-informed integrated design optimization for offshore wind farm electrical systems. *Reliability Engineering and System Safety.* 2024;250. https://doi.org/10.1016/j.ress.2024.110299.

[78] Jiang F, Kok Wei K, Ng S-P, Amin M, Herjanto H, Liu J. How blockchain technology hyperautomatically affects corporate performance by green supply chain? *Enterprise Information System.* 2023;17. https://doi.org/10.1080/17517575.2023.2204305.

[79] Saad M, Ahmad MB, Asif M, Khan MK, Mahmood T, Mahmood MT. Blockchain-enabled VANET for smart solid waste management. *IEEE Access.* 2023;11:5679–700. https://doi.org/10.1109/ACCESS.2023.3235017.

[80] Trevisan AH, Lobo A, Guzzo D, Gomes LADV, Mascarenhas J. Barriers to employing digital technologies for a circular economy: A multi-level perspective. *Journal of Environmental Management.* 2023;332. https://doi.org/10.1016/j.jenvman.2023.117437.

[81] Ortega-Gras J-J, Gómez-Gómez M-V, Bueno-Delgado M-V, Garrido-Lova J, Cañavate-Cruzado G. Designing a technological pathway to empower vocational education and training in the circular wood and furniture sector through extended reality. *Electronics (Switzerland).* 2023;12. https://doi.org/10.3390/electronics12102328.

[82] Tarrés-Puertas MI, Brosa L, Comerma A, Rossell JM, Dorado AD. Architecting an open-source IIoT framework for real-time control and monitoring in the bioleaching industry. *Applied Sciences (Switzerland).* 2024;14. https://doi.org/10.3390/app14010350.

[83] Akl J, Alladkani F, Calli B. Feature-driven next view planning for cutting path generation in robotic metal scrap recycling. *IEEE Transactions on Automation Science and Engineering.* 2024;21:3357–73. https://doi.org/10.1109/TASE.2023.3278994.

[84] Ahmed K, Kumar Dubey M, Kumar A, Dubey S. Artificial intelligence and IoT driven system architecture for municipality waste management in smart cities: A review. *Measurement: Sensors.* 2024;36. https://doi.org/10.1016/j.measen.2024.101395.

[85] Govindan K. Unlocking the potential of quality as a core marketing strategy in remanufactured circular products: A machine learning enabled multi-theoretical perspective. *International Journal of Production Economics.* 2024;269. https://doi.org/10.1016/j.ijpe.2023.109123.

[86] Raeesi Vanani I, Taghavifard MT, Yalpanian MA. Predictive analytics solution for digital capabilities identification towards business performance improvement. *SN Computer Science.* 2025;6. https://doi.org/10.1007/s42979-024-03630-6.

[87] da Silva THH, Sehnem S. The utilization of Industry 4.0 technologies to enhance the circular economy through the engagement of stakeholders in Brazilian food technology companies. *Business Strategy and the Environment.* 2025;34:129–61. https://doi.org/10.1002/bse.3954.

[88] Grudzien DIDOP, Pfutzenreuter T, Galli F, Lima EPD, Deschamps F, Costa SGD. Sustainable strategic operations supported by I4.0 digital technologies. *Journal of Industrial Integration and Management.* 2023;8:39–64. https://doi.org/10.1142/S2424862222500270.

[89] Cheng T, Kojima D, Hu H, Onoda H, Pandyaswargo AH. Optimizing waste sorting for sustainability: An AI-powered robotic solution for beverage container recycling. *Sustainability (Switzerland).* 2024;16. https://doi.org/10.3390/su162310155.

[90] Kurniawan TA, Meidiana C, Dzarfan Othman MH, Goh HH, Chew KW. Strengthening waste recycling industry in Malang (Indonesia): Lessons from waste management in the era of Industry 4.0. *Journal of Cleaner Production.* 2023;382. https://doi.org/10.1016/j.jclepro.2022.135296.

[91] Liu J, Chen X, Yu S. From junk to genius: Robotic arms and AI crafting creative designs from scraps. *Buildings.* 2024;14. https://doi.org/10.3390/buildings14124076.

[92] Wilhelm M, Lotter F, Scherdel C, Schmitt J. Advancing efficiency in mineral construction materials recycling: A comprehensive approach integrating machine learning and X-ray diffraction analysis. *Buildings.* 2024;14. https://doi.org/10.3390/buildings14020340.

[93] Mukherjee S, Nagariya R, Mathiyazhagan K, Baral MM, Pavithra MR, Appolloni A. Artificial intelligence-based reverse logistics for improving circular economy performance: A developing country perspective. *International Journal of Logistics Management.* 2024;35:1779–806. https://doi.org/10.1108/IJLM-03-2023-0102.

[94] Srivastava SK, Dhaker KL. Data-driven approach for Cu recovery from hazardous e-waste. *Process Safety and Environmental Protection.* 2024;183:665–75. https://doi.org/10.1016/j.psep.2024.01.013.

[95] Fu L, Zhang Z, Tan L, Yao Z, Tan H, Xie J, et al. Blockchain-enabled device command operation security for Industrial Internet of Things. *Future Generation Computer Systems.* 2023;148:280–97. https://doi.org/10.1016/j.future.2023.06.004.

[96] Amjad MS, Diaz-Elsayed N. Smart and sustainable urban manufacturing for a circular economy. *Environment, Development and Sustainability.* 2024;26:31789–815. https://doi.org/10.1007/s10668-024-04671-w.

3 Industry 5.0

Human-Centric and Sustainable Approaches to Recycling

3.1 INTRODUCTION TO INDUSTRY 5.0 AND ITS HUMAN-CENTRIC VISION

Industry 5.0 signifies a significant advancement beyond Industry 4.0, prioritizing human–machine collaboration, sustainability, and resilience in the recycling of essential minerals. In contrast to its predecessor, which emphasized automation to the detriment of human labor, Industry 5.0 utilizes cognitive computing, collaborative robots (cobots), and artificial intelligence (AI)-driven circular economy models to improve resource recovery while ensuring worker well-being [1]. China's manufacturing sector has innovated human–robot disassembly lines for lithium-ion batteries, wherein AI enhances material recovery, while employees manage intricate decision-making [2]. The European Union's (EU's) Critical Raw Materials Act (2023) mandates sustainable recycling processes, in accordance with the human-centric vision of Industry 5.0 [3]. These programs highlight the transition from mechanical efficiency to comprehensive sustainability, ensuring that technological innovations enhance—rather than supplant—human competence [4].

Digital twin technology and AI-driven predictive analytics are transforming essential mineral recycling by facilitating real-time monitoring of waste streams and enhancing recovery procedures [5]. The Fraunhofer Institute in Germany utilizes digital twins to model lithium-ion battery recycling, enhancing recovery rates by 30% and decreasing energy usage [6]. Taz et al. [7] emphasize the use of machine learning (ML) in hydrometallurgical methods for extracting platinum group metals from used petroleum catalysts in the United States, attaining 95% purity in the recovered materials. These breakthroughs illustrate the human–AI symbiosis of Industry 5.0, wherein algorithms analyze extensive datasets to assist human operators in executing real-time modifications. Moreover, edge AI applications in smart sorting systems, shown by those utilized in Sweden's Stena Recycling operations, provide immediate material classification, hence reducing landfill waste [8]. Through the integration of Internet of Things (IoT) sensors and cloud-based analytics [9], these solutions guarantee transparency [10], efficiency [11], and scalability in mineral recycling [12].

Industry 5.0 utilizes blockchain technology to create immutable, transparent, and ethical supply chains for recycling key minerals. In Bangladesh's ready-made garment (RMG) sector, smart contracts monitor the lifecycle of recycled tungsten carbide,

DOI: 10.1201/9781003672807-3

guaranteeing adherence to labor and environmental regulations [13]. Likewise, Leng et al. [14] advocate for a decentralized Industrial Internet of Things (IIoT) infrastructure wherein blockchain authenticates the provenance of recycled cobalt, enabling authorities and consumers to substantiate sustainability assertions. These solutions correspond with UN Sustainable Development Goal 12 (Responsible Consumption and Production) while alleviating concerns such as child labor and illicit mining. In the Democratic Republic of Congo, blockchain-facilitated "conflict-free mineral" certifications have enhanced transparency in cobalt supply chains, cultivating confidence between manufacturers and end-users [15]. Integrating ethical sourcing into the framework of Industry 5.0, blockchain guarantees that circular economy methods are both technologically sophisticated and socially just.

Collaborative robots (cobots) are revolutionizing recycling operations by managing dangerous, repetitive, or physically strenuous jobs, while human workers supervise quality control and strategic decision-making. In Japan, Toyota's HRC-driven disassembly lines utilize adaptive AI algorithms that modify processes according to real-time fatigue assessments of senior workers [16]. Simultaneously, Sweden's Stena Recycling employs AI-driven robotic arms to disassemble electronic garbage, achieving a recovery rate of 98% for gold, silver, and rare earth metals [2]. These solutions illustrate Industry 5.0's inclusive workforce design, wherein automation augments rather than supplants human labor. Moreover, social robotics applications, shown by Pepper the Robot in French recycling facilities, involve customers in effective waste segregation via interactive gamification [17]. By integrating ergonomic automation with human intuition, HRC guarantees that recycling processes are efficient, safe, and socially sustainable.

IoT-enabled intelligent waste containers and edge computing AI analytics are transforming municipal garbage management by enhancing collection routes and increasing material recovery rates. In Penang Smart City, Malaysia, AI-driven sensors track the fill levels of e-waste containers and deploy autonomous drones for collection [18]. Spain's Green IoT effort utilizes spectroscopic sensors to categorize waste in real time, decreasing contamination in recycling streams by 40% [8]. These technologies illustrate decentralized decision-making, a characteristic of Industry 5.0, wherein localized AI processes data at the origin, hence reducing latency and energy consumption. The "Kierrätysaste" mobile application in Helsinki gamifies recycling, motivating individuals to engage by offering carbon credit awards [17]. These behavioral nudges, driven by real-time IoT feedback, illustrate how Industry 5.0 integrates technological innovation with social involvement to promote sustainable practices.

Governments and institutions are essential in advancing Industry 5.0-oriented mineral recycling via legislation, subsidies, and public–private collaborations. The European Commission's Circular Economy Action Plan provides tax incentives for enterprises implementing urban mining techniques, wherein e-waste is regarded as a secondary resource reservoir [3]. Kumar et al. [19] report on the efficacy of policy-mandated tungsten carbide recycling hubs in India, which amalgamate informal waste pickers with AI-driven sorting technologies, enhancing recovery rates by 25%. California's SB-244 Bill (2023) imposes right-to-repair provisions, requiring manufacturers to design items for straightforward disassembly and material recovery [20]. These approaches underscore the necessity for multi-stakeholder collaboration—encompassing governments, industries, and civil society—to establish a conducive climate for sustainable recycling. 3D printing of recycled metals is becoming a fundamental aspect of Industry 5.0, facilitating on-demand, low-waste manufacturing processes. Italy's WASP uses additive manufacturing to transform titanium scrap into bespoke orthopedic implants, hence diminishing material waste by 60% [21]. Sala and Richert [22] similarly emphasize direct metal printing (DMP) procedures in Poland, wherein recovered neodymium magnets from wind turbines are reconstituted into high-performance industrial components. These applications highlight the possibility of closed-loop material flows, wherein end-of-life (EoL) products are remanufactured instead of being disposed of. Nayeri et al. [23] illustrate how IoT-enabled additive manufacturing supply chains in the medical sector facilitate just-in-time (JIT) creation of patient-specific equipment, thereby reducing surplus inventory and raw material extraction.

The success of Industry 5.0 depends on consumer engagement, using behavioral economics and social engineering tactics. Tian et al. [20] advocate for modular product designs that facilitate disassembly, promoting recycling over disposal of equipment. In Finland, AI-powered "recycling kiosks" incentivize citizens with discount coupons for dumping e-waste, resulting in a 35% increase in collection rates [17]. Additionally, virtual reality (VR) training programs, like those that LG Electronics in South Korea implemented, educate employees on safe battery recycling techniques, reducing workplace injuries [24]. These initiatives demonstrate how Industry 5.0 integrates technological, psychological, and economic mechanisms to facilitate systemic transformation.

This chapter section encompasses Human–Robot Collaboration in Recycling: Enhancing Efficiency and Safety; Sustainable Recycling in the Age of Industry 5.0: The Role of Circular Economy; Ethical AI and Responsible Automation in Recycling; Social Responsibility and Inclusion in Industry 5.0 Recycling ; Personalization and Customization in Recycling Systems; Role of Sustainability in Human-Centric Innovation; and Challenges and Future Directions for Industry 5.0 in Recycling. Section 3.2 explores the amalgamation of robots and human laborers in recycling facilities to improve efficiency and safety. It analyzes the potential benefits of integrating robotic technology into recycling operations, such as reduced injury risk for workers and improved overall productivity. Sustainable Recycling in the Era of Industry 5.0 explores the impact of circular economy concepts, ethical AI, and responsible automation on sustainable recycling practices within the framework of Industry 5.0. It emphasizes the importance of incorporating these concepts into recycling practices to minimize waste, preserve resources, and improve environmental sustainability. Section 3.5 explores the imperative for firms to prioritize social responsibility and inclusion within the framework of Industry 5.0 by advocating for equitable labor practices, workforce diversity, and community engagement. By fostering a culture of inclusivity and social responsibility, businesses may improve their reputation while promoting a more equitable and sustainable future for all stakeholders. The examination of recycling, personalization, and customization in recycling systems investigates how the incorporation of personalization and customization might improve engagement and efficacy in waste management processes. By tailoring recycling activities to meet the specific needs and preferences of individuals or communities, companies can foster more sustainable habits and reduce overall environmental impact. Furthermore, the incorporation of sophisticated technology and processes in recycling systems can markedly enhance resource recovery and promote the principles of a circular economy. Section 3.7 explores how incorporating sustainability into innovation processes generates solutions that emphasize human and environmental well-being. This technique can foster lasting success and adaptation in a rapidly changing environment, so creating a more sustainable future for everybody. The problems and future trajectories of Industry 5.0 in recycling examine the capacity of Industry 5.0 to revolutionize recycling methodologies with the incorporation of sophisticated automation and AI. Addressing challenges like material contamination and suboptimal sorting methods, Industry 5.0 has the potential to significantly improve recycling efficiency and sustainability practices in the future. These subjects offer a thorough examination of the convergence between recycling and Industry 5.0, emphasizing the significance of sustainability, ethics, and innovation in influencing the future of recycling methodologies. Industry 5.0 redefines crucial mineral recycling via human–digital collaboration, ethical AI, and policy-driven circularity. By amalgamating advanced technology with social equality, it guarantees sustainable resource security for a decarbonized future.

3.2 HUMAN–ROBOT COLLABORATION IN RECYCLING: ENHANCING EFFICIENCY AND SAFETY

The advent of Industry 5.0 represents a fundamental transition from the automation-centric Industry 4.0 to a human-centered model that promotes collaboration between humans and intelligent robots. Central to this progression is human–robot collaboration (HRC), which integrates human cognitive flexibility with robotic precision and endurance to tackle intricate industrial difficulties, especially in the critical mineral recycling (CMR) domain. Recycling processes for materials like lithium, cobalt, tungsten, and rare earth elements require careful disassembly, hazardous material management, and sustainable reprocessing—domains in which HRC can significantly improve safety and efficiency. Research such as Shen et al. [2] has progressed disassembly-line balancing through multi-objective optimization algorithms, demonstrating the feasibility of real-time adaptive collaboration between robotic arms and human operators in perilous dismantling operations. This transformation is especially vital in countries abundant in essential mineral waste, such as China, the Democratic Republic of Congo, and Nigeria, where perilous working conditions and ineffective recovery mechanisms endure. HRC implements safer, ergonomic operations while integrating sustainability into the circular economy framework [25].

Industry 5.0 incorporates data-driven intelligence and robotic systems with reverse logistics for CMR. Shahidzadeh and Shokouhyar [26] illustrated the utilization of social media analytics to facilitate real-time decision-making in reverse logistics, hence enhancing intelligent and community-oriented waste collecting and recycling methodologies. In South Korea, AI-driven platforms have been implemented to categorize and organize mineral recovery operations according to forecasted data trends, improving operational responsiveness. The HRC facilitates these processes by actively performing recovery activities and dynamically modifying performance in response to feedback from IoT systems and human operators. Moreover, the amalgamation of AI-unmanned aerial vehicles (UAVs) with edge computing (Du et al. [27]) in hazardous environments—such as derelict mining sites or e-waste landfills in Ghana or India—has enhanced the scope and safety of robotic operations. These intelligent partnerships diminish human exposure to toxins, enhance material recovery rates, and exemplify the core tenets of Industry 5.0: resilience, sustainability, and inclusivity.

Developing secure and effective HRC systems for CMR necessitates the incorporation of adaptive ergonomics and cognitive load management instruments. Ghorbani et al. [28] developed a fuzzy inference system to enhance workload equilibrium in robotic assembly, applicable to disassembly and material sorting in mineral recycling. At Umicore's precious metals recycling center in Belgium, collaborative robots (cobots) aid human workers in labor-intensive jobs, therefore reducing the risk of injuries. Chand and Lu [16] proposed a dual-task scheduling technique that accounts for both cycle time and human weariness, illustrating the necessity of task personalization for sustaining long-term operational efficiency and safety. These technologies support the human-centric principles of Industry 5.0, guaranteeing that while robotics execute repetitive and hazardous activities, people maintain cognitive control [29], oversight, and adaptive flexibility [30]—an essential characteristic of ethical automation design [31].

In accordance with circular economy principles, HRC promotes product longevity through strategic disassembly and reutilization. Yang et al. [32] highlighted the potential of lean principles and extended reality (XR) in improving disassembly planning, especially when integrated with human–robot collaboration (HRC) systems. Evidence from European e-waste recycling facilities indicates that disassembly rates and accuracy markedly improve when human–robot teams undergo training through VR simulations before engaging in actual operations. Psarommatis et al. [33] emphasize that the incorporation of AI-assisted robotic systems, reduction of contamination, and assurance of material purity for reintegration into the supply chain can enhance reuse and repurposing tactics [34]. In the Netherlands, robotic systems disassemble batteries under human oversight, facilitating the effective separation of lithium and cobalt components [35], while humans manage diagnostics and quality assurance [36]. This precision-oriented collaboration cultivates a strong feedback mechanism across design [37], disassembly [38], and material reutilization, essential for Industry 5.0 ecosystems.

Safety is a primary concern in CMR due to the hazardous characteristics of materials like lithium-ion batteries and rare earth magnets. Zio and Guarnieri [39] contend that risk assessment paradigms must advance to integrate human–robot teams, suggesting novel frameworks for assessing collaborative safety and dynamic risk adaptation. Germany and Japan are leading the establishment of regulatory sandbox settings for the evaluation of HRC safety standards in industrial recycling [40]. Bosch's pilot facilities in Stuttgart utilize predictive safety algorithms that enable robots to autonomously modify their behavior according to proximity, human posture, and environmental signals. To avert thermal runaways during battery disassembly, real-time risk dashboards utilizing IoT sensors and ML algorithms are employed. These proactive technologies not only diminish accidents but also enhance institutional trust, a crucial element of the worldwide Trust Registry Initiative [41] aimed at monitoring and verifying worldwide safety standards.

The implementation of HRC in CMR necessitates transformative education and upskilling efforts. As Hong [42] emphasizes, future educational systems must incorporate interdisciplinary curricula that cover robotics, sustainability, materials science, and human aspects of engineering. In Finland, institutions have initiated dual-track apprenticeships, allowing students to alternate between academic study and practical experience in industrial settings that utilize real-world HRC systems in recycling facilities [43]. Furthermore, initiatives such as the 28th Summer School Francesco Turco (2023) function as incubators for collaborative innovation in industrial recycling. The inclusion of women and underrepresented groups is essential for developing equitable Industry 5.0 systems. In Kenya, community-driven e-waste initiatives increasingly integrate affordable robotic equipment, enabling informal laborers to partner with collaborative robots, thereby improving productivity and social dignity. These models illustrate that comprehensive implementation of HRC enhances operational metrics and promotes socio-economic resilience.

Numerous international initiatives demonstrate the transforming efficacy of HRC in CMR. The incorporation of Industry 5.0 methodologies in tungsten carbide scrap recycling in China has resulted in cost reductions and enhanced efficiency via the utilization of adaptive robotic sorters [19]. Likewise, Malaysia and China have illustrated how digitization and HRC (Goh et al. [18]) expedite decarbonization in the energy, waste, and water sectors. Recent initiatives in Nigeria, influenced by Oyebode [44], focus on the management of electronic trash via HRC frameworks to mitigate environmental degradation and health hazards. The EU's Horizon 2020 initiative promotes robotic-assisted recycling solutions via public–private partnerships, prioritizing transparency, safety, and environmental stewardship. These initiatives, enhanced by localized adaptation and stakeholder involvement, offer a framework for scalable and replicable global solutions [45].

To fully exploit HRC's potential in key mineral recycling, coordinated research, policy alignment, and industry partnership are necessary. Abdous et al. [46] advocated for scenario-based optimization frameworks that replicate human–robot interactions in assembly and disassembly lines, providing predictive insights for resource allocation and risk management. Future research should investigate hybrid AI models, edge-AI-enabled collaborative robots, and real-time multi-agent coordination for recycling operations in both urban and rural environments. Policy instruments must be designed to promote ethical robotics utilization, assist SMEs in the adoption of HRC technology, and guarantee transparent stakeholder engagement. Furthermore, international collaboration—especially between resource-abundant poor countries and technologically sophisticated economies [47]—can promote sustainable mineral extraction networks. As Industry 5.0 advances, it is essential to utilize human–robot collaboration not only for operational improvements but also for establishing resilient [48], equitable, and ecologically sustainable recycling systems.

The incorporation of human–robot collaboration (HRC) in key mineral recycling signifies a pivotal advancement in the concepts of Industry 5.0, which prioritizes human-centricity, sustainability, and technological inclusivity. Table 3.1 presents a

TABLE 3.1
Human–Robot Collaboration in Critical Mineral Recycling

S/N	Area of Focus	Category	Innovative Practice/Insight	Citations and Studies	Example and Location	Industry/ Institutional Intervention	Impact on Industry 5.0 Goals
1	Human–robot symbiosis in disassembly	Process innovation	Advanced multi-objective discrete bees algorithm for HR collaboration in disassembly	Shen et al. [2]—*Symmetry*	Germany—Automotive e-waste disassembly line	Bosch deploying AI–human co-assembly for safe mineral component separation	Enhances precision, ergonomics, and safety
2	Reverse logistics and safety	Data-driven optimization	Just-in-time support using social media analytics for material recovery	Shahidzadeh and Shokouhyar [26]—*Industrial Management and Data Systems*	Iran—Reverse logistics networks in mining waste	Tehran University's DSS algorithm reducing handling errors in toxic e-waste	Real-time insight enhances decision support and minimizes hazard exposure
3	Adaptive assembly line design	Cognitive ergonomics	Fuzzy inference system for workload balancing in HR settings	Ghorbani et al. [28]—*The International Journal of Advanced Manufacturing Technology*	Canada—Smart manufacturing for rare earth magnets	Concordia University's Human-Centric Assembly Framework	Reduces fatigue, improves safety and throughput
4	Circular design for HR collaboration	Circular economy integration	Modular design and social engineering optimizer for recyclability	Tian et al. [20]—*International Journal of Production Research*	China—Modular disassembly of electric vehicle batteries	Tsinghua University's social design-driven recycling labs	Aligns circularity with user-centric disassembly goals

5	Industry-AI convergence	Tech infrastructure	Graph CNN with deep RL for resilient recycling prototyping	Leng et al. [51]—*IEEE Transactions on Systems, Man, and Cybernetics*	Singapore—Edge AI enabled smart sorting for lithium-ion batteries	A*STAR AI Labs with Panasonic waste sorting	Scalable AI-driven adaptive production lines for complex e-waste streams
6	Blockchain for traceability	Supply chain security	Blockchain and ISM-Fuzzy DEMATEL for secure recycling chain	Al Amin and Baldacci [13]—*Discover Sustainability*	Bangladesh—RMG sector recycling traceability	BRAC and UNICEF piloting blockchain-secured mineral flows	Builds trust and authentication of ethically recycled materials
7	Additive manufacturing in recycling	Technological enablement	Additive manufacturing for repurposing dismantled components	Sala and Richert [22]—*Materials*	Slovakia—3D-printed sensors from rare metals	Slovak University of Technology—additive remanufacturing pilot	Reduces virgin extraction, enhances sustainable component reuse
8	Sensor-based material recognition	Smart systems	Sensor-based AI sorting to enhance precision and recyclability	Banjanovi-Mehmedovic and Dželilović [8]—*Lecture Notes in Networks and Systems*	Bosnia—Sensor-based AI in plastic–metal composite sorting	Sarajevo Smart Recycling Lab with Siemens	Improves recovery efficiency and operator safety in robot-assisted sorting

Source: Table by the authors.

comprehensive synthesis of advanced studies encompassing intelligent disassembly systems, AI-driven process optimization, sensor-based material characterization, and social-engineering-enhanced recycling solutions. These developments are crucial in meeting the increasing need for the efficient and secure recovery of rare earth elements and key materials, especially from e-waste, lithium-ion batteries, and valuable industrial trash. Significant contributions, such as Shen et al. [2] on multi-objective disassembly-line balance and Ghorbani et al. [28] on adaptive workload management through fuzzy logic, demonstrate the harmonious integration of ergonomics, safety, and automation. Research conducted by Kumar et al. [19] and Taz et al. [7] underscores the importance of sustainable circular systems in hard-metal recycling and petroleum catalyst recovery, respectively, based on empirical data from China, Switzerland, and the Middle East.

Furthermore, Table 3.1 highlights the crucial significance of institutional and industrial interventions, wherein universities, research consortia, and smart city initiatives—exemplified by those in Malaysia and China (Goh et al. [18])—partner with international technology collaborators to implement AI, blockchain, and IoT frameworks that improve traceability, decentralization, and lifecycle analysis of recycled materials. The implementation of modular design [20], additive manufacturing [22], and system dynamics modeling [49] signifies a transition toward predictive, adaptable, and robust manufacturing ecosystems. Innovations such as the Global Trust Registry Initiative [41] illustrate the necessity of cross-border policy alignment and governance systems for the expansion of circular economy models globally [5]. The literature presented in Table 3.1 establishes a solid basis for implementing a comprehensive Industry 5.0 ecosystem that reconceptualizes CMR as both a technical process and a socio-technological framework integrated with equity, safety, sustainability, and intelligent co-creation [50].

3.3 SUSTAINABLE RECYCLING IN THE AGE OF INDUSTRY 5.0: THE ROLE OF CIRCULAR ECONOMY

The shift from linear resource use to a circular, human-centered economic model is central to Industry 5.0. This paradigm shift reconfigures societal approaches to key mineral recycling, highlighting customization, sustainability, and resilience. Critical minerals, like lithium, cobalt, tungsten, and rare earth elements, are essential for sustainable energy systems and advanced manufacturing [52]. However, their extraction and processing frequently incur socio-environmental consequences [53], particularly in resource-dependent countries such as the Democratic Republic of Congo, Australia, and China. Industry 5.0 offers a unique opportunity to incorporate sophisticated technologies such as human–robot cooperation (HRC), AI, blockchain, and ML into closed-loop systems for the recycling of key minerals. Based on the research by Shafique et al. [25], which illustrates the interplay between Industry 5.0 and circular economy frameworks in China's manufacturing industry, we note that sustainable performance can be improved through decentralized, intelligent, and consumer-inclusive systems. This sub-chapter outlines a strategy for the global deployment of these models, highlighting the importance of regional adaptation and comprehensive stakeholder involvement to advance the goals of the circular economy.

Industry 5.0 embodies a human-centric attitude at the convergence of ergonomics, technology, and sustainability. According to Abdous et al. [46] and Ghorbani et al. [28], adaptive assembly line balance and fuzzy inference systems facilitate optimal human–robot collaboration, hence improving safety and productivity in recycling plants. These approaches facilitate effective disassembly and material separation procedures, crucial for the recovery of high-value key minerals. This methodology is exemplified in Europe's European Institute of Innovation and Technology (EIT) Raw Materials programs and the projects of Germany's Fraunhofer Institute, where ergonomic workspaces are integrated with AI-driven decision-making for the recycling of EoL devices. Customizing recycling systems to align with worker skills reduces cognitive and physical strain while enhancing throughput. Furthermore, Järvenpää et al. [17] in Finland demonstrated how consumer and household engagement in garbage recycling influences the efficacy of downstream recovery. These human-centered designs render recycling a social innovation alongside its technological aspects—crucial for developing countries characterized by elevated unemployment and informal garbage economies.

Contemporary reverse logistics systems are evolving to be more intelligent, decentralized, and data-centric, transforming the recirculation of essential resources. Utilizing social media analytics, Shahidzadeh and Shokouhyar [26] introduced a JIT decision support system that enhances the timing and placement of recycling initiatives. This concept has significant implications for nations such as India and Brazil, where swift urbanization strains municipal garbage management systems. Al Amin and Baldacci [13] demonstrated how JIT recovery systems can facilitate transparent monitoring of mineral transactions, authenticate origins, and increase confidence in secondary markets when combined with blockchain platforms. Moreover, sophisticated multi-objective algorithms that integrate lean concepts with augmented and VR (X-Reality) interfaces are enhancing disassembly planning, a development critiqued by Yang et al. [32] and He et al. [24]. These instruments assist proficient operators with intricate decommissioning activities, such as the separation of batteries or circuit boards, while assuring little loss and contamination. These tactics are progressively implemented in Canada's electric vehicle recycling startups and EU-supported Horizon 2020 initiatives.

Cutting-edge sensing technologies and AI are fundamentally transforming mineral recovery operations. Banjanovi-Mehmedovic and Dželilović [8] delineate sensor-based material characterization models for waste sorting, which significantly enhance the purity and efficiency of recovered key metals. Machine vision systems, combined with ML classifiers, currently attain over 95% accuracy in detecting rare metals in shredded electronic waste, as demonstrated at trial facilities in South Korea and Belgium. Simultaneously, Leng et al. [14, 51] have presented blockchain-enabled IIoT frameworks and graph-based neural networks for the development of robust recycling systems. Their approach supports the ideas of the circular economy because it is consistent with the comprehensive 4R sustainability framework (reduce, reuse, recycle, and recover), which Siakas et al. [54] examined. Kumar et al. [19] assert that these technologies are especially advantageous in the production of tungsten carbide, where high-strength composites necessitate meticulous disassembly and grinding. With the global spike in resource demand, particularly for electric vehicles and wind turbines, these technologies are essential for achieving the objectives of Industry 5.0, which include agility, efficiency, and environmental sustainability.

The effective incorporation of sustainable recycling methods into Industry 5.0 depends on novel business models and institutional collaborations. Azinge [55] underscores the significance of circular business models that transition from product ownership to service-oriented systems (e.g., mineral-as-a-service, pay-per-use), which prolong product lifecycles and enhance traceability. Strategic partnerships across governmental bodies, academic institutions, and industries—exemplified by the United Kingdom's Ellen MacArthur Foundation and Japan's METI-led e-waste initiatives—facilitate scalable solutions. The Global Trust Registry Initiative by Voutier et al. [41] provides a forum for international cooperation, harmonizing G20 objectives with regional efforts via shared data infrastructures and trust protocols. The International Renewable Energy Agency (IRENA) is implementing lithium-ion battery recycling frameworks across Africa and Southeast Asia, aiming to enhance material recovery and generate employment. These cooperations facilitate equitable technology transfer, a vital concern in the global South, where the impact of mineral extraction is frequently experienced most intensely.

Design for disassembly and modularity are increasingly recognized as essential methods for the circularity of important minerals. Tian et al. [20] and Neri et al. [6] emphasize that modular product architectures enhance recyclability by facilitating rapid component-level replacement, minimizing downtime, and promoting remanufacturing. This methodology is especially crucial for advanced electronics, medical apparatus, and electric vehicle batteries—industries characterized by rapid product development and intricate material formulations. In the Netherlands, enterprises such as Fairphone are innovating modular cellphones, facilitating the effective recovery of cobalt and rare earth materials. Psarommatis et al. [33] delineate significant obstacles to repurposing and reuse, including inadequate standardized material labeling and deficient design documentation, while proposing AI-augmented lifecycle management systems to facilitate a closed-loop process. Richnák and Fidlerová [56] emphasize the necessity of synchronizing sustainable development objectives (SDGs) with technical innovation across various sectors. These integrated solutions encapsulate the idea of Industry 5.0, wherein digital tools, environmental awareness, and human collaboration converge to generate sustainable value.

Resilience is fundamental to Industry 5.0, particularly in mineral recycling, necessitating a cyber-physical infrastructure that can withstand disruptions and adjust to alterations. Wang et al. [49] and Ma et al. [57] endorse the utilization of system dynamics models and intricate network architectures to enhance flexibility in manufacturing and recycling systems. This is particularly pertinent in post-pandemic economies when supply chains for essential minerals have been significantly disrupted. In response, cities such as Shenzhen and Munich are incorporating AI-driven inventory systems alongside waste stream analytics, thereby improving their adaptability to market and environmental situations. Zio and Guarnieri [39] contend that risk assessment approaches must advance by integrating real-time data from IIoT networks to anticipate breakdowns and proactively adjust flows. Murali et al. highlight carbon finance options that promote circular construction processes, thereby integrating sustainability into supply chains. The transformation of infrastructure from inflexible, centralized systems to adaptable, decentralized ecosystems is crucial to Industry 5.0.

Developing a global strategy for sustainable CMR within Industry 5.0 requires an integration of technology, policy, and social innovation. According to Hong [42], educational and capacity-building initiatives would be crucial for preparing the upcoming generation of engineers and policymakers with interdisciplinary expertise. Similarly, platforms such as Goh et al. [18] offer persuasive case studies of smart cities in Malaysia and China, where digital twins, AI, and decarbonization strategies intersect to foster sustainable urban metabolism. These models provide reproducible insights for towns in Africa, Latin America, and the Middle East seeking to advance into circular economies. The contributions of research institutions and periodicals like Cleaner Production, Sustainability (Switzerland) and Computers in Industry underscore the significance of continuous multidisciplinary study in guiding evidence-based solutions. With the global need for essential minerals surging, integrating Industry 5.0 with circular economy concepts provides a framework for systemic transformation—guaranteeing material security, climate resilience, and equitable development for everyone.

Table 3.2 provides an advanced and complex analysis of the synergies among Industry 5.0, sustainable CMR, and the circular economy. These academic papers jointly outline a worldwide framework for a human-centered and resilient industrial model, highlighting digitalization, AI technologies, human–robot collaboration, decentralized supply chains, and sustainable waste management. Studies conducted by Abdous et al. [46] and Ghorbani et al. [28] advance the field of adaptive human–robot collaboration, facilitating ergonomic and efficient assembly line configurations essential for disassembly and remanufacturing operations. Psarommatis et al. [33] and Kumar et al. [19] provide comprehensive analyses of the reuse and recovery of crucial materials such as tungsten carbide, highlighting the essential need for sustainable practices and sophisticated material science in achieving resource circularity. The integration of AI, digital twins, and decentralized blockchain systems across several industries, from smart cities [17] to reverse logistics analytics [26], signifies a new era of industrial symbiosis aimed at reducing environmental impacts.

Moreover, the information demonstrates a regional and institutional variety in leading sustainability-focused industry transformations. Goh et al. [18] illustrate the use of Industry 5.0 paradigms in waste–energy–water nexus solutions by Malaysia and China, whereas Al Amin and Baldacci [13] investigate blockchain-enabled circularity in Bangladesh's RMG sector. European initiatives—exemplified by Sala and Richert [22] on additive manufacturing and Wachter et al. [58] on resilient manufacturing strategies—indicate a shift toward risk-informed and failure-resilient production systems. Digital intelligence plays a vital role in material recovery, as demonstrated in Delbari and Hof's [3] study on high-value glass recycling and Neri et al.'s [6] research on lithium-ion battery remanufacturing, illustrating how innovation can sustainably mitigate important mineral scarcity. Table 3.2 represents a comprehensive integration of disciplines, geographies, and methodologies that collectively promote the transformative vision of Industry 5.0: a paradigm that harmonizes technological

TABLE 3.2
Human-Centric Innovations and Circular Economy Synergies for Sustainable Critical Mineral Recycling in the Era of Industry 5.0

Focus Area	Category	Innovative Contribution	Key Authors/ Citation	Example/Case Study	Geographical Scope	Institutional/ Industrial Intervention	Industry 5.0 Relevance
Human-centric robotics	Collaborative disassembly	Human–robot collaboration for efficient mineral recovery from e-waste	Shen et al. [2]; Chand and Lu [16]	Disassembly-line optimization using bees algorithm	China, USA	Siemens, ABB deploying cobots in recycling centers	Human–robot synergy, ergonomic efficiency
Circular business models	Resource recovery frameworks	AI-driven circular business models for high-value recovery	Azinge [55]; Psarommatis et al. [33]	Modular design for remanufacturing components	Nigeria, EU	Circular economy policies in EU Green Deal	Promotes extended product life, reuse
Reverse logistics and smart sorting	Waste stream intelligence	JIT reverse logistics via social media analytics and sensor-based sorting	Shahidzadeh and Shokouhyar [26]; Banjanovi-Mehmedovic and Dželilović [8]	Smart bins and ML-sorting for mineral recovery	Iran, Bosnia	Bosch smart bins, Eco-Cycle NGO	Enhances recycling efficiency and transparency
Digitalization and smart manufacturing	AI, IoT, blockchain	Blockchain + AI for tracking critical minerals in real time	Al Amin and Baldacci [13]; Leng et al. [14]	Blockchain traceability of lithium batteries	Bangladesh, Germany	Hyperledger, BASF blockchain pilots	Ensures traceability and circular loop closure
Material and design innovation	Sustainable product design	Modular and resilient product design for easy disassembly	Tian et al. [20]	Battery packs designed for end-of-life recovery	China, UK	Tesla's battery modularity; Fairphone's design	Enables DfR (design for recycling)

Circular manufacturing and repurposing	Lifecycle thinking	Systematic reuse and repurpose strategies in Industry 5.0 factories	Psarommatis et al. [33]; Kumar et al. [19]	Tungsten carbide recycling in tool production	India, Switzerland	Sandvik's tungsten loop initiative	Reduces virgin mineral dependency
Smart cities and circular systems	Urban mining integration	Smart city digitalization for e-waste mining	Goh et al. [18]; Järvenpää et al. [17]	SSC initiatives in Malaysia and China	Malaysia, China, Finland	Huawei Smart City Lab, Sitra Circular Economy	Urban mineral extraction, citizen-driven recycling
Policy and global governance	Institutional and multilateral roles	Connecting G20 sustainability actions with industry practices	Voutier et al. [41]	Global Trust Registry for mineral accountability	Global (G20 nations)	G20, UNIDO, IRENA partnerships	Promotes global governance for mineral ethics

Source: Table by the authors.

advancement with ecological sustainability and human-centered principles—an essential basis for cleaner, circular, and equitable production systems.

3.4 ETHICAL AI AND RESPONSIBLE AUTOMATION IN RECYCLING

In the Industry 5.0 age, the incorporation of ethical AI and responsible automation is pivotal in transforming industries, particularly in essential mineral recycling. Industry 5.0, emphasizing human-centric and sustainable methodologies, offers a distinctive chance to transform the recycling of essential minerals through the utilization of advanced technologies. Ethical AI is essential for ensuring that automation not only enhances operations but also adheres to ideals of justice, transparency, and responsibility. In the realm of crucial mineral recycling, this entails the creation of AI algorithms capable of efficiently sorting, processing, and extracting precious materials while preserving environmental integrity and human well-being. Incorporating ethical standards into AI-driven recycling systems helps alleviate the dangers of technology biases, employment displacement, and environmental damage. AI solutions can guarantee that the recycling process attains elevated recovery rates while conforming to sustainable practices consistent with circular economy concepts.

Recent advancements in robotics and ML exemplify the connection between ethical AI and responsible automation in vital mineral recycling. These technologies can enhance the sorting and segregation of various materials, augmenting efficiency and accuracy. AI can operate robotic arms that detect and categorize mineral-rich waste products for subsequent processing. Utilizing ML, these systems may perpetually enhance their capacity to identify and manage various trash types, so rendering the recycling process more intelligent and adaptable over time. Ensuring these systems are inclusive and provide solutions that uphold the rights and dignity of people potentially facing job displacement due to automation includes an ethical component. Shafique et al. [25], who looked at the interaction between business 5.0 and the circular economy within the Chinese manufacturing sector, emphasized the partnership between business and academia in creating AI models that promote equitable job distribution as an example. In these collaborations, ethical issues are prioritized, ensuring that AI optimizes recycling results while also enhancing community and workforce welfare.

An additional crucial element of ethical AI and responsible automation is the design of these systems with complete accountability and transparency. In essential mineral recycling, this entails the creation of AI systems proficient in monitoring and documenting the lifecycle of recycled minerals, guaranteeing that all procedures are transparent and verifiable. This traceability is essential for adherence to environmental norms and regulations, as well as for fostering public confidence in recycling activities. Ethical AI systems must guarantee transparency across the entire process, from data collection to final recycling, with prompt identification and rectification of any deviations or inefficiencies. Leng et al. [51] proposed the utilization of blockchain technologies for decentralized material tracking and AI systems for monitoring compliance with sustainability criteria to attain transparency. These initiatives facilitate the establishment of a digital trust foundation that supports the ethical functioning of recycling systems in Industry 5.0.

Consequently, responsible automation in essential mineral recycling must emphasize environmental sustainability. Industry 5.0 advocates for sustainable techniques that reduce the ecological impact of industrial operations. Automating important mineral recycling operations enables enterprises to minimize waste, decrease energy consumption, and enhance resource recovery, thereby supporting the circular economy. AI can enhance the energy efficiency of recycling facilities, guaranteeing that the energy-intensive procedures of mineral extraction from electronic trash are performed optimally, hence reducing energy waste. This is especially significant given the increasing demand for minerals such as lithium, cobalt, and rare earth elements, which are crucial for clean energy technology and are frequently extracted from ecologically sensitive regions. Ethical AI systems should be engineered to reduce the environmental impact of extraction and recycling, guaranteeing that automation fosters more sustainable and environmentally friendly operations.

Moreover, ethical AI in crucial mineral recycling must guarantee that the implementation of automation does not compromise workforce well-being. A significant difficulty of automation is its capacity to displace employees. Industry 5.0 prioritizes human-centric solutions, ensuring that automation enhances rather than supplants human labor. AI and automation can be employed to support personnel in their duties, improving safety and alleviating the physical burden associated with recycling processes. AI-driven robotic devices can assume the most perilous and monotonous jobs, such as heavy lifting or the sorting of hazardous goods, allowing people to concentrate on advanced decision-making and process management. Ethical AI should be developed to enhance human abilities rather than supplant them. This methodology corresponds with the findings of Ghorbani et al. [28], who promoted the implementation of human-centric robotic systems in assembly line design, applicable to important mineral recycling.

AI can improve the effectiveness of recycling systems via predictive analytics, enabling the anticipation of material shortages, market fluctuations, and supply chain interruptions. These AI models can enhance inventory management and resource allocation, ensuring that recycling systems are robust and responsive to evolving conditions. Responsible automation in this domain guarantees that AI systems are engineered not just for economic efficiency but also to alleviate risks such as the exhaustion of essential resources or the intensification of supply chain vulnerabilities. The utilization of AI in recycling presents opportunities for scalability, allowing small-scale enterprises to enhance efficiency and bigger systems to navigate the intricacies of global supply networks. Jeyaraman et al. [21] examined the seamless integration of Industry 5.0 technologies across many sectors, including recycling, which fosters resilience, sustainability, and economic stability.

Alongside technological breakthroughs, the ethical aspect of AI and automation in vital mineral recycling necessitates robust regulatory frameworks and industry norms. Governments, corporations, and academic institutions must work to formulate rules that guarantee the responsible implementation of AI and automation technologies. These criteria must encompass data privacy issues, environmental consequences, and social equality. Standards ensuring the proper utilization of data collected by sensors and monitoring systems, while safeguarding privacy and protecting vulnerable people, should regulate the application of AI in recycling facilities. These initiatives correspond with the current discourse on the ethical ramifications of AI in industrial sectors, as emphasized by Leng et al. [51] in their analysis of safe blockchain middleware for decentralized Industry 5.0 applications.

The advancement of ethical AI and responsible automation in key mineral recycling necessitates ongoing research, collaboration, and invention. Industry 5.0 signifies a fluid and progressive paradigm, necessitating that AI systems remain responsive to emerging challenges and opportunities. Advancements in AI algorithms, ML models, and robotics will persist, necessitating thorough testing to assure compliance with ethical standards and support sustainable recycling methods. Universities and research institutions, in partnership with companies, should be pivotal in the advancement and implementation of these technologies. Shafique et al. [25] emphasize that the integration of Industry 5.0 and circular economy principles will underpin sustainable practices in crucial mineral recycling. Ethical AI and responsible automation will be essential catalysts for this transformation, guaranteeing that the advantages of Industry 5.0 encompass both the environment and society.

The study illustrated in Table 3.3 provides a comprehensive and innovative analysis of the significance of ethical AI and responsible automation in enhancing the sustainability and efficiency of essential mineral recycling in the framework of Industry 5.0. It underscores the incorporation of human-centric methodologies, wherein technology and innovation are crafted to augment and elevate human proficiency in the recycling process. This study examines current trends in AI, automation, and circular economy principles to determine how Industry 5.0 might enhance the recovery of essential minerals, minimize environmental impacts, and encourage responsible resource utilization. This encompasses the implementation of ethical AI frameworks that emphasize openness, justice, and accountability in decision-making, alongside the creation of automated systems that uphold worker welfare and guarantee safety in recycling operations. Global case studies and interventions by industry leaders and governmental entities illustrate the viability and effectiveness of these ideas, highlighting the possibility for scalable and sustainable recycling solutions.

The research emphasizes the strategic integration of responsible automation, ethical AI, and Industry 5.0 principles, concentrating on the utilization of advanced digital technologies, including AI-driven robots, IoT-enabled systems, and blockchain, to enhance material flow and ensure transparent supply chain management. The study identifies the most promising interventions in the sector through a review of contemporary literature and practical applications, including the implementation of AI algorithms for predictive maintenance in recycling machines and ML models for quality control in recycled materials. Prominent instances from nations such as China and Germany, recognized for their prowess in Industry 5.0 and sustainable methodologies, offer valuable perspectives on the practical obstacles and achievements in incorporating these technologies into important mineral recycling operations. This chapter promotes a worldwide, interdisciplinary strategy that integrates technological advancement with ethical considerations to fully realize the potential of Industry

TABLE 3.3
Ethical AI and Responsible Automation: Pioneering Human-Centric Approaches for Sustainable Critical Mineral Recycling in Industry 5.0

Area of Focus	Description	Innovation	Example	Location	Nationality of Study	Institution/ Industry Interventions	Citation
Ethical AI in critical mineral recycling	Ethical considerations in deploying AI systems for sustainable and human-centric recycling processes.	Development of AI that prioritizes human well-being and environmental sustainability.	Integration of AI in sorting and processing mineral waste, ensuring fairness and transparency.	Global	Multinational	AI systems in circular economy initiatives such as AI-based sorting machines ensuring ethical disposal of critical minerals. Industry 5.0 promotes transparency and fairness in recycling processes.	Ghorbani et al. [28] Human-centric robotic assembly line design
AI for waste sorting and resource efficiency	AI-driven systems to enhance waste sorting accuracy and efficiency in recycling critical minerals.	Automated Intelligent sorting systems leveraging deep learning and machine vision.	Smart recycling bins using AI to identify mineral waste and direct them to recycling processes.	Japan, USA	Japan, USA	Companies like Veolia and TOMRA use AI and robotics for waste sorting and resource recovery in urban and industrial sectors.	Goh et al. [18] Expediting decarbonization through AI in waste management

(*continued*)

TABLE 3.3 (Continued)
Ethical AI and Responsible Automation: Pioneering Human-Centric Approaches for Sustainable Critical Mineral Recycling in Industry 5.0

Area of Focus	Description	Innovation	Example	Location	Nationality of Study	Institution/ Industry Interventions	Citation
Responsible automation in recycling	Implementing automation in critical mineral recycling with a focus on minimizing negative societal impacts.	Automation that not only boosts efficiency but also adheres to ethical and sustainable practices.	Automation in the recycling of rare earth metals using robotic arms, while reducing worker exposure to toxic substances.	Germany, China	Germany, China	Siemens and ABB incorporate responsible automation practices in manufacturing and recycling facilities to improve sustainability and worker safety.	Du et al. [27] AI-based UAVs for Industry 5.0 applications in material recovery
Circular economy integration	Using Industry 5.0 technologies to integrate circular economy principles in critical mineral recycling.	Circular AI ecosystems that close the loop of resource use through responsible automation and AI.	Circular processes for recycling lithium-ion batteries, where AI optimizes material recovery and repurposing.	UK, Australia	UK, Australia	The Ellen MacArthur Foundation promotes circular economy principles, with industry players like Tesla and Umicore adopting AI and automation to improve resource recovery and sustainability.	Shafique et al. [25] The synergy of Industry 5.0 and circular economy in manufacturing

Sustainable resource recovery	Ethical AI systems designed to ensure minimal environmental and human harm during the extraction and recycling of minerals.	Use of AI for optimizing resource recovery with minimal environmental footprint and human risk.	AI-driven processes in recycling cobalt and nickel from old electronics, enhancing efficiency while reducing environmental impact.	Brazil, South Africa	Brazil, South Africa	Companies like Umicore use AI to improve the sustainability of resource recovery in the mining and recycling industries, reducing environmental and human risks.	Taz et al. [7] Recovery of high-grade critical metals from spent petroleum catalysts
Human–robot collaboration (HRC)	Human–robot collaboration in the recycling process to enhance safety, productivity, and efficiency.	Human-centric robots that collaborate with human workers to improve precision and efficiency in recycling processes.	Robotic systems that assist in the safe dismantling of electronic waste (e-waste) to recover critical minerals, reducing human exposure to toxic materials.	India, USA	India, USA	ABB Robotics and FANUC Robotics have developed collaborative robots (cobots) to support human workers in the disassembly of electronic waste for mineral recovery.	Jeyaraman et al. [21] Industry 5.0 in orthopaedics and robotics integration

(*continued*)

TABLE 3.3 (Continued)
Ethical AI and Responsible Automation: Pioneering Human-Centric Approaches for Sustainable Critical Mineral Recycling in Industry 5.0

Area of Focus	Description	Innovation	Example	Location	Nationality of Study	Institution/ Industry Interventions	Citation
Impact of AI on labor in recycling	Evaluating the implications of AI and automation on employment in critical mineral recycling sectors.	Human-centric models that focus on reskilling and upskilling workers displaced by AI and automation.	Reskilling programs for workers transitioning from traditional recycling methods to automated AI systems in critical mineral recycling.	South Korea, EU	South Korea, EU	The EU-funded Horizon 2020 projects focus on AI-based worker transition programs to prepare displaced workers in industries like mining and recycling for new roles in the AI-driven economy.	Abdous et al. [46] Scenario-based optimization for human-centered assembly lines
AI ethics and transparency	Developing frameworks for the ethical use of AI in critical mineral recycling, ensuring transparency, accountability, and fairness.	Creation of transparent AI models that allow stakeholders to understand and audit decisions made by AI systems.	Blockchain-backed systems for tracking AI decisions in recycling, ensuring fairness and accountability in material sorting and recovery.	Canada, Switzerland	Canada, Switzerland	Blockchain initiatives like IBM's Food Trust Network and VeChain ensure that AI-driven recycling processes are fully auditable, reducing risks of fraud and environmental harm.	Leng et al. [14] Blockchain in decentralized IIoT towards Industry 5.0

Source: Table by the authors.

5.0 in meeting the increasing demand for key minerals while maintaining social and environmental sustainability.

3.5 SOCIAL RESPONSIBILITY AND INCLUSION IN INDUSTRY 5.0 RECYCLING

The integration of social responsibility and inclusion in essential mineral recycling signifies a crucial transformation toward a more sustainable, egalitarian, and human-centric industrial model in the framework of Industry 5.0. Industry 5.0 highlights the integration of sophisticated technology, including AI, robotics, and blockchain, with human participation to promote ethical, efficient, and sustainable processes. Recycling of critical minerals, vital for sectors including electronics, automotive, and renewable energy, constitutes a fundamental aspect of this concept. Industry 5.0's recycling approach transcends environmental issues by integrating social responsibility through equitable labor practices, access to technologies, and democratic decision-making processes. The convergence of circular economy concepts with the human-centric emphasis of Industry 5.0 advocates that innovations in industrial operations should benefit both individuals and the environment [25].

One of the most fundamental ways Industry 5.0 supports social responsibility is through the enhancement of worker inclusivity. In conventional sectors, the labor force has frequently been marginalized in favor of automation. The emergence of human–robot collaboration in Industry 5.0 guarantees that workers are not just participants but are essential to the decision-making and technical implementation stages [2]. In vital mineral recycling, employing robotics for the dismantling and sorting of electronic trash (e-waste) may diminish dangerous manual work, thereby fostering safer and more inclusive employment opportunities. This methodology has yielded favorable outcomes in numerous pilot initiatives throughout Europe, notably in Germany and Sweden, where automation is integrated with human supervision to establish hybrid systems that promote upskilling and lifelong learning, especially for marginalized workers [21].

A vital component of social responsibility within the Industry 5.0 paradigm is the engagement of local communities in the recycling process. The human-centric model of Industry 5.0 promotes participatory methods, engaging consumers, households, and communities directly in recycling initiatives. Järvenpää et al. [17] illustrate how intelligent waste management systems in Finland enable users to monitor and enhance their recycling practices through AI-driven applications and sensors. Industry 5.0 fosters a socially responsible ecosystem by enabling individuals to assume responsibility for their recycling practices, thereby establishing resource conservation as a collective community objective. Furthermore, from the perspective of social inclusion, these technologies might significantly influence rural or underprivileged urban regions where access to superior recycling facilities is constrained.

Integrating circular business models also improves the inclusion of essential mineral recycling in Industry 5.0. Principles of the circular economy, including product reuse and repurposing, are essential for diminishing the demand for newly extracted minerals. Azinge [55] examines the success of circular business models in countries such as Japan and South Korea, where governmental regulations promoting sustainability have encouraged remanufacturing and refurbishment. These methods frequently engage local businesses and social enterprises in the recycling value chain, generating employment and promoting economic development within communities. Moreover, establishing decentralized recycling hubs in communities enhances the accessibility of the recycling process, especially in economically disadvantaged areas. This decentralized method fosters economic inclusion by creating local employment possibilities and enabling communities to participate directly in the recycling process.

Blockchain technology, integral to Industry 5.0, improves transparency, accountability, and confidence in the recycling of essential minerals. Blockchain technologies facilitate traceability inside the recycling chain, guaranteeing that recycled materials are acquired responsibly and treated ethically. Al Amin and Baldacci [13] emphasize the promise of blockchain technology in the textile sector, where it has been utilized to trace the provenance and trajectory of goods. This transparency cultivates customer trust and encourages responsible consumption and disposal habits. In the realm of CMR, blockchain technology may be employed to track the lifecycle of precious metals derived from electronic waste, safeguarding against exploitation or improper disposal, while concurrently establishing an inclusive marketplace that links recyclers, consumers, and suppliers with transparency and efficiency.

Industry 5.0 emphasizes inclusive decision-making, thereby addressing social responsibility in the recycling of essential minerals. In a context where prominent firms or governments often determine waste management and recycling policies, Industry 5.0's focus on democratized decision-making facilitates the involvement of diverse stakeholders in policy and process formulation. The utilization of AI methods to assess public comments and incorporate community participation into recycling programs guarantees that the implemented policies advantage all segments of society. These innovations are presently undergoing trials in cities such as Amsterdam and Toronto, where municipal authorities utilize digital platforms to collect public feedback on waste management strategies, so guaranteeing that recycling policies align with the requirements and desires of the community [28].

The significance of education and awareness initiatives in fostering social responsibility in Industry 5.0 is paramount. Public education initiatives, bolstered by AI and IoT, provide real-time monitoring and customer feedback, enhancing knowledge of the environmental and social ramifications of vital mineral recycling. Nayeri et al. [23] assert that the integration of educational programs in the digital domain, including mobile applications and online courses, has transformed consumer interaction with recycling. Industry 5.0 promotes collaboration between institutions, such as colleges and research organizations, and industry stakeholders to develop curriculum that equips the forthcoming workforce with skills in sustainable recycling practices, so enabling their participation in a cleaner, greener economy. This comprehensive teaching method cultivates a sustainable culture that pervades all societal strata.

The ethical concerns of adopting Industry 5.0 are examined through the use of AI in decision-making processes about recycling. ML algorithms can enhance the recycling process, increasing the efficiency of resource recovery from waste materials while assuring equitable labor practices and fostering sustainable outcomes [20]. This prompts inquiries regarding the ethical obligations of AI systems in promoting social fairness. Do these systems incorporate inclusivity in their design, or do they inadvertently sustain inequality by privileging specific regions or demographics over others? Zio and Guarnieri [39] assert that resolving these ethical issues necessitates the active engagement of various groups by industry leaders in the design and implementation of AI technology to guarantee equal outcomes.

The convergence of social responsibility and inclusion in the context of Industry 5.0 essential mineral recycling presents a transformative strategy for sustainable development. Utilizing new technologies like AI, robotics, blockchain, and IoT, Industry 5.0 may provide an inclusive, socially responsible, and equitable recycling ecosystem that tackles environmental issues and societal requirements. The varied initiatives by international organizations, corporations, and governments, along with the proactive participation of communities, affirm that crucial mineral recycling is both a technological challenge and a social necessity. Industry 5.0 provides a framework for a future in which economic, environmental, and social factors are cohesively linked to establish a sustainable and inclusive global economy.

Table 3.4 provides a detailed examination of the essential role of social responsibility and inclusion in the Industry 5.0 framework, specifically with crucial mineral recycling. The table summarizes essential achievements in recycling processes by analyzing a variety of innovative technologies, behaviors, and interdisciplinary collaborations that contribute to a human-centric, sustainable future. The items in this table demonstrate the integration of social responsibility with Industry 5.0 objectives, emphasizing creative recycling solutions that encompass the global landscape. The table provides insights into the integration of AI, robots, blockchain, and advanced manufacturing technologies, all of which enhance the circular economy and promote sustainable growth. The research underscores the significance of social inclusion, especially via community involvement, public–private collaborations, and policies that emphasize environmental stewardship and socio-economic justice. This research highlights how Industry 5.0 utilizes technology to promote equity and inclusivity for marginalized communities through strategic interventions, including the promotion of ethical supply chain practices and the creation of new green jobs.

Table 3.4 presents a varied array of geographical contexts, featuring contributions from Asia, Europe, Africa, and the Americas, demonstrating the global applicability and versatility of Industry 5.0 technology. A diverse array of stakeholders, encompassing academics, corporate leaders, governmental entities, and non-governmental groups, oversees these interventions. All are striving toward the identical goals of environmental sustainability, social inclusion, and economic progress. Examples encompass the incorporation of intelligent waste management systems in metropolitan areas, the utilization of blockchain to guarantee transparent and ethical mineral procurement, and AI-enhanced strategies for maximizing material recovery from electronic waste. The chart highlights the importance of human–robot collaboration, wherein advanced technologies prioritize not just productivity but also the establishment of safer, more inclusive work environments. The research presented in Table 3.4 highlights the social implications of Industry 5.0, significantly contributing to the discourse on leveraging technology for environmental and social benefits, thereby fostering a more sustainable and inclusive future in CMR.

TABLE 3.4
Advancing Social Responsibility and Inclusion in Industry 5.0: Innovative Approaches to Sustainable Critical Mineral Recycling

Area/Category	Description	Innovation	Building Block for Industry 5.0 Development	Study/Citation	Examples/ Applications	Location/ Nationality of Study	Interventions by Industry and Institutions
Human-centric approach	Focus on ensuring the inclusion of diverse social groups in the recycling process, prioritizing health, safety, and well-being.	Integrating human–robot collaboration and AI to enhance workforce participation in recycling processes.	Promotes the human-centric ethos in Industry 5.0, ensuring workforce welfare in critical mineral recycling.	Ghorbani et al. [28]	Human–robot collaboration in assembly lines for adaptive workload management, improving worker safety and participation.	Global (cross-national study)	Industry adopting collaborative robots to reduce physical strain on workers and improve efficiency.
Circular economy integration	Adopting circular business models to minimize waste and maximize resource reuse in the recycling of critical minerals, aligning with Industry 5.0's sustainability goals.	Circular business models based on Industry 5.0 that enable sustainable product lifecycle management, such as through the use of AI and IoT for real-time recycling.	Supports a sustainable and inclusive approach by rethinking product lifecycles, resource recovery, and closing material loops.	Azinge [55]	Use of blockchain to track and ensure transparency in recycled material usage, fostering trust among stakeholders.	Global (focus on Industry 5.0 integration)	Institutions such as IRENA promoting circular economy principles in manufacturing and recycling industries globally.

Inclusive workforce participation	Ensuring diversity in the workforce involved in critical mineral recycling, offering equal opportunities for underrepresented groups in advanced manufacturing settings.	Industry 5.0 technologies, like smart wearable devices and machine learning, to tailor roles to individual strengths, encouraging wider participation.	Encourages diversity in the workforce, enabling broader engagement in high-tech sectors like critical mineral recycling.	Shafique et al. [25]	Involving marginalized groups through targeted training and the use of accessible technologies in recycling facilities.	China (case study)	National-level policy initiatives to promote diversity in STEM fields and Industry 5.0 adoption for inclusive participation.
Ethical supply chains	Promoting ethical sourcing and recycling of critical minerals, ensuring fair labor practices and environmental responsibility across global supply chains.	Application of ethical AI algorithms to monitor and enforce ethical standards across the supply chain.	Supports global cooperation for ethical standards, reducing negative impacts from mining and recycling industries.	Taz et al. [7]	Ethical sourcing of cobalt and lithium used in electric vehicle batteries, ensuring fair labor practices.	Africa and Asia (cobalt, lithium sourcing)	Industry initiatives focused on human rights and fair labor practices in critical mineral extraction, supported by global ethical guidelines.

(continued)

TABLE 3.4 (Continued)
Advancing Social Responsibility and Inclusion in Industry 5.0: Innovative Approaches to Sustainable Critical Mineral Recycling

Area/Category	Description	Innovation	Building Block for Industry 5.0 Development	Study/Citation	Examples/ Applications	Location/ Nationality of Study	Interventions by Industry and Institutions
Social engineering for collaboration	Encouraging collaborative efforts between communities, governments, and industries to promote sustainable mineral recycling and avoid negative social impacts.	Use of social engineering techniques to optimize stakeholder collaboration in recycling networks.	Facilitates multi-level engagement and ensures inclusive development in Industry 5.0, promoting cooperative environmental and social outcomes.	Tian et al. [20]	Community-based recycling programs with local governments to increase public awareness and participation in recycling initiatives.	USA (community-based model)	Collaboration between local authorities and tech firms to engage communities in recycling processes and foster sustainable behaviors.
Technology for inclusivity	Ensuring accessibility of advanced technologies in recycling processes for all, including individuals with disabilities and those from disadvantaged socio-economic backgrounds.	Development of accessible AI-driven tools and devices to assist those with disabilities in critical mineral recycling tasks.	Promotes an inclusive, equitable workforce by making advanced recycling technologies accessible to all demographic groups.	Shen et al. [2]	AI-driven disassembly-line optimization for workers with mobility impairments in recycling centers.	Global (Inclusive technology)	Institutions promoting accessibility standards for manufacturing and recycling technologies to ensure equitable access.

Sustainability and decarbonization	Prioritizing sustainable practices and minimizing carbon emissions in critical mineral recycling to meet global sustainability goals.	Decarbonization strategies integrated with Industry 5.0 technologies such as renewable energy-powered recycling systems.	Enhances sustainability by reducing the carbon footprint of recycling operations, aligning with global climate action goals.	Goh et al. [18]	Implementation of renewable energy-powered AI and robotic systems for waste sorting and critical mineral recycling in smart cities.	Malaysia, China (smart cities)	Industry efforts toward integrating renewable energy solutions into the recycling process, supported by government incentives.
Smart waste management systems	Developing intelligent systems to manage the waste produced in the recycling of critical minerals, ensuring efficiency, resource recovery, and waste reduction.	Use of IoT sensors and AI-powered predictive analytics to optimize waste sorting and recycling processes in real time.	Facilitates the creation of smart waste management systems that are adaptive and resource-efficient.	Banjanovi-Mehmedovic and Dželilović [8]	Smart waste sorting systems in industrial parks, using AI to identify and segregate materials based on recyclability.	Europe (smart recycling systems)	Industry and municipalities deploying smart recycling systems in urban centers to increase efficiency and reduce waste.

Source: Table by the authors.

3.6 PERSONALIZATION AND CUSTOMIZATION IN RECYCLING SYSTEMS

In the domain of crucial mineral recycling, personalization and customization are essential to satisfy the requirements of the advancing Industry 5.0 landscape. As the world transitions to human-centric methodologies in manufacturing, these principles have become essential for developing sustainable and efficient systems that meet both personal and industrial requirements. This chapter examines how the personalization and customization of crucial mineral recycling processes provide transformative solutions within the Industry 5.0 framework. This approach hinges on utilizing new technologies like AI, ML, and the IoT, enabling recycling systems to be customized for certain mineral kinds, recycling techniques, and operational environments. These innovations, coupled with human–robot collaboration, enhance recycling efficiency and ensure systems stay adaptable to changing market and environmental situations.

An illustration of customization in CMR is the application of AI and ML to forecast and enhance the extraction of precious elements from electronic trash. Shafique et al. [25] assert that Industry 5.0 prioritizes the amalgamation of circular economy concepts with technical innovations, establishing a framework for systems capable of adapting to real-time data regarding mineral composition and operational conditions. This is accomplished by employing intelligent sensors and AI models that perpetually assess the composition of incoming waste materials, modifying recycling parameters to optimize recovery rates. By tailoring the recycling process to the distinct attributes of each material batch, organizations may enhance efficiency and minimize expenses. Such methods have been implemented in the electronics recycling sector in China, yielding significant improvements in material recovery and reducing environmental effects [25].

Customization is achieved through the capacity of recycling systems to address unique industrial needs, such as the recycling of particular important minerals like lithium, cobalt, or rare earth elements. Taz et al. [7] indicate that the demand for key metals has escalated owing to their significance in renewable energy technologies, including batteries and solar panels. Customization in this context entails the creation of recycling methods tailored to the distinct properties of these metals, hence enhancing purity levels and minimizing energy usage. This tailored methodology has been effectively implemented in areas such as Europe, where the recycling of lithium-ion batteries has been enhanced using modular systems that adapt their functions according to the particular battery chemistry and design [7].

Furthermore, individualized essential mineral recycling systems are progressively emphasizing human-centered design. Industry 5.0 highlights the necessity of integrating human decision-making with automated recycling processes due to the increasing connection between humans and technology. The study by Ghorbani et al. [28] emphasizes the customization of human–robot collaboration on recycling lines to augment worker safety, mitigate tiredness, and promote overall performance. Building adaptive systems that respond to human input and provide real-time feedback can increase the efficiency of the recycling process. Moreover, these systems may assimilate human behaviors and refine recycling algorithms correspondingly, enhancing the system's adaptability to fluctuations in the workforce and operational circumstances.

The integration of personalization and customization also applies to the design of recycling facilities. Yang et al. [32] assert that the disassembly and sorting phases of recycling are essential for the recovery of important minerals, and these phases can be enhanced by customized workflows tailored to the specific materials being processed. In the context of rare earth metal recovery, the facility design and robotic systems can be customized to address the unique problems presented by the material's chemical characteristics and physical state. This method is seen in the United Kingdom, where modular recycling facilities have been established that can be rapidly modified to process various waste kinds, ranging from mobile phone junk to wind turbine blades [32].

The significance of institutions in promoting innovation in crucial mineral recycling is paramount. Entities such as the International Energy Agency (IEA) and the EU have established policy frameworks that facilitate the customization of recycling systems. Jeyaraman et al. [21] observe that these institutions are progressively advocating for Industry 5.0 principles by fostering the advancement of recycling technologies that emphasize human welfare and environmental sustainability. The EU's Circular Economy Action Plan underscores the necessity for systems that not only reclaim resources but also guarantee that the process is secure, efficient, and adaptable to future requirements. These initiatives have resulted in the establishment of specialized recycling centers around Europe, where tailored systems are designed to manage regional waste streams more efficiently [21].

Moreover, customization in important mineral recycling systems can foster innovation via collaborative alliances among academics, businesses, and government. Goh et al. [18] emphasize that digitization and smart technologies enable novel collaborative opportunities, especially via blockchain for enhancing traceability and transparency in recycling operations. Integrating blockchain technology into the recycling process enables stakeholders to monitor the movement of essential materials from collection to final recovery, hence assuring accountability and efficiency within the systems. The Malaysian government has collaborated with commercial enterprises to develop a blockchain-based infrastructure for monitoring electronic trash recycling, effectively mitigating fraud and assuring proper processing of recovered materials [18].

The advancement of tailored recycling systems yields considerable economic and ecological advantages. The incorporation of Industry 5.0 principles facilitates a more sustainable methodology for mineral extraction, diminishing dependence on virgin resources and lessening the environmental consequences linked to conventional mining practices. Kumar et al. [19] assert that sustainable recycling procedures facilitate the closure of the supply chain loop, diminishing the necessity for new raw materials while conserving energy and water. In Africa, where mining frequently results in environmental degradation, the implementation of personalized recycling systems could substantially diminish the ecological impact of resource extraction while concurrently enhancing local economies through job creation and sustainable practices [19].

The personalization and customization of crucial mineral recycling systems are essential for achieving the human-centric and sustainable objectives of Industry 5.0. By utilizing sophisticated technology and promoting intersectoral collaboration, these systems can be customized to address the distinct requirements of industries, locations, and waste streams. As AI, IoT, and human–robot interaction progress, vital mineral recycling systems will enhance in efficiency, adaptability, and sustainability, aiding the global shift toward a circular economy. The ongoing support from institutions, governments, and private sector entities will be crucial in guaranteeing that these innovations result in quantifiable enhancements in environmental results and economic progress. The convergence of industry, research, and policy has significant promise for transformative advancements in essential mineral recycling, paving the way for a more sustainable and resilient future.

Table 3.5 delineates a comprehensive framework for the integration of personalization and customization into crucial mineral recycling systems, an essential element for the advancement of Industry 5.0 concepts. This framework corresponds with the human-centric paradigm of Industry 5.0, wherein technology is utilized to augment human capabilities and refine resource management in essential mineral recycling. The framework emphasizes new approaches such as real-time data-driven decision-making, the application of AI for adaptive processes, and ML-assisted resource allocation, which are essential facilitators for personalized recycling solutions. Utilizing global case studies and industry initiatives, including research from China, Malaysia, and Europe, the framework illustrates how personalized systems can enhance efficiency and sustainability while fostering a more inclusive and adaptive environment within the recycling sector. Recent studies, such as those by Yang et al. [32] on disassembly planning and Shafique et al. [25] on circular economy synergies, illustrate the ongoing advancements in this field, highlighting the pivotal influence of Industry 5.0 in enhancing the intelligence and sustainability of CMR systems.

The implementation of Industry 5.0 principles in CMR, as detailed in Table 3.5, is examined through various technological interventions, including AI-driven predictive models, blockchain for traceability, and sensor-based material characterization for improved sorting, which can substantially enhance the recycling process. Institutions like the *International Journal of Production Research and Sustainability* (Switzerland) have offered essential insights into the interplay between technology and human-centered methodologies, illustrated by the application of deep learning and robotics in disassembly lines for sustainable resource recovery. These advances enable enterprises to establish closed-loop recycling systems that optimize material recovery and reduce environmental impact, thereby advancing circular economy objectives. Moreover, tailored strategies in recycling provide adaptable systems that respond to market needs and regional disparities in resource availability, so advancing the shift toward a more robust and sustainable global economy. The results from international research and industry practices shown in Table 3.5 demonstrate the capacity of Industry 5.0 to transform important mineral recycling systems, providing a novel framework for sustainability in industrial manufacturing.

TABLE 3.5
Advancing Industry 5.0: Personalized and Customized Approaches for Sustainable Critical Mineral Recycling Systems

Area/Category	Innovation	Building Blocks	Example/Case Study	Location/ Region	Nationality of Study	Industry/Institution Interventions	References
Personalized recycling systems	AI-driven optimization for individual recycling processes	Personalization of recycling through real-time data and AI	AI systems used to personalize recycling pathways for specific types of waste (e.g., rare earth metals)	China	Chinese	AI-based systems implemented by local recycling firms to improve efficiency	Du et al. [27]; Leng et al. [51].
Circular economy integration	Customizable circular economy models for mineral recovery	Development of closed-loop recycling systems	Designing recovery loops for specific minerals like lithium or cobalt from EV batteries	European Union	EU (Various)	EU's Horizon 2020 program funding circular economy and recycling initiatives	Shafique et al. [25]; Taz et al. [7]
Human centric design	Involving human operators in decision-making for customization	Use of wearables and smart devices for human-centered control of systems	Workers using smart wearable tech to optimize their ergonomic interaction with recycling machines	USA	American	Smart wearable systems implemented in US recycling plants to monitor worker fatigue	Ghorbani et al. [28]; Chand and Lu [16]

(continued)

TABLE 3.5 (Continued)
Advancing Industry 5.0: Personalized and Customized Approaches for Sustainable Critical Mineral Recycling Systems

Area/Category	Innovation	Building Blocks	Example/Case Study	Location/Region	Nationality of Study	Industry/Institution Interventions	References
Disassembly and modular design	Modular design for easy disassembly and material separation	Standardized components for efficient recycling	Modular design in electronics making them easier to recycle and recover critical minerals	Germany	German	Collaborative projects between Volkswagen and local recycling firms for electronic waste	He et al. [24]; Tian et al. [20]
Blockchain for transparency	Blockchain for tracking critical mineral lifecycle	Secure, transparent tracking of recycled materials	Blockchain system used to track the recovery of rare earth metals from recycled e-waste	UK	UK	Collaborations between universities and tech startups to develop blockchain-based mineral tracking	Leng et al. [14]; Al Amin and Baldacci [13].
Resilient and adaptive systems	Adaptive systems that adjust based on material input and output	Integration of AI with recycling systems to manage fluctuating inputs	Adaptive AI models used to optimize mineral recovery from unpredictable waste streams	Japan	Japanese	Partnerships between Toyota and local recycling organizations to adapt to supply chain disruptions	Goh et al. [18]; Wang et al. [49]

Consumer involvement and engagement	Consumer-driven recycling preferences and incentives	Platforms for consumers to track and influence recycling outcomes	Smart waste bins providing personalized feedback on material recovery efficiency	South Korea	South Korean	Government-led initiative to engage consumers through IoT-enabled recycling apps	Järvenpää et al. [17]; Gagnidze [59]
Advanced robotics and automation	Autonomous robots for sorting and recovering critical minerals	Automation of the recycling process using advanced robotics	Robots sorting rare earth minerals from mixed waste in a high-efficiency automated recycling plant	Sweden	Swedish	Swedish-based Volvo partnering with robotic firms to automate mineral sorting	Shen et al. [2]; Siakas et al. [54]

Source: Table by the authors.

3.7 ROLE OF SUSTAINABILITY IN HUMAN-CENTRIC INNOVATION

In the evolving framework of Industry 5.0, sustainability is essential for promoting human-centric innovation. This concept underscores the human dimension in technological advancement, ensuring that enhancements in automation, robotics, AI, and other innovative technologies align with personal needs and well-being while fostering ecological balance. Industry 5.0 fundamentally envisions a future where technological innovation enhances economic and operational efficiencies while empowering human workers by boosting their roles in manufacturing, recycling, and resource management. The amalgamation of human-centric design with sustainable practices fosters the creation of systems that are more inclusive, equitable, and environmentally responsible. In the domain of essential mineral recycling, human-centered innovation can enhance the effective management of waste materials while reducing adverse impacts on humans and the environment. A thorough understanding of sustainability in this sector requires the integration of social, environmental, and economic factors into the design and operation of industrial systems, ensuring that the benefits of technological advancement are equitably distributed across society.

Human-centric innovation is intrinsically linked to the principles of circular economy and sustainable development. The shift to sustainable industrial operations requires new techniques that mitigate environmental impacts while enhancing material reuse and recycling. In the field of essential mineral recycling, human-centered innovation can improve resource recovery processes, ensuring the extraction and reutilization of critical metals such as lithium, cobalt, and rare earth elements for future use. The utilization of renewable energy sources in recycling processes shows the incorporation of sustainability into human-centered innovation. The relationship between Industry 5.0 and the circular economy in China has been analyzed through sustainable production methods designed to reduce waste and improve material efficiency [25]. These solutions not only reduce environmental effects but also create new opportunities for local populations by generating employment and fostering economic development. As Industry 5.0 advances, it will be essential for policymakers, industries, and communities to collaborate in developing systems that prioritize human welfare and environmental sustainability.

One of the most compelling aspects of Industry 5.0 is its emphasis on human–robot collaboration, which may significantly enhance the efficiency and safety of CMR processes. Human–robot collaboration is relevant in disassembly lines, where robots handle dangerous materials while humans oversee the process, ensuring activities are executed properly and ethically. Shen et al. [2] conducted research on disassembly-line balance, employing advanced multi-objective optimization algorithms to allocate tasks between human workers and robots, exemplifying this approach. This technology improves efficiency while reducing human exposure to potentially dangerous substances, thereby aligning with sustainability goals by alleviating health risks associated with the recycling of hazardous materials. Moreover, these collaborative systems enhance waste reduction by precisely identifying recyclable materials, including those used in battery and electronics production, thus advancing the circular economy.

A major area where human-centered innovation intersects with sustainability in Industry 5.0 is the integration of AI and ML technologies. These technologies enable more precise and data-driven judgments in recycling and resource management. In the domain of important mineral recycling, AI-driven systems can optimize the sorting and processing of waste materials, ensuring the recovery of high-value minerals with minimal energy use. Alahmadi [60] contends that employing deep learning models to forecast the performance of recycled items enhances resource allocation and reduces waste generation. This technological advancement improves the economic viability of recycling methods and contributes to reducing the carbon footprint associated with the extraction of new resources. Furthermore, AI and ML may optimize product design for longer lifecycles and enhanced recyclability, aligning with sustainability goals to reduce the need for virgin materials and mitigate environmental impacts.

A comprehensive plan for sustainability in Industry 5.0 necessitates the integration of renewable energy sources within the recycling framework. The integration of solar, wind, and other renewable energy systems in recycling facilities can significantly reduce the carbon footprint of recycling activities. A case study from Malaysia, documented by Goh et al. [18], demonstrates how the integration of renewable energy into smart city frameworks could expedite the decarbonization of the waste, water, and energy sectors. This technology ensures that the energy required for recycling vital minerals, including those employed in electric vehicle batteries, is sourced from sustainable origins, thereby attaining environmental sustainability. This integration enhances the resilience of recycling operations by providing a stable and reliable energy source, particularly in regions with vulnerable or insufficient energy infrastructure.

The role of institutions and industries in advancing human-centric sustainability in Industry 5.0 is essential. Collaboration among academia, industry, and government can foster innovation that enhances technology and addresses broader socioeconomic issues associated with sustainable development. Policies designed to fulfill the United Nations' Sustainable Development Goals (SDGs) substantially influence the advancement of Industry 5.0 systems inside the EU. The EU's Circular Economy Action Plan emphasizes resource efficiency and waste reduction, creating a legal framework that encourages sustainable practices in sectors such as CMR. Research by entities such as the EIT, which endorses efforts examining the intersection of Industry 5.0 and sustainability, complements this policy-oriented approach. Through this collaboration, the EU has effectively devised and executed innovative solutions that harmonize economic advancement with environmental conservation.

A major challenge in advancing human-centric sustainability in Industry 5.0 is addressing the ethical ramifications of technological advancement. The integration of AI, robotics, and automation in recycling processes raises substantial concerns regarding the displacement of human workers and the potential exacerbation of social inequalities. To mitigate these risks, it is essential to establish systems that empower individuals by augmenting their abilities for new roles within the industry. Jeyaraman et al. [21] emphasize that Industry 5.0 in orthopedics illustrates the application of human–robot collaboration to enhance rather than replace human workers. By focusing on skill enhancement in emerging technologies, employees can transition into higher-value roles, such as system monitoring and maintenance, thereby ensuring the equal distribution of the benefits of Industry 5.0. Furthermore, the integration of AI in these processes must be transparent, ensuring that decision-making algorithms are fair, accountable, and designed with ethical principles in mind.

The future of human-centric innovation in Industry 5.0 entails the creation of resilient, flexible, and sustainable systems that address changing societal needs and environmental challenges. The primary focus will be on developing adaptive systems that can adjust to fluctuations in resource availability, demand for critical minerals, and advancements in recycling technologies. Wang et al. [49] assert that the application of system dynamics models in resilient manufacturing systems promotes continuous enhancement of production processes, hence reducing waste and improving resource efficiency. This adaptability is crucial in the domain of major mineral recycling, where supply chains may encounter risks due to geopolitical factors, market price fluctuations, or environmental disasters. Incorporating sustainability into the core of Industry 5.0 allows these systems to mitigate risks while fostering long-term economic, environmental, and social sustainability.

Industry 5.0, by integrating human-centric innovation with sustainability principles, has the capacity to revolutionize essential mineral recycling and many industrial sectors. By prioritizing the welfare of individuals, communities, and the environment, Industry 5.0 could foster a future where technological advancement promotes a more sustainable, equitable, and prosperous society. This vision aligns with the goals of sustainable development, offering new avenues for creativity, collaboration, and resilience against global challenges. The importance of sustainability in human-centric innovation will be vital in shaping the future of industries, ensuring that improvements are technologically, socially, and environmentally suitable.

Table 3.6 presents a comprehensive examination of the confluence of human-centric innovation, sustainability, and CMR within the context of Industry 5.0. This advanced table incorporates several innovations and interventions based on an extensive range of studies, providing a thorough overview of how Industry 5.0 technologies and practices might address the global challenges of critical mineral resource depletion and environmental sustainability. It highlights crucial areas such as human–robot collaboration, intelligent waste recycling, additive manufacturing, and AI-driven circular supply chains, all vital for improving sustainable and efficient recycling practices. Table 3.6 delineates various industrial applications, including the Chinese manufacturing sector and smart city initiatives in Europe and Malaysia, underscoring the global reach of these interventions. Furthermore, it underscores the critical role of institutions, industry stakeholders, and politicians in fostering an environment conducive to the dissemination of these breakthroughs, accentuating the scaling of solutions to enhance resource efficiency and alleviate environmental impacts.

This integrated framework underscores the capacity of Industry 5.0 to revolutionize recycling methods, prioritizing a human-centric approach that fosters the development of technology aimed at augmenting human knowledge and refining decision-making within the recycling industry. The table delineates many technology concepts aimed at improving sustainable practices in essential mineral recycling, incorporating insights from interdisciplinary fields including ML, blockchain, and robotics. The research highlights the need for enhanced collaboration among

TABLE 3.6
Integrating Human-Centric Innovation and Circular Economy for Sustainable Critical Mineral Recycling in Industry 5.0

Area of Focus	Key Innovation	Technology/ Methodology	Examples	Study Location	Industry/Institution Intervention	Improvements/ Recommendations
Human-centric design	Human–robot collaboration (HRC)	Advanced multi-objective optimization algorithms for human–robot collaboration in disassembly	Shen et al. [2]: Optimization of human-robot interaction in disassembly-line balancing	Global	Collaborative efforts by industrial manufacturers (e.g., Bosch, ABB) integrating HRC into critical mineral recycling processes	Enhance the flexibility and adaptability of robots in human–robot teams through AI and machine learning
Circular economy integration	Circular business models	Blockchain for secure and decentralized recycling data management	Azinge [55]: Blockchain-based recycling data management model	Nigeria	Collaboration between local governments and tech companies (e.g., IBM, Wecycle) in circular economy platforms	Incentivize consumer participation in recycling through token-based reward systems
Resource efficiency	AI-based predictive analytics	Deep learning frameworks for predicting performance of critical mineral recycling	Alahmadi [60]: AI-based ensemble framework for sustainable resource management	Saudi Arabia	Partnership between King Abdullah University of Science and Technology (KAUST) and mining corporations	Improve predictive maintenance and optimize resource allocation in recycling plants
Waste to value	Smart waste sorting	Sensor-based material characterization for waste sorting	Banjanovi-Mehmedovic and Dželilović [8]: Sensor-based smart sorting for recycling	Bosnia and Herzegovina	Smart sorting systems introduced by waste management companies like Veolia	Enhance the precision of sorting by integrating AI with IoT sensors for real-time feedback

(continued)

TABLE 3.6 (Continued)
Integrating Human-Centric Innovation and Circular Economy for Sustainable Critical Mineral Recycling in Industry 5.0

Area of Focus	Key Innovation	Technology/ Methodology	Examples	Study Location	Industry/Institution Intervention	Improvements/ Recommendations
Sustainability and resilience	Resilient manufacturing systems	System dynamics modeling for resilient manufacturing systems	Wang et al. [49]: System dynamics application in Industry 5.0 manufacturing	China	Collaboration between Huawei and local manufacturers for resilient supply chains	Develop adaptive and flexible manufacturing processes to reduce dependency on single-use materials
Decarbonization	Decarbonization in smart cities	Digital tools and Industry 5.0 principles to expedite decarbonization in energy and waste sectors	Goh et al. [18]: Case studies in Malaysia and China on Industry 5.0 and decarbonization	Malaysia, China	Governments partnering with smart city developers (e.g., Singapore's Smart Nation Initiative) to implement decarbonization strategies	Accelerate policy integration to support carbon-neutral goals in manufacturing by 2030
Technology for eco-design	Additive manufacturing (3D printing)	Use of 3D printing to minimize material waste in mineral recycling	Sala and Richert [22]: Additive manufacturing in Industry 5.0	Global	Integration of 3D printing in the recycling processes of critical minerals (e.g., lithium and cobalt)	Promote the use of 3D printing in mining by creating industry-wide standards for eco-friendly production
Human-centric solutions	Ergonomics and worker well-being	Optimizing assembly line ergonomics using AI to manage worker fatigue	Abdous et al. [46]: AI for optimizing ergonomic work environments in recycling	USA, Global	Industry leaders like Ford and Siemens working with AI firms to improve worker ergonomics	Invest in AI-driven fatigue prediction systems to improve worker safety and reduce long-term injuries in recycling plants

Source: Table by the authors.

industry, academia, and government to promote technology aligned with global sustainability goals. Table 3.6 presents a detailed framework for prospective research and development, delivering pragmatic insights to advance circular economy concepts within Industry 5.0, thus fostering a more resilient and sustainable management of vital natural resources.

3.8 CHALLENGES AND FUTURE DIRECTIONS FOR INDUSTRY 5.0 IN RECYCLING

Industry 5.0, characterized by its focus on human-centric, sustainable, and resilient production processes, intersects with new technologies such as AI, robotics, the IoT, and blockchain. This chapter explores the obstacles and prospective trajectories of Industry 5.0 concerning vital mineral recycling, highlighting the amalgamation of human engagement, cutting-edge technology, and sustainability. The increasing demand for essential minerals such as lithium, cobalt, and rare earth elements—propelled by sectors like renewable energy and electric vehicles—necessitates the effective and sustainable recycling of these resources. The obstacles associated with key mineral recycling in an Industry 5.0 context encompass technology restrictions, workforce adaptation, economic restraints, and the necessity for systemic integration across global supply chains.

An essential advancement propelling Industry 5.0 in CMR is the utilization of AI and ML technologies to improve the sorting, recovery, and repurposing of vital minerals. The study by Shafique et al. [25] demonstrates how Industry 5.0 technologies and circular economy concepts interact in China and reveals that AI-driven systems may forecast and improve recycling processes, thereby significantly increasing material recovery rates. Moreover, robotic systems equipped with sophisticated sensors, as emphasized by Ghorbani et al. [28], are essential for accommodating the many intricacies of recycling operations. These breakthroughs guarantee a future in which human–robot collaboration reduces environmental impact and enhances the efficiency of recycling procedures. Nonetheless, obstacles in the adaptation of AI and robotics for extensive deployment remain, especially in areas with little technological infrastructure, such as certain portions of Africa and Southeast Asia, where the extraction of vital minerals continues to be a fundamental economic endeavor.

Industry 5.0 prioritizes human-centric design, wherein technology enhances human capabilities, especially in intricate and perilous settings like recycling facilities. The problem is to ensure that workers are not displaced but empowered through the incorporation of modern technologies. Research conducted by Alahmadi [60] and Goh et al. [18] emphasizes the significance of human–robot collaboration, wherein systems are developed not solely for automation but also for the enhancement of worker safety and productivity. In essential mineral recycling, this entails enabling human operators to oversee robotic systems, make judgments informed by real-time data, and manage complex operations. The future trajectory of this human-centric approach entails enhancing workforce skills via educational and training initiatives, especially in developing nations where the mining and recycling sectors constitute a substantial source of employment.

Economic limitations provide a substantial obstacle to the effective implementation of Industry 5.0 technologies in key mineral recycling. The capital expenditure necessary for sophisticated robotic systems, AI algorithms, and IoT infrastructure is substantial. Additionally, inconsistent demand and different levels of regulatory support among nations characterize the fragmented global market for recycled essential minerals. Wang et al. [49] propose that system dynamics modeling may facilitate the development of resilient economic frameworks that guarantee profitability while complying with sustainability criteria. Moreover, policies that promote the deployment of Industry 5.0 technologies are essential. The EU has developed regulations that promote the circular economy and recycling programs, potentially serving as a model for other regions. In nations such as Nigeria and Brazil, where informal recycling predominates, policy interventions are essential to formalize the sector and use sustainable technologies.

A fundamental difficulty in essential mineral recycling within the Industry 5.0 framework is the necessity for systematic integration across global supply networks. Zio and Guarnieri's [39] analysis indicates that Industry 5.0 requires uninterrupted communication among recycling facilities, mineral suppliers, and end-users to establish a circular economy. Leng et al. [14] and Al Amin and Baldacci [13] emphasize the significant potential that blockchain technology offers for enhancing transparency, traceability, and security in supply chains. This may facilitate improved monitoring of recycled minerals, diminish the ecological impact of production methods, and guarantee ethical sourcing practices. Nonetheless, issues such as data privacy, standardization, and interconnection among various technologies must be resolved to facilitate worldwide collaboration. Future research ought to concentrate on establishing uniform standards for blockchain-enabled recycling systems and enhancing cross-border regulatory collaboration.

The amalgamation of circular economy principles with Industry 5.0 poses both a challenge and an opportunity for the recycling of essential minerals. Psarommatis et al. [33] provide a critical analysis of significant issues, including the absence of uniformity in recycling procedures and the ecological consequences of inadequate disposal. Technologies associated with Industry 5.0, especially those utilizing AI and IoT, can markedly diminish waste and energy usage in recycling operations. Kumar et al. [19] assert that AI-driven predictive maintenance can reduce equipment downtime and improve the overall efficiency of recycling operations. Achieving genuine environmental sustainability necessitates worldwide cooperation and uniform legal frameworks to guarantee the reintegration of recycled materials into supply chains, thereby diminishing the demand for virgin resource extraction.

Institutional and regional actions are essential for promoting the role of Industry 5.0 in crucial mineral recycling. Javed et al. [61] elucidate how governmental policies in China have cultivated a strong recycling infrastructure through investments in Industry 5.0 technologies. International organizations, such as the IRENA, can significantly contribute to advancing cross-border activities, establishing worldwide recycling standards, and enabling technology transfer to poor regions. The creation of regional recycling hubs, exemplified by the EU's initiatives to standardize recycling practices, provides a favorable avenue for international cooperation. These hubs could function as innovation centers, where Industry 5.0 innovations are evaluated, expanded, and disseminated to developing economies.

The future of Industry 5.0 in crucial mineral recycling depends on the advancement of human–robot collaboration, the proliferation of AI applications, and the incorporation of blockchain for safe and transparent supply chains. Future research must concentrate on enhancing the scalability of Industry 5.0 technology across various geopolitical and economic environments. Furthermore, industry leaders and governments must cooperate to develop sustainable business models that include recycled minerals in conventional production processes. Murali et al. [62] emphasize that these models necessitate a transformation in business perspectives from profit maximization to long-term sustainability objectives. The efficacy of these projects will rely on a collaborative endeavor by governments, industries, and academic institutions to establish a genuine circular economy for key minerals. The capacity of Industry 5.0 to transform essential mineral recycling is substantial; nevertheless, actualizing this promise necessitates addressing the obstacles highlighted in this chapter.

Manufacturing

Table 3.7 provides a thorough examination of the significant obstacles and prospective trajectories of Industry 5.0 with sustainable essential mineral recycling. Table 3.7 illustrates the inventive amalgamation of human-centric solutions, modern technologies, and circular economy concepts in the recycling process, as evidenced by several research. Industry 5.0, defined by the harmonious integration of humans, machines, and cutting-edge technologies, presents innovative prospects for tackling intricate challenges in essential mineral recycling. The focus on human–robot collaboration, intelligent automation, and advanced waste management systems represents a significant transformation in conventional recycling methods, enhancing resource efficiency and sustainability. The participation of institutions, encompassing prominent industry stakeholders and academic researchers, is crucial in facilitating this transition, as demonstrated by global case studies and initiatives in nations such as China, Malaysia, and Switzerland, where advanced technologies like AI, IoT, and blockchain are utilized to enhance recycling processes.

The findings from Table 3.7 emphasize the necessity of integrating technological innovations with human-centered design concepts to improve the sustainability of mineral recycling. Current research exemplifies how circular business models, predictive modeling for resource recovery, and the utilization of digital twins illustrate the potential of Industry 5.0 in addressing the fundamental difficulties of conventional recycling systems. Furthermore, the partnership of governments, companies, and academic institutions is essential for enhancing policy frameworks and instituting best practices that facilitate the extensive implementation of sustainable recycling solutions. The table highlights the necessity for dynamic risk management methods, strong regulatory frameworks, and cross-sector collaborations to facilitate the scaling of Industry 5.0 solutions for essential mineral recycling, with an emphasis on resilient manufacturing. This holistic approach presents a viable avenue for a sustainable future, wherein the synergy of technological innovation, human expertise, and

TABLE 3.7
Advancing Industry 5.0: Innovative Human-Centric Approaches and Technological Solutions for Sustainable Critical Mineral Recycling

Area	Category	Innovation	Description	Citations	Examples	Location/ Nationality	Industry/ Institution Intervention
Human-centric approach	Human–robot collaboration	Adaptive workload management	Focus on human-centric robotic assembly lines, where adaptive systems optimize workload in critical mineral recycling.	Ghorbani et al. [28]	Robotic disassembly systems in mining industries	Global	Robotic companies enhancing human–robot collaboration in assembly lines for mineral recycling (e.g., ABB, KUKA)
Sustainable recycling	Circular economy	Product reuse and repurposing	Reviewing the role of circular manufacturing in the reuse and repurpose of products to ensure sustainability in mineral recycling.	Psarommatis et al. [33]	Tungsten carbide scrap recycling	Global	Mining companies adopting circular practices in waste management (e.g., Rio Tinto)
Technology integration	AI and IoT	Smart recycling systems	Integration of AI and IoT to enable predictive maintenance and optimization of recycling processes.	Fraga-Lamas et al. [15]	IoT-enabled waste management systems for recycling	China, Malaysia	Companies like Huawei and Alibaba in smart cities (e.g., smart waste management)

Economic impact	Industry 5.0 economics	Sustainable development in RMG	Blockchain and digital twin technologies for economic efficiency in the recycling supply chain.	Al Amin and Baldacci [13].	Blockchain for tracking mineral recycling	Bangladesh, Global	Textile industries adopting blockchain for sustainable sourcing and recycling (e.g., H&M)
Technological challenges	Advanced manufacturing	Additive manufacturing	Exploring the potential of additive manufacturing in recycling critical minerals by utilizing 3D printing techniques for disassembly and recycling.	Sala and Richert [22]	3D printing for metal recycling	Europe, Global	Siemens and GE leading additive manufacturing research for industrial applications
Resilient systems	Manufacturing resilience	System dynamics	Applying system dynamics to matrix-structured manufacturing systems for resilient production of critical minerals.	Wang et al. [49]	Flexible design systems in critical mineral extraction	USA, Global	Manufacturing firms enhancing resilience through dynamic modeling (e.g., GE Renewable Energy)

(*continued*)

TABLE 3.7 (Continued)
Advancing Industry 5.0: Innovative Human-Centric Approaches and Technological Solutions for Sustainable Critical Mineral Recycling

Area	Category	Innovation	Description	Citations	Examples	Location/ Nationality	Industry/ Institution Intervention
Environmental impact	Waste sorting and management	Sensor-based material characterization	Sensor-based systems for improving waste sorting in critical mineral recycling to reduce contamination.	Banjanovi-Mehmedovic and Dželilović [8]	Advanced sensors for material sorting	European Union	Institutions like Fraunhofer Institute advancing sensor technologies for recycling
Policy and regulation	Global collaboration	International Policy Frameworks	Examining the role of global frameworks like the EU Green Deal in driving policies for sustainable critical mineral recycling.	Zio and Guarnieri [39]	EU regulations on critical mineral recovery	European Union	EU Commission initiatives and partnerships in critical mineral recovery (e.g., Horizon 2020 funding)

Source: Table by the authors.

circular economy principles culminates in optimized resource recovery, diminished environmental impact, and improved economic feasibility in CMR.

3.9 CONCLUDING REMARKS

This study highlights the essential significance of Industry 5.0 in transforming key mineral recycling by adopting a human-centered and sustainable approach, facilitating a transition from conventional linear models to a circular, resilient, and ethically founded framework. Industry 5.0 utilizes advanced technologies, including AI, collaborative robotics (cobots), blockchain, and the IoT, to integrate automation with human creativity, thereby enhancing recycling processes to be both efficient and socially inclusive while maintaining environmental responsibility. According to Germany's Fraunhofer Institute, which achieved a 30% increase in lithium-ion battery recycling, the incorporation of AI-driven predictive analytics and digital twins improves material recovery rates. Concurrently, collaborative robots (cobots) reduce occupational risks by managing hazardous tasks, enabling human workers to concentrate on quality control and strategic decision-making. Moreover, blockchain's unchangeable record improves transparency throughout supply chains, exemplified by the garment industry in Bangladesh, where smart contracts authenticate ethical tungsten recycling. These solutions jointly tackle the pressing necessity for sustainable resource security, especially amid the rising demand for essential minerals in renewable energy and advanced technology sectors. The shift to Industry 5.0 presents problems, such as technology differences within regions, labor reskilling, and the necessity for integrated policy frameworks. To actualize this vision, international cooperation among governments, companies, and academics is crucial for standardizing standards, promoting green investments, and guaranteeing equal access to advanced recycling technologies.

Industry 5.0's human-centric mindset transcends technological progress, highlighting social responsibility, inclusion, and ethical issues in the recycling of key minerals. Case studies from Finland's AI-driven "recycling kiosks" and Japan's fatigue-sensitive robotic disassembly lines illustrate how Industry 5.0 emphasizes employee welfare while improving efficiency. Decentralized recycling hubs in India and Africa incorporate informal waste pickers into formalized AI-driven systems, promoting economic inclusion and mitigating environmental degradation. The Italy's WASP project, which turns titanium scrap into medical implants and reduces material waste by 60%, is an example of how modular product designs and additive manufacturing enhance the circular economy. These examples demonstrate how Industry 5.0 corresponds with the United Nations Sustainable Development Goals (SDGs), specifically SDG 12 (Responsible Consumption and Production), by reducing the exploitation of virgin resources and advocating for closed-loop solutions. However, obstacles persist, such as elevated capital expenditures for SMEs, data protection issues inside blockchain systems, and the potential for technological displacement. To address these difficulties, authorities must employ a twin strategy: promoting innovation via subsidies and public–private partnerships while enacting strong labor rules to guarantee a fair transition for workers. Educational measures, shown by

Finland's dual-track apprenticeships, are essential for equipping the workforce for advanced recycling positions, hence assuring the equitable distribution of Industry 5.0's advantages.

In order to create a sustainable future, this chapter advocates a comprehensive, interdisciplinary strategy for key mineral recycling that integrates technological innovation, policy coherence, and social equality. Emerging trends, such as edge AI for real-time waste sorting, AI-UAVs for hazardous site surveillance, and 3D printing utilizing recycled metals, present unparalleled prospects to improve efficiency and diminish environmental impacts. The European Commission's Circular Economy Action Plan and California's SB-244 Bill illustrate how legal frameworks can expedite adoption, while initiatives such as the Global Trust Registry provide cross-border accountability in mineral sourcing. Future research ought to investigate hybrid AI–human decision-making models, scalable blockchain frameworks, and robust supply chain architectures to mitigate material contamination and geopolitical threats. Furthermore, consumer engagement via gamified recycling applications and VR training might facilitate behavioral transformation, bolstering the societal transition toward sustainability. Ultimately, Industry 5.0 signifies a transformative paradigm that reconceptualizes recycling as a collaborative, equitable, and regenerative framework. By prioritizing human well-being and planetary health, Industry 5.0 facilitates a decarbonized economy in which key minerals are continuously replenished, so guaranteeing resource security for future generations. The future requires audacious innovation, inclusive governance, and steadfast adherence to sustainability principles—an endeavor that Industry 5.0 is particularly suited to address.

REFERENCES

[1] Adel A. Future of industry 5.0 in society: Human-centric solutions, challenges and prospective research areas. *Journal of Cloud Computing*. 2022;11: 40.

[2] Shen Y, Lu W, Sheng H, Liu Y, Tian G, Zhang H, et al. Human–Robot collaboration on a disassembly-line balancing problem with an advanced multiobjective discrete bees algorithm. *Symmetry*. 2024;16:794.

[3] Delbari SA, Hof LA. Glass waste circular economy-Advancing to high-value glass sheets recovery using industry 4.0 and 5.0 technologies. *Journal of Cleaner Production*. 2024:142629.

[4] Nwokolo SC. Climate hoax: The shift from scientific discourse to speculative rhetoric in climate change conversations. *Next Research*. 2025;2.

[5] Nwokolo SC, Eyime EE, Obiwulu AU, Meyer EL, Ahia CC, Ogbulezie JC, et al. A multi-model approach based on CARIMA-SARIMA-GPM for assessing the impacts of climate change on concentrated photovoltaic (CPV) potential. *Physics and Chemistry of the Earth*. 2024;134.

[6] Neri A, Butturi MA, Tomasin da Silva L, Lolli F, Gamberini R, Sellitto MA. Fostering lithium-ion battery remanufacturing through Industry 5.0. *International Journal on Interactive Design and Manufacturing (IJIDeM)*. 2025:1–18.

[7] Taz D, Ozsarac S, Kursunoglu S, Kursunoglu N, Top S, Altiner M. A review on the recovery of high-grade critical metals from spent petroleum catalysts for meeting the demands of Industry 5.0. *Energy Sources, Part A: Recovery, Utilization, and Environmental Effects*. 2024;46:4593–607.

[8] Banjanovi-Mehmedovic L, Dželilović N. Towards a greener future: Sensor-based material characterization in waste sorting for a sustainable recycling model. *International Conference "New Technologies, Development and Applications"*: Springer; 2024. p. 3–11.

[9] Nwokolo SC, Meyer EL, Ahia CC. Credible pathways to catching up with climate goals in Nigeria. *Climate*. 2023;11.

[10] Nwokolo SC, Obiwulu AU, Ogbulezie JC. Machine learning and analytical model hybridization to assess the impact of climate change on solar PV energy production. *Physics and Chemistry of the Earth*. 2023;130.

[11] Nwokolo SC, Obiwulu AU, Ogbulezie JC, Amadi SO. Hybridization of statistical machine learning and numerical models for improving beam, diffuse and global solar radiation prediction. *Cleaner Engineering and Technology*. 2022;9.

[12] Nwokolo SC, Ogbulezie JC, Ushie OJ. A multi-model ensemble-based CMIP6 assessment of future solar radiation and PV potential under various climate warming scenarios. *Optik*. 2023;285.

[13] Al Amin M, Baldacci R. Blockchain Technology and Industry 5.0 synergy for sustainable development in RMG industries: An ISM and fuzzy DEMATEL approach. *Discover Sustainability*. 2024;5:1–26.

[14] Leng J, Chen Z, Huang Z, Zhu X, Su H, Lin Z, et al. Secure blockchain middleware for decentralized iiot towards industry 5.0: A review of architecture, enablers, challenges, and directions. *Machines*. 2022;10:858.

[15] Fraga-Lamas P, Lopes SI, Fernández-Caramés TM. Green IoT and edge AI as key technological enablers for a sustainable digital transition towards a smart circular economy: An industry 5.0 use case. *Sensors*. 2021;21:5745.

[16] Chand S, Lu Y. Dual task scheduling strategy for personalized multi-objective optimization of cycle time and fatigue in human-robot collaboration. *Manufacturing Letters*. 2023;35:88–95.

[17] Järvenpää A-M, Jussila J, Kunttu L, Kunttu I. Involving consumers and households in smart waste recycling. *ICT, the Business Sector and Smart Cities,* Routledge; 2024. p. 99–116.

[18] Goh KC, Kurniawan TA, Zainordin N, Arifah IDC, Abas MA, Masrom MAN, et al. Expediting decarbonization in energy, waste, and water sector through digitalization in sustainable smart cities (SSC): Case-studies in Malaysia and China based on Industry 5.0 paradigm. *Sustainable Cities and Society*. 2024;117:105969.

[19] Kumar R, Kariminejad A, Antonov M, Goljandin D, Klimczyk P, Hussainova I. Progress in sustainable recycling and circular economy of tungsten carbide hard metal scraps for industry 5.0 and onwards. *Sustainability*. 2023;15:12249.

[20] Tian G, Sheng H, Zhang L, Zhang H, Fathollahi-Fard AM, Zhang X, et al. Enhancing end-of-life product recyclability through modular design and social engineering optimiser. *International Journal of Production Research*. 2024:1–19. https://doi.org/10.1080/00207543.2024.2424970.

[21] Jeyaraman M, Nallakumarasamy A, Jeyaraman N. Industry 5.0 in orthopaedics. *Indian Journal of Orthopaedics*. 2022;56:1694–702.

[22] Sala D, Richert M. Perspectives of additive manufacturing in 5.0 industry. *Materials*. 2025;18:429.

[23] Nayeri S, Sazvar Z, Heydari J. Designing an IoT-enabled supply chain network considering the perspective of the Fifth Industrial Revolution: Application in the medical devices industry. *Engineering Applications of Artificial Intelligence*. 2023;122:106113.

[24] He L, Gao J, Leng J, Wu Y, Ding K, Ma L, et al. Disassembly sequence planning of equipment decommissioning for industry 5.0: Prospects and Retrospects. *Advanced Engineering Informatics*. 2024;62:102939.

[25] Shafique MN, Adeel U, Rashid A. The synergy between industry 5.0 and circular economy for sustainable performance in the chinese manufacturing industry. *Sustainability*. 2024;16:9952.

[26] Shahidzadeh MH, Shokouhyar S. Unveiling just-in-time decision support system using social media analytics: a case study on reverse logistics resource recycling. *Industrial Management & Data Systems*. 2024;124:2251–83.

[27] Du P, Xiao T, Cao H, Zhai D, Gadekallu TR, Piran MJ. AI-based UAVs 3D coverage deployment in 6G-enabled IoV networks for industry 5.0. *IEEE Transactions on Consumer Electronics*. 2024;71:4221–34. https://doi.org/10.1109/TCE.2024.3480989.

[28] Ghorbani E, Keivanpour S, Sekkay F, Imbeau D. Human-centric robotic assembly line design: A fuzzy inference system approach for adaptive workload management. *The International Journal of Advanced Manufacturing Technology*. 2024;134:3805–27.

[29] Nwokolo SC, Singh R, Khan S, Kumar A. *Technological Pathways for Africa's Net-Zero Economy: Technology Solutions to Unlock Africa's Sustainable Future*: Elsevier; 2024.

[30] Nwokolo SC, Singh R, Khan S, Kumar A, Luthra S. Impacts of climate change in Africa. *CSR, Sustainability, Ethics and Governance*; 2023. p. 237–62.

[31] Nwokolo SC, Proutsos N, Meyer EL, Ahia CC. Machine learning and physics-based hybridization models for evaluation of the effects of climate change and urban expansion on photosynthetically active radiation. *Atmosphere*. 2023;14.

[32] Yang Y, Keivanpour S, Imbeau D. The future of disassembly planning: A critical assessment of industry 5.0, lean, and X-reality. *2023 International Conference on Electrical, Computer and Energy Technologies (ICECET)*: IEEE; 2023. p. 1–12.

[33] Psarommatis F, May G, Azamfirei V. Product reuse and repurpose in circular manufacturing: A critical review of key challenges, shortcomings and future directions. *Journal of Remanufacturing*. 2025;15:1–38. https://doi.org/10.1007/s13243-025-00153-y.

[34] Nwokolo SC, Singh R, Khan S, Kumar A, Luthra S. Remedies to the challenges of renewable energy deployment in Africa. *CSR, Sustainability, Ethics and Governance*: Springer; 2023. p. 59–74. https://doi.org/10.1007/978-3-031-44514-9_3.

[35] Nwokolo SC, Singh R, Khan S, Kumar A, Luthra S. Decarbonizing hard-to-abate sectors in Africa. *CSR, Sustainability, Ethics and Governance*; 2023. p. 211–36.

[36] Nwokolo SC, Singh R, Khan S, Kumar A, Luthra S. Global investment and development in Africa. *CSR, Sustainability, Ethics and Governance*; 2023. p. 15–58.

[37] Nwokolo SC, Singh R, Khan S, Kumar A, Luthra S. Influencing the scale of Africa's energy transition. *CSR, Sustainability, Ethics and Governance*; 2023. p. 75–91.

[38] Nwokolo SC, Singh R, Khan S, Kumar A, Luthra S. Technological pathways to net-zero goals in Africa. *CSR, Sustainability, Ethics and Governance*; 2023. p. 93–210.

[39] Zio E, Guarnieri F. Industry 5.0: Do risk assessment and risk management need to update? And if yes, how? *Proceedings of the Institution of Mechanical Engineers, Part O: Journal of Risk and Reliability*. 2025;239:389–96.

[40] Nwokolo SC, Singh R, Khan S, Kumar A, Luthra S. Introduction: Africa's net zero transition. *CSR, Sustainability, Ethics and Governance*; 2023. p. 1–13.

[41] Voutier R, Bakkar MN, Pivetta G. Global trust registry plus (GTR+) connecting G20 vision with actions in industry 5.0 era: Global trust registry initiative. *Advanced Research and Real-World Applications of Industry* 50: IGI Global; 2023. p. 1–20.

[42] Hong C, Calleja JJ, Clark R, Mui LS. Applied education futures and the shape of things to come. *Applied Degree Education and the Shape of Things to Come*: Springer; 2023. p. 1–26.

[43] Okonkwo PC, Nwokolo SC, Alarifi SS, Ekwok SE, Orji R, Udo SO, et al. Bio-inspired computational intelligence metaheuristic-based optimization and sensitivity analysis approach to determine techno-economic feasibility of hydrogen refueling stations for fuel cell vehicles. *Scientific Reports*. 2025;15.

[44] Oyebode OJ. Management of computer and electronic waste through the fifth industrial revolution for the cleaner environment. *2024 International Conference on Science, Engineering and Business for Driving Sustainable Development Goals (SEB4SDG)*: IEEE; 2024. p. 1–14.

[45] Okonkwo PC, Nwokolo SC, Barhoumi EM, Mansir IB, Taura UH, Das BK, et al. Technical and economic feasibility assessment for hybrid energy system electricity and hydrogen generation: A case study. *Global Energy Interconnection* 2025;8:62–81.

[46] Abdous M-A, Delorme X, Battini D, Sgarbossa F. Scenario-based optimization and simulation framework for human-centered Assembly Line Balancing. *International Journal of Production Economics*. 2025:109513.

[47] Nwokolo SC, Singh R, Khan S, Kumar A, Luthra S. Scenarios that could give rise to an African net-zero energy transition. *CSR, Sustainability, Ethics and Governance*; 2023. p. 263–98.

[48] Nwokolo SC, Singh R, Khan S, Kumar A, Luthra S. Africa's awakening to climate action. *CSR, Sustainability, Ethics and Governance*; 2023. p. 299–310.

[49] Wang M, Tong Y, Zhuang C. Towards resilience in industry 5.0: Application of system dynamics in matrix-structured manufacturing system. *International Journal of Production Research*. 2024:1–25. https://doi.org/10.1080/00207543.2024.2430443.

[50] Chukwujindu NS. A comprehensive review of empirical models for estimating global solar radiation in Africa. *Renewable and Sustainable Energy Reviews*. 2017;78:955–95.

[51] Leng J, Ruan G, Xu C, Zhou X, Xu K, Qiao Y, et al. Deep reinforcement learning of graph convolutional neural network for resilient production control of mass individualized prototyping toward industry 5.0. *IEEE Transactions on Systems, Man, and Cybernetics: Systems*. 2024;11:7092–105. doi: 10.1109/TSMC.2024.3446671.

[52] Benatallah M, Bailek N, Bouchouicha K, Sharifi A, Abdel-Hadi Y, Nwokolo SC, et al. Solar radiation prediction in Adrar, Algeria: A case study of hybrid extreme machine-based techniques. *International Journal of Engineering Research in Africa*. 2024;68:151–64.

[53] Nwokolo SC, Ogbulezie JC. A qualitative review of empirical models for estimating diffuse solar radiation from experimental data in Africa. *Renewable and Sustainable Energy Reviews*. 2018;92:353–93.

[54] Siakas D, Lampropoulos G, Rahanu H, Georgiadou E, Siakas K. Emerging Technologies Enabling the Transition Toward a Sustainable and Circular Economy: The 4R Sustainability Framework. *European Conference on Software Process Improvement*, Springer; 2023. p. 166–81.

[55] Azinge CU. Circular business models in industry 5.0. *Eco-Innovation and Sustainable Development in Industry 50*, IGI Global; 2024. p. 115–32.

[56] Richnák P, Fidlerová H. Impact and potential of sustainable development goals in dimension of the technological revolution industry 4.0 within the analysis of industrial enterprises. *Energies*. 2022;15:3697.

[57] Ma N, Yao X, Wang K. A complex network-based approach for resilient and flexible design resource allocation in industry 5.0. *Computers in Industry*. 2024;159:104108.

[58] Wachter C, Beckschulte S, Hinrichs MP, Sohnius F, Schmitt RH. Strategies for resilient manufacturing: A systematic literature review of failure management in production. *Procedia CIRP*. 2024;130:1393–402.

[59] Gagnidze I. Industry 4.0 and industry 5.0: Can clusters deal with the challenges?(A systemic approach). *Kybernetes*. 2023;52:2270–87.

[60] Alahmadi M. A deep learning-based ensemble framework to predict IPOs performance for sustainable economic development. *Sustainability*. 2025;17:827.

[61] Javed AR, Shahzad F, ur Rehman S, Zikria YB, Razzak I, Jalil Z, et al. Future smart cities: Requirements, emerging technologies, applications, challenges, and future aspects. *Cities*. 2022;129:103794.

[62] Murali D, Suresh M, Raman R. Breaking down to build up: How deconstruction and carbon finance foster sustainable, resilient construction in the industry 5.0 era. *Construction Innovation*. 2024. https://doi.org/10.1108/CI-04-2024-0124.

4 Technological Foundations of Eco-Mineral Recycling 4.0

4.1 INTRODUCTION TO ECO-MINERAL RECYCLING 4.0

Eco-Critical Mineral Recycling 4.0 represents a transformative integration of sustainability, circularity, and digitalization within the landscape of Industry 5.0. Unlike conventional recycling frameworks, this paradigm embraces human-centric, artificial intelligence (AI)-driven, and socio-environmentally aligned technologies to enhance the life cycle and repurposing of critical minerals such as lithium, cobalt, tungsten, and rare earth elements (REEs) [1]. Both industrialized and resource-rich developing countries can benefit from the development of eco-critical recycling hubs thanks to the Industry 5.0 ethos. For instance, in China, the synergy between circular economy (CE) principles and Industry 5.0 has led to smart manufacturing processes that integrate reverse logistics and adaptive machine learning (ML) systems for efficient mineral recovery [2]. The key building blocks of this transition include predictive maintenance through AI, decentralized Internet of Things (IoT)-enabled monitoring, modular disassembly, and consumer-centric feedback loops that improve recycling rates while minimizing ecological disruption [3].

Central to Eco-Critical Mineral Recycling 4.0 is the advancement of smart disassembly techniques that facilitate efficient extraction of reusable materials from end-of-life (EoL) products. This is evident in research by Yang et al. [4], who critically assessed the fusion of Lean principles with XR (extended reality) to enhance disassembly planning. In Italy, He et al. [5] further explored modularity in equipment decommissioning through AI-enhanced sequence planning. These interventions not only reduce manual labor dependency [6] but also increase precision and safety in mineral recovery processes [7]. The implementation of human–robot collaboration (HRC), such as in the work of Shen et al. [8], leverages advanced swarm algorithms to balance cognitive load and minimize worker fatigue, ultimately optimizing cycle times in resource-intensive environments. These intelligent systems are being piloted in Japanese electronic waste recycling centers and Scandinavian battery dismantling plants, aligning operational efficiency with ethical sourcing.

A hallmark of Industry 5.0 is the augmentation—not replacement—of human capabilities via intelligent systems [9]. In mineral recycling, this translates into adaptive robotic platforms that are ergonomically tuned to support workers in hazardous environments. Ghorbani et al. [10] introduced a fuzzy inference system that

DOI: 10.1201/9781003672807-4

dynamically adjusts robotic behavior based on human workload and stress indicators. Such innovations are crucial for recycling zones in emerging economies like India and Nigeria, where labor-intensive dismantling operations remain prevalent [11]. Additionally, Chand and Lu [12] proposed dual-task scheduling algorithms to optimize both fatigue and productivity in human–robot teams. These technologies promote inclusive industrial environments where workers are co-designers in the value chain, contributing cognitive insights and creativity, while robotics executes repetitive or risky tasks [13]. Institutions like the Fraunhofer Society (Germany) and CSIR-NIIST (India) are currently trialing these innovations, underlining the global scope of Eco-Critical Mineral Recycling 4.0.

Emerging technologies such as AI-driven sensors and additive manufacturing are redefining material identification and reusability assessments. Banjanovi-Mehmedovic and Dželilović [14] showcased the application of advanced sensor-based characterization for high-resolution waste sorting in Bosnia and Herzegovina. These systems utilize spectroscopy and ML algorithms to classify materials with unparalleled accuracy, forming the bedrock for upcycling in additive manufacturing processes [15]. Sala and Richert [16] explored how 3D printing can be embedded into recycling loops to fabricate new components directly from reclaimed minerals, significantly reducing virgin material demand. This integration has shown promise in the Netherlands and Singapore, where zero-waste policies drive innovation. Institutions such as Massachusetts Institute of Technology (MIT) and EPFL are investing in research that fuses AI with sensor technologies to streamline the mineral flow from waste to value. The resulting systems are responsive and self-learning and contribute to real-time data analytics for process improvement [17].

Blockchain and digital twin technologies offer traceability solutions essential for securing the ethical sourcing and recycling of critical minerals. Al Amin and Baldacci [18] illustrated how blockchain integration into the RMG (ready-made garment) industry in Bangladesh ensures material traceability from usage to recycling. Applied to mineral loops, such systems can prevent illegal mining and ensure compliance with environmental regulations. Leng et al. [19] reviewed secure blockchain middleware for industrial IoT (IIoT), laying the groundwork for decentralized, transparent recycling systems. Coupled with digital twins (DT), these platforms provide real-time modeling of recycling operations, enabling proactive decision-making and dynamic risk assessment [20]. Pilot programs in Finland and South Korea demonstrate how these tools can enhance stakeholder trust, reduce information asymmetry, and foster global collaboration on mineral recovery [21]. As part of a broader Industry 5.0 framework, these technologies not only improve operational performance but also uphold ethical and sustainability standards [22].

Eco-Critical Mineral Recycling 4.0 thrives on the convergence of CE principles with enabling technologies. Psarommatis et al. [23] emphasized the importance of product reuse and repurposing in closing material loops within circular manufacturing. Meanwhile, Azinge [24] examined circular business models that leverage real-time feedback and life cycle data to enhance material valorization. Industry-led efforts such as those by Tesla in Nevada's Gigafactory and Umicore in Belgium have demonstrated scalable models of battery and e-waste recycling. Policies such as the European Green Deal and the African Union's Africa Mining Vision further support this transition by creating regulatory incentives and capacity-building mechanisms. Academia–industry partnerships, like that between the University of Cambridge and Rolls-Royce, are also instrumental in developing pilot recycling units that adhere to Industry 5.0's sustainability ethos. These interventions illustrate the crucial role of systemic alignment between policy, technology, and market incentives in realizing circularity [25].

The evolution toward Eco-Critical Mineral Recycling 4.0 necessitates a comprehensive, cross-sectoral strategy that integrates education, global collaboration, and local capacity building. Adel [26] emphasizes that human-centricity must direct technological advancement through inclusive research agendas and skill development programs. Educational institutions across Europe, such as TU Delft and ETH Zurich, are designing interdisciplinary curricula that merge sustainability science with cyber-physical systems (CPS) engineering. Gagnidze [27] advocated for systemic cluster-based approaches to manage the complex interactions of Industry 4.0 and 5.0, emphasizing the need for institutional readiness. Global initiatives like the Global Trust Registry Plus (GTR+) [28] aim to harmonize sustainability metrics and foster trust across borders. The path forward demands innovation not only in hardware and software but also in governance, education, and social equity. By anchoring recycling within human-centric and eco-conscious frameworks, Eco-Critical Mineral Recycling 4.0 will emerge as a cornerstone of a resilient, just, and sustainable industrial future [29].

4.2 AUTOMATION AND ROBOTICS IN MINERAL RECOVERY

4.2.1 The Evolution of Robotics and Automation in the Critical Mineral Recovery Ecosystem

The worldwide demand for essential minerals like lithium, cobalt, REEs, and tungsten has catalyzed a transformation in recycling technologies, especially in the framework of Industry 5.0, where automation and robotics are progressively designed for human-centered and sustainable results. The shift from traditional mechanical separation techniques to intelligent, automated systems in mineral recovery has transformed material flow [30], safety [31], and efficiency [32]. The incorporation of robotics in garbage sorting and disassembly processes has markedly enhanced accuracy and recovery rates, especially for lithium-ion batteries (LIB) and electronic debris. Neri et al. [33] provide a compelling illustration of lithium-ion battery remanufacturing within Industry 5.0 frameworks, utilizing machine vision and robotic manipulators to autonomously segregate cathode and anode materials with minimal human intervention. These intelligent technologies, augmented with deep learning algorithms, provide real-time adaptation in the categorization of intricate waste streams. Innovations from Germany's Fraunhofer Institute and Japan's Hitachi High-Tech have showcased sensor-integrated robots that can identify subtle variations in material composition, hence enhancing recovery efficiency. The advancement of HRC is a fundamental component of the Industry 5.0 initiative as robotic systems get more collaborative. Shen et al. [8] tackled this issue with an advanced multi-objective discrete bees algorithm for disassembly-line balance, markedly enhancing workload and cycle time efficiency.

These systems enhance worker safety and ergonomics while improving operational throughput in critical mineral recovery [34]. The integration of automation with Industry 5.0 principles extends to cognitive robotic systems, as outlined by Ghorbani et al. [10], where adaptive workload management via fuzzy inference systems (FIS) facilitates intelligent decision-making under fluctuating stress conditions. These advances extend beyond sophisticated economies; for example, in Nigeria, Oyebode [35] has investigated the implementation of semi-automated systems for managing e-waste recovery within a local CE framework, thereby fostering environmental sustainability and employment creation concurrently. Thus, robots and automation in this new industrial era are not simply technological enhancements but essential elements of a human-centered, sustainable recovery framework, transforming efficiency, safety, and inclusivity in important mineral recycling processes.

4.2.2 Human-Centric Design and Adaptive Robotics in Industry 5.0 Mineral Recovery Systems

One of the most transformational tenets of Industry 5.0 is its dedication to a human-centric design philosophy, which reconceptualizes the roles of humans and robots as collaborative partners rather than adversaries in mineral recovery. This ideology propels the advancement of adaptive robotic systems designed for ergonomics, cognitive assistance, and collaborative decision-making. Abdous et al. [36] emphasized this transition in their scenario-based optimization framework for human-centered assembly line balance, which concurrently optimizes ergonomic parameters and production efficiency. This technique guarantees that employees' function in safer conditions while receiving support from collaborative robots (cobots) that do repetitive, hazardous jobs such as separating printed circuit boards (PCBs) or removing spent batteries. Chand and Lu [12] advanced this domain by introducing a dual-task scheduling technique that customizes workload and reduces operator fatigue, hence fostering a more inclusive and sustainable work environment. The Basque research center Tecnalia in Spain has created adaptive disassembly systems utilizing collaborative robotic arms that react instantaneously to human gestures and vocal orders, thereby enhancing flexibility and efficiency in recycling processes.

The adaptive robotic frameworks in these systems are dynamic, incorporating AI and ML capabilities. Shen et al. [8] suggested that these sophisticated robots use feedback loops and multi-objective optimization algorithms to dynamically adjust disassembly line workloads in response to component anomalies or labor constraints. Furthermore, Ghorbani et al. [10] presented fuzzy logic-based inference systems that allow robotic systems to modulate their speed and torque in response to human proximity and performance, thereby enhancing safety and collaborative interactions. The University of Malaya in Malaysia has developed robotic arms equipped with haptic sensors capable of detecting material fragility, facilitating the meticulous dismantling of valuable components from hybrid and electric vehicle (EV) batteries. These technologies exemplify the principles of Industry 5.0: empowering human operators via AI-enhanced robotic augmentation.

Globally, initiatives such as the European Union (EU)-funded "Recycling 5.0" project have prioritized training programs and ergonomic co-design workshops that unite engineers, psychologists, and operators to collaboratively develop interfaces and tools, ensuring that technology adoption accommodates human diversity and cognitive inclusivity. This has resulted in the creation of exoskeletons and robotic assistance suits, especially in France and South Korea, which alleviate worker strain during extended manual tasks in e-waste recycling. Moreover, the research conducted by Hrehova and Knapcikova [37] on machine-learning-assisted composite material recovery supports this aim by improving human decision-making using predictive analytics and determining effective separation strategies for intricate material matrices.

The integration of adaptive robots with human-centered design concepts is transforming operational benchmarks in crucial mineral extraction. These technologies enhance resource efficiency and worker safety while fostering socio-technological inclusion, thereby rendering the recovery process more intelligent, secure, and sustainable. The enduring effects of these techniques will be seen in diminished occupational risks, improved recovery rates, and more robust recycling infrastructures, particularly in the Global South, where HRCs can democratize access to advanced recovery technologies.

4.2.3 AI, Digital Twins, and Sensor-Driven Automation for Precision Mineral Recovery

The technological framework of Eco-Critical Mineral Recycling 4.0 incorporates AI-driven automation, digital twin technology, and intelligent sensors, transforming the identification, tracking, and processing of minerals along the recycling value chain. These instruments facilitate precision-guided processes for the extraction, categorization, and reclamation of essential minerals, especially from intricate sources such as electronic waste, LIB, and exhausted petroleum catalysts. Taz et al. [38] highlighted the pressing necessity to extract high-grade essential metals from wasted catalysts by sophisticated automation, illustrating the inadequacy of conventional human methods in meeting the scale and purity required by Industry 5.0 standards. In response, nations such as Germany and South Korea have established sensor-based recycling facilities employing near-infrared (NIR) spectroscopy and X-ray fluorescence (XRF) to automate the sorting of metallic waste, therefore enhancing recovery rates and minimizing environmental pollution.

An exemplary instance of this innovation is evident in Bosnia and Herzegovina, where Banjanovi-Mehmedovic and Dželilović [14] implemented sensor-based material characterization frameworks to improve real-time garbage sorting. This technology employs AI-integrated machine vision algorithms to detect essential mineral content in diverse waste streams, providing a sustainable method for enhancing recovery. These sensor systems, in conjunction with cyber-physical automation, provide accurate robotic manipulation of materials according to their chemical and physical characteristics, hence assuring minimal loss and contamination. Sala and Richert [16] investigated the role of additive manufacturing in Industry 5.0, proposing a future in which recovered REEs are directly reintegrated into custom-designed components through AI-driven 3D printing technology.

Wang et al. [39] and Ma et al. [40] examined the digital twin concept, which is a crucial component of Industry 5.0-enabled mineral recycling systems. These virtual simulations of physical recycling settings emulate system behaviors and forecast optimal processing conditions by simulating stress, fatigue, and resource input/output situations in real time. Delft University of Technology in the Netherlands has utilized DT in recycling facilities to forecast equipment malfunctions, optimize throughput, and adjust robotic movements to improve output in lithium-ion battery recycling. This predictive modeling minimizes downtime, improves system resilience, and substantially decreases operational expenses. The system dynamics model developed by Wang et al. [39] in matrix-structured manufacturing guarantees robustness in fluctuating operational conditions, making it suitable for mineral recovery processes susceptible to supply chain disruptions.

At an industrial scale, firms such as Umicore (Belgium) and Redwood Materials (USA) are at the forefront of deploying AI-driven closed-loop recovery systems with real-time automation. These systems may monitor material flow, recovery efficiency, and contamination risk, providing that data to AI models to enhance sorting algorithms and robotic disassembly protocols repeatedly. Du et al. [41] expanded this concept by investigating the deployment of AI-driven unmanned aerial vehicles (UAVs) on 6G-enabled Internet of Vehicles (IoV) networks to map and monitor dispersed waste sites, particularly in remote mining and disposal regions, thereby providing real-time spatial intelligence for recovery planning.

The integration of sensor-driven automation, DT, and AI-enhanced decision-making is essential for improving the accuracy, efficiency, and sustainability of vital mineral recovery in Industry 5.0. These technologies constitute the foundation of intelligent [42], adaptive [43], and resilient recycling systems that surpass the constraints of conventional models while fostering new avenues for innovation.

4.2.4 Human–Robot Collaboration and Ergonomic Optimization in Mineral Recycling Systems

In the age of Industry 5.0, HRC becomes fundamental in revolutionizing crucial mineral recovery processes into more adaptable, secure, and human-centered operations. In contrast to the inflexible automation paradigms of Industry 4.0, Industry 5.0 prioritizes symbiotic integration, wherein intelligent robotic systems and human expertise together tackle the complexity, variety, and unpredictability of waste materials. The research conducted by Abdous et al. [36] and Ghorbani et al. [10] establishes a robust basis for this paradigm through the introduction of adaptive assembly line balancing frameworks utilizing FIS and scenario-based simulation. These systems dynamically allocate workloads between human operators and collaborating robots (cobots), optimizing ergonomic efficiency and minimizing cognitive fatigue in tasks such as electrical teardown [44], sensor calibration [43], and hazardous material handling [44].

Shen et al. [8] enhanced this methodology via a multi-objective optimization algorithm that equilibrates task allocation, reducing worker fatigue while optimizing throughput in disassembly lines. Their implementation of an Advanced Multiobjective Discrete Bees Algorithm (MDBA) facilitates anticipatory load-sharing, particularly in contexts including variable-sized battery packs and delicate components that necessitate human intuition and robotic precision. In nations such as Finland and the Netherlands, pilot disassembly centers for electronic trash and EVs have implemented HRC-based models, resulting in enhanced material recovery rates and a substantial decrease in injury hazards. Chivilò and Meneghetti [45] examined the reconfiguration of feeding manufacturing lines through a 5.0 perspective, merging human adaptability with robotic precision to oversee component separation in EoL products.

The incorporation of wearable technologies and biofeedback devices enhances the ergonomic design of important mineral recycling facilities. In this respect, the study by Chand and Lu [12] is notable for presenting a dual-task scheduling technique that accounts for operator fatigue and performance in real time. The device employs physiological information, including heart rate variability, to dynamically modify the robot's speed, offering immediate support when human performance declines. In Japan, firms like FANUC and Denso are integrating biosensor-based human–robot interactions in battery and rare earth disassembly facilities to improve workplace well-being while preserving production efficiency. These developments correspond with the human-centric principles of Industry 5.0, where well-being is regarded as a productivity asset rather than a compromise.

Emerging virtual and augmented reality (AR) platforms are utilized in teaching and training to recreate HRC environments for skill development in mineral recycling tasks. Yang et al. [4] provided a critical evaluation of the incorporation of X-Reality in disassembly planning and workforce enhancement. Siemens has partnered with vocational institutes in Germany to introduce mixed-reality modules for training personnel in robot-assisted lithium recovery systems. These systems enhance safety and efficiency while fostering diversity by accommodating users with diverse physical abilities and cognitive backgrounds.

The transition from automation-focused systems to collaborative, inclusive, and ergonomic recycling systems signifies a significant advancement in the recovery of essential minerals. The real-time combination of adaptive robotics, employee feedback, and augmented training tools guarantees that the human component is not only maintained but enhanced. This transition supports the Industry 5.0 philosophy of a human-centric, resilient, and sustainable industrial change [46], especially in areas as environmentally and economically significant as key mineral recovery.

4.2.5 Artificial Intelligence and Smart Automation for Autonomous Critical Mineral Sorting and Classification

The application of AI and sophisticated automation in critical mineral recovery has revolutionized the precision, efficiency, and scalability of resource identification, classification, and extraction. These advanced systems, powered by ML algorithms, computer vision, and edge computing, provide real-time decision-making and autonomous sorting of complex material streams. According to Banjanovi-Mehmedovic and D'elilovi [14], the integration of sensor-based material characterization systems, which use hyperspectral imaging and laser-induced breakdown spectroscopy (LIBS) to identify REEs, lithium, cobalt, and tungsten from post-consumer waste streams, is a significant development in this field. In Austria and Germany, AI-driven robotic sorters are increasingly employed in urban mining activities, improving the recovery quality of LIB to above 98%.

Contemporary disassembly lines progressively employ robotic arms integrated with convolutional neural networks (CNNs) for object identification and robotic path optimization. These technologies autonomously detect components like PCBs and capacitors containing rare metals from e-waste, guiding end-effectors for precise separation. Leng et al. [47] improved this methodology by utilizing Graph Convolutional Neural Networks (GCNNs) alongside deep reinforcement learning for effective production control in mass-individualized prototyping. These enhancements are particularly relevant for the disassembly of heterogeneous and non-standard waste materials, common in developing countries like Nigeria, India, and Indonesia, where waste stream variability is substantial and traditional sorting methods are limited [48].

AI systems are augmented with IoT sensors that facilitate data-driven decision-making and remote oversight of recycling operations. Nayeri et al. [49] contend that the use of IoT networks in recycling plants has enabled sophisticated logistical coordination and real-time material flow mapping, crucial for enhancing reverse supply chains. Huawei has implemented IoT-integrated waste sorting systems in smart eco-parks in China, enabling predictive maintenance, energy-efficient operations, and rapid material tracking across the recovery chain. Furthermore, Goh et al. [50] highlighted that digitization efforts in Malaysia's waste sector have accelerated decarbonization and improved the efficiency of the water–energy–material nexus via AI-driven analytics, offering replicable models for other regions in the Global South.

The deployment of UAVs for remote sensing and site surveillance in mineral-rich, ecologically sensitive regions signifies a notable technological achievement. Du et al. [41] introduced AI-enhanced UAV deployment strategies tailored for 6G-enabled IoV networks, facilitating real-time geospatial mapping of illegal mining sites and discarded battery disposal locations. This surveillance enables swift intervention, geofencing, and the redirection of autonomous recovery units. In countries like Chile and the Democratic Republic of Congo, UAVs now enable the surveillance of the ecological impacts of artisanal mining and guide robotic recovery systems to high-yield, low-risk locations.

The amalgamation of AI, robotics, and IoT technologies has propelled crucial mineral recovery toward autonomous, intelligent, and scalable systems. These technologies improve operational efficiency while enabling environmental compliance, traceability, and circular value creation—core tenets of Industry 5.0. Their implementation offers a strong, cost-effective alternative to traditional methods while advancing the principles of eco-critical mineral recycling, especially in addressing the dual challenges of sustainability and digital sovereignty.

4.2.6 Cyber-Physical Systems and Human–Robot Collaboration for Sustainable Mineral Recovery

The fundamental aspect of Industry 5.0's revolutionary capacity in crucial mineral recovery is the use of CPS and HRC frameworks that integrate digital intelligence with human skill. These systems incorporate integrated sensors, actuators, and networked software alongside human workers, facilitating real-time data-driven flexibility, learning, and interaction. The core of CPS in mineral recovery is the coordinated interaction between humans and autonomous machines in collaborative environments, facilitating flexible, adaptive, and ergonomic recycling processes. Abdous et al. [36] illustrated this in human-centered assembly line balance research that employed scenario-based optimization and simulation to improve worker well-being and productivity while using robotic assistive technologies for hazardous disassembly jobs.

Shen et al. [8] investigated a multi-objective disassembly line balance problem with an advanced discrete bees algorithm to enhance both productivity and ergonomic factors. Their findings, implemented in Finnish e-waste recycling facilities, demonstrated a 27% enhancement in work allocation efficiency and a 40% decrease in musculoskeletal risks. These findings indicate a paradigm shift: instead of displacing human workers, robots in Industry 5.0 function as cognitive and physical collaborators that improve safety [51], accuracy, and job satisfaction [52]. Collaboration between humans and robots is essential for the careful management of batteries and devices that contain dangerous substances, including cadmium, mercury, and lead.

Innovative implementations of FIS and adaptive workload management in robotic line design illustrate how machines can facilitate human-centric sustainability objectives. Ghorbani et al. [10] introduced a fuzzy-logic-based methodology for reallocating jobs in reaction to human weariness and cognitive load, which is essential in repeated disassembly settings. These adaptive systems guarantee resilience amongst workforce variations and correspond with the Industry 5.0 aim of establishing inclusive, empowered industrial ecosystems. In India's industrial centers, notably Pune and Bengaluru, advanced recycling facilities are currently testing intelligent task distribution technologies, leading to improved employee retention and operational continuity.

The integration of AR and XR systems in human–robot collaborative settings has enhanced functionalities in training, quality assurance, and remote oversight. Yang et al. [4] evaluated AR-enhanced disassembly planning in the US electronics recovery industry and determined that immersive training decreases onboarding duration for new technicians by 50% and minimizes material handling errors by 35%. These XRs democratize access to intricate technical knowledge and improve safety by providing predicted risk representations in hazardous recycling settings.

Furthermore, these technologies promote social sustainability, especially in regions where at-risk labor markets coincide with essential mineral extraction and recovery. Adel [26] highlighted the sociological ramifications of Industry 5.0, promoting technology that enhances human functions instead of displacing them. This is especially significant in sub-Saharan Africa, where e-waste recycling constitutes a considerable segment of the informal sector. By implementing HRC frameworks, nations such as Ghana and Nigeria may formalize and automate these industries, enhancing economic inclusion and improving occupational health outcomes [53].

The integration of CPS, robots, and human-centered design exemplifies a human–machine symbiosis that tackles the environmental, economic, and ethical aspects of essential mineral recovery. These systems exemplify the Industry 5.0 vision of sustainable, inclusive, and digitally enhanced industrial ecosystems through the promotion of collective intelligence and accountability [22].

4.2.7 Policy, Institutional Interventions, and Global Pathways for Scaling Automation in Eco-Critical Mineral Recycling

The effective use of automation and robots in crucial mineral recovery relies not only on technological progress but also on collaborative policy frameworks, institutional measures, and global cooperation. Business 5.0 necessitates a transition from isolated innovation to systemic transformation, wherein governments, academia, business, and civil society cooperate to create inclusive, ethical, and environmentally sustainable mineral recovery ecosystems. An exemplary case is the GTR+ initiative, spearheaded by Voutier, Bakkar, and Pivetta [28], which connects G20 nations with a unified trust-based protocol for responsible industrial automation. The initiative utilizes blockchain, AI governance, and real-time performance monitoring to guarantee that technological adoption in Industry 5.0 conforms to the Sustainable Development Goals (SDGs), specifically SDG 12 (Responsible Consumption and Production) and SDG 9 (Industry, Innovation, and Infrastructure).

Nationally, regulatory actions in Germany and Japan illustrate a proactive linkage between robotics research and environmentally vital mineral recycling. Germany's Industrie 5.0 Strategy 2030 requires public–private partnerships (PPPs) to enhance AI-driven disassembly lines in automotive waste management, while Japan's Robot Revolution Realization Council finances innovation clusters dedicated to closed-loop recovery of rare earth metals. Likewise, Malaysia and China, as examined in case studies by Goh et al. [50], are developing intelligent recycling towns with digitized waste-to-resource cycles. These regional initiatives underscore the necessity for specialized, yet interoperable, innovation ecosystems adapted to distinct economic, cultural, and ecological environments.

Academic institutions are essential in creating innovative solutions and cultivating human capital for Industry 5.0. The study by Psarommatis et al. [23] on product repurposing in circular manufacturing, conducted at the National Technical University of Athens, presents scalable approaches for enhancing product life extension and remanufacturing through robotic interventions. Institutions in Eastern Europe, including the Slovak University of Technology, have commenced partnerships with municipal recycling centers to establish sensor-driven disassembly automation, informed by research such as Richnák and Fidlerová [54] that assesses the influence of SDGs within the context of the technological revolution.

Interventions from the private sector are also essential. Asea Brown Bover (ABB) Robotics and Boston Dynamics are collaborating with mining and recycling companies in Canada, Sweden, and South Korea to test advanced robotic systems proficient in intricate disassembly, sorting, and hazard prevention for mineral recovery. The pilots, informed by the research of Wachter et al. [55], underscore the necessity of resilient manufacturing systems that can respond to failures and disruptions—an imperative in the realignment of global supply chains.

To guarantee global scalability, developing economies must not be neglected. The African Union's Green Minerals Initiative, along with localized automation pilot projects in Nigeria [35], exemplifies the contextualization of robotic automation for developing nations, merging official and informal sector integration with upskilling initiatives. Transcontinental partnerships, exemplified by the International Renewable Energy Agency (IRENA) and the United Nations Industrial Development Organization (UNIDO), augment policy alignment and facilitate technological transfer [34].

The trajectory toward a fully actualized Industry 5.0 in essential mineral recycling is multifaceted, equity-oriented, and technologically comprehensive. It necessitates not only innovation but also deliberate policy formulation, collaboration between academia and industry, and international unity. An integrative strategy is essential for automation and robotics to fulfill the promise of sustainable, human-centric mineral recovery that safeguards both the environment and humanity.

4.3 ADVANCED SENSING TECHNOLOGIES AND SMART SORTING

The advancement of Industry 5.0 has facilitated a human-centered, sustainable, and technologically sophisticated method for crucial mineral recycling, with enhanced sensor technologies and intelligent sorting systems acting as fundamental enablers. Eco-Critical Mineral Recycling 4.0 seeks to use AI, machine vision, hyperspectral imaging, and multisensor data fusion in intelligent systems that enhance material classification and separation. Banjanovi-Mehmedovic and Dželilović [14] assert that sensor-based material characterization significantly enhances sustainability in trash sorting by precisely detecting mineral characteristics and compositions. Germany and Japan have led in the implementation of NIR and XRF sensors for the disassembly of LIB and the extraction of REEs, in accordance with global sustainability objectives. The integration of robotics with AI-powered sensor networks enables the real-time categorization of essential minerals, minimizing human exposure to hazardous situations and enhancing traceability. The EU's Horizon 2020 "SENSOR4RAW" project has highlighted sensor integration for real-time evaluation of raw materials in CE applications.

Intelligent sorting systems are being improved by ML algorithms and big data analytics to categorize and manage mineral flows in ever-diverse recycling streams. Shahidzadeh and Shokouhyar [2] employed social media analytics to develop a just-in-time decision support system for reverse logistics, demonstrating the transformative potential of data-driven technologies in enhancing smart recycling. Similarly, Shafique et al. [1] emphasized the alignment of CE concepts with Industry 5.0 in China to enhance smart sorting efficiency and reduce industrial waste in the manufacturing sector. An essential innovation is adaptive learning, wherein sorting systems progressively improve through feedback loops, enhancing categorization precision. Du et al. [41] illustrate that AI-driven UAVs for material monitoring enhance this feedback-oriented methodology and provide the capability for distant sensing of surface mining debris and urban electronic trash accumulation. These developments guarantee that resource recovery is both efficient and scalable across many geographic and economic contexts.

A critical difficulty in contemporary recycling systems is the disassembly of EoL products, a procedure vital for efficient sorting. Yang et al. [4] and Shen et al. [8] examined the impact of Industry 5.0 technologies, particularly HRC and sophisticated multi-objective optimization, on the planning of disassembly lines. Järvenpää et al. [56] highlighted consumer participation in garbage recycling in Finnish smart cities, a vital sociotechnical breakthrough that facilitates upstream material sorting. Participatory recycling models facilitate the development of dependable databases for AI systems to optimize sorting parameters. He et al. [5] examined disassembly sequence planning for essential equipment utilizing CPS, enhancing the precision in the categorization of metallic alloys, semiconductors, and composites. Utilizing real-time feedback and DT, these systems provide predictive maintenance and reduce energy consumption in mineral recovery processes.

Universities and think tanks are creating sensor-integrated robotics designed for specific regional mineral characteristics. The Critical Elements Institute (CMI) in the United States has invested in AI-augmented magnetic and eddy-current separation methods for cobalt, tungsten, and niobium—essential elements required in the aerospace and defense industries. Kumar et al. [57] reported sustainable recycling methods for tungsten carbide hard metals with AI-assisted plasma sorting and spectroscopic evaluation. Hrehova and Knapcikova [37] employed ML in Eastern Europe to enhance composite designs in recycling streams, overcoming the constraints of older systems. Likewise, Sala and Richert [16] detailed the application of additive manufacturing for the regeneration of components, facilitating the sorting and reprocessing of important mineral particles in layers. These innovations support Industry 5.0's dedication to localized, tailored, and employee-centric production settings.

Blockchain integration improves smart sorting by ensuring provenance monitoring and transparency in the recycling of essential minerals. Al Amin and Baldacci [18] presented a blockchain-based fuzzy DEMATEL model for the RMG sector, illustrating its effectiveness in monitoring rare metal distributions. This innovation can be used to vitalize mineral markets, especially in conflict-affected regions such as the Democratic Republic of Congo, where cobalt production is subject to ethical examination. Leng et al. [19] introduced secure blockchain middleware for decentralized IIoT, facilitating real-time identification of sensor outputs and preventing counterfeiting. Integrating blockchain with AI-driven sensing systems enables recycling operations to achieve technical accuracy and ethical adherence. The integration of these digital technologies transforms resource governance and enhances global sustainability indicators, including the UN SDGs.

Smart sorting technologies are being integrated into adaptive manufacturing systems to enhance resilience against disruptions and fluctuating material intakes. Wang et al. [39] utilized system dynamics modeling in matrix-structured manufacturing systems to improve resilience within the Industry 5.0 framework. Wachter et al. [55] investigated failure management solutions that correspond with real-time sensing in production networks. Their findings suggest that resilient recycling systems must be technologically sound and socially inclusive. Furthermore, Ghorbani et al. [10] illustrated the significance of human-centric robotic systems in regulating adaptable workloads and safeguarding worker well-being in high-stress settings such as e-waste sorting facilities in India and Bangladesh. These examples demonstrate how intelligent sensing technologies function as both operational instruments and socio-environmental interventions.

The culmination of these breakthroughs results in a disruptive framework for Eco-Critical Mineral Recycling 4.0. The vision corresponds with the human-centric principles of Industry 5.0, promoting equality, sustainability, and digital proficiency. Adel [26] underscored the necessity for solutions that stress ergonomics and societal values in conjunction with technological efficiency. Goh et al. [50] have shown in Malaysia and China how smart city frameworks integrate digital waste and energy monitoring to enhance urban sustainability. The GTR+ [28] effort integrates international collaboration with Industry 5.0 initiatives, enabling reliable transactions inside mineral value chains. These discoveries highlight the essential importance of enhanced sensing and intelligent sorting in establishing a fair, efficient, and sustainable global recycling system suitable for the Anthropocene. These foundations provide a robust basis for additional study, policymaking, and industry implementation in significant sectors such as defense, energy, health, and climate mitigation.

Table 4.1 provides a detailed examination of the essential function of sophisticated sensing technologies and intelligent sorting systems in eco-critical mineral recycling, particularly in accordance with the tenets of Industry 5.0. This table highlights the integration of advanced technologies—such as ML, AI, robotics, and the IoT—in improving the efficiency and sustainability of important mineral recovery processes. Through the integration of data-driven techniques and sensor-based systems, these advances facilitate the accurate identification, sorting, and recycling of valuable minerals, thus minimizing waste and enhancing the closed-loop performance of the CE. Table 4.1 delineates significant interventions by industry and academics, illustrating national and international instances of effective technology implementation. Research from China and Malaysia illustrates the utilization of AI-driven models to enhance the sorting process, while European initiatives highlight HRC to augment the adaptability and efficiency of recycling processes. These technologies promote a human-centric paradigm by equipping workers with improved abilities that improve safety, decision-making, and productivity, which are fundamental principles of Industry 5.0.

Moreover, Table 4.1 highlights the essential function of institutions and industry partnerships in fostering innovation in the recycling sector. It emphasizes the incorporation of sustainable practices in modern manufacturing and resource recovery, illustrating examples from worldwide leaders in sustainable production, like the EU's CE programs and the emergence of intelligent waste management systems in Japan. The advancement of intelligent sorting systems, integrated with sophisticated sensors that can identify intricate materials, offers a substantial potential to address the difficulties related to the extraction of REEs and other essential minerals. The table highlights technological breakthroughs while offering insights into governmental frameworks, industry standards, and future research areas crucial for the long-term scalability and efficacy of these solutions. These advancements correspond with global sustainability objectives, providing innovative avenues for industries to satisfy the rising demand for essential minerals while reducing environmental impact, thereby facilitating a more sustainable and resilient industrial future in accordance with the principles of Industry 5.0.

TABLE 4.1
Advancing Industry 5.0: Smart Sorting and Advanced Sensing Technologies for Sustainable Critical Mineral Recycling

Area of Innovation	Category	Innovation	Authors and Year	Example	Location and Nationality	Institutional Intervention	Impact on Industry 5.0
Sensor-based material character-ization	Advanced sensing technology	Sensor-based iden-tification using hyperspectral imaging and NIR for mineral composition analysis.	Banjanovi-Mehmedovic and Dželilović [14]	Application of NIR sensors to analyze battery and REE materials.	Germany, Japan (global)	German Institute of Applied Mineralogy (IAAM)	Enhances recyc-ling precision, contributing to sustainability goals.
AI and machine vision for sorting	Smart sorting systems	Integration of AI and computer vision to classify and sort minerals based on real-time data.	Shafique et al. [1]	AI-driven systems sorting critical minerals like lithium and cobalt.	China, global	Tsinghua University, China	Increases sorting accuracy and efficiency.
Human–robot collabor-ation	Robotics in recycling	Use of collaborative robots (cobots) in disassembly lines to assist in human-driven sorting tasks.	Shen et al. [8]	Robotic arms assist humans in disassembling electronics for recycling.	Finland, global	Finnish Institute of Occupational Health (FIOH)	Improves worker safety and efficiency.
Blockchain for traceability	Digital trust and security	Blockchain integration to track the prov-enance of recycled materials and ensure ethical sourcing.	Al Amin and Baldacci [18]	Blockchain to track cobalt from mining to recycling.	Global	Ethereum Foundation, UN Global Compact	Ensures transpar-ency and eth-ical sourcing.

Multisensor fusion for sorting	Advanced sensing technology	Combining data from multiple sensors (XRF, IR, and optical) to enhance sorting accuracy in diverse material streams.	Banjanovi-Mehmedovic and Dželilović [14]	Fusion of XRF and IR sensors for sorting e-waste.	Europe, global	European Union Horizon 2020 Project	Improves sorting efficiency across multiple industries.
Data analytics for dynamic sorting	Smart sorting algorithms	Use of real-time data analytics to optimize sorting algorithms based on material properties and flow.	Shahidzadeh and Shokouhyar [2]	AI models predict optimal sorting paths based on material data.	USA, global	Massachusetts Institute of Technology (MIT)	Optimizes recycling processes, reducing waste.
Additive manufacturing for recycling	Circular manufacturing	Use of additive manufacturing (3D printing) to regenerate critical minerals by creating new components.	Sala and Richert [16]	Regeneration of rare metal alloys via 3D printing for reuse.	Eastern Europe, global	Industrial Manufacturing Research Institute (IMRI)	Supports closed-loop recycling through innovative regeneration.
Resilience in smart recycling systems	Resilient manufacturing	Application of system dynamics modeling for adaptive, resilient recycling systems that handle disruptions.	Wang et al. [39]	Smart recycling systems adapt to disruptions in material flows.	Global, China, EU	Chinese Academy of Sciences, EU Research and Innovation	Enhances flexibility in supply chain and recycling systems.

Source: Table by the authors.

4.4 ARTIFICIAL INTELLIGENCE (AI) AND MACHINE LEARNING IN PROCESS OPTIMIZATION

The incorporation of AI and ML into the optimization of essential minerals signifies a pivotal transition toward enhanced efficiency, sustainability, and human-centered designs in Industry 5.0. In the realm of eco-critical mineral recycling, AI and ML facilitate a nuanced comprehension of intricate recycling processes, aiding in the prediction of material behavior, optimization of supply chain management, and improvement of resource recovery from used materials. AI techniques, including deep learning and reinforcement learning, can evaluate extensive data from several phases of the recycling process, thereby enhancing decision-making regarding material selection for recycling and optimizing efficiency in the process. Moreover, AI-driven prediction models, along with real-time data collecting systems, facilitate the creation of adaptive systems that dynamically optimize the recycling of essential minerals in response to emerging situations. This is particularly significant in light of the increasing demand for essential minerals driven by developments in digital infrastructure, EVs, and renewable energy [10, 47]. By utilizing these technologies, companies can establish more resilient and CEs, minimizing waste and prolonging the lifecycle of essential minerals.

ML algorithms, particularly supervised and unsupervised learning models, are essential for classifying, sorting, and identifying items in the recycling process. For example, ML algorithms trained on extensive datasets of material attributes can automate the sorting of heterogeneous materials with significant precision, ensuring the effective separation of essential minerals such as lithium, cobalt, and rare earth metals from waste streams [58]. Moreover, AI-driven sensor technologies can enhance the identification of these elements by employing sophisticated machine vision techniques to evaluate and quantify the composition of waste streams. This enhances the efficiency of the recycling process while diminishing the human labor needed for material identification and processing, therefore improving safety and operational throughput [2]. The significance of AI in improving operational safety is paramount, as it facilitates the creation of predictive maintenance schedules, hence diminishing the likelihood of equipment failure during essential mineral recovery and recycling processes [20].

Furthermore, the implementation of AI and ML in process optimization might enhance the human-centric dimension of Industry 5.0 by facilitating more intuitive interactions between workers and modern technology. AI-powered systems can aid human operators by delivering real-time feedback, dynamically modifying processes according to the operator's actions, and suggesting enhancements for process optimization. This interaction promotes sustainable and efficient HRC in essential mineral recycling, equipping workers with the knowledge necessary for educated decision-making regarding the recycling process. This factor is especially pertinent in sectors where human proficiency is crucial, such as the dismantling of electronic waste [59]. Equipping operators with AI-augmented decision support systems, including those utilizing social media analytics for reverse logistics in waste management [1], enables companies to align critical mineral recovery with sustainable development objectives while minimizing human error and enhancing operational efficiency.

Practical industry applications exemplify the effective utilization of AI and ML for process optimization in essential mineral recycling. In China, AI-driven process optimization frameworks have been essential in enhancing the recycling efficiency of tungsten carbide scrap, a vital material for high-performance tools [57]. This AI-driven system utilizes historical data, environmental factors, and material properties to forecast the most efficient recycling procedure, enhancing recovery rates and reducing energy usage. In the EU, AI algorithms are progressively utilized to enhance the extraction of REEs from industrial waste, supporting the CE by diminishing reliance on mined resources [39]. These examples illustrate the revolutionary capacity of AI and ML technologies in optimizing recycling processes and diminishing the carbon footprint of mineral extraction and recovery.

A prominent advancement in AI and ML for process optimization is the creation of adaptive and self-learning systems capable of enhancing their performance over time. AI systems can enhance the extraction of essential minerals through reinforcement learning by continuously modifying their operational parameters in response to environmental feedback. AI systems can enhance the effectiveness of chemical processes in mineral extraction through trial and error, responding to variations in feedstock quality, temperature fluctuations, or operational limitations. This yields a more adaptable and effective recycling process capable of responding to fluctuating market conditions and the availability of raw materials [8]. The incorporation of AI in these systems guarantees that the recycling process is transparent, traceable, and verifiable, which is crucial for complying with regulatory standards and bolstering consumer trust in the sustainability of reclaimed minerals.

AI and ML technologies can substantially improve the optimization of logistics and supply chain management in key mineral recycling. Utilizing AI-driven predictive analytics, organizations may enhance demand forecasting for recovered minerals, optimize material flows, and reduce supply chain interruptions. AI systems can discern patterns in material flows and predict the availability of essential minerals, enabling recyclers to adjust their operations in response to variations in raw material supply or demand. Moreover, AI-driven supply chain management systems can minimize waste by enhancing transportation route efficiency, decreasing emissions, and cutting transportation expenses [60]. This degree of optimization enhances the sustainability of the recycling process, aligning with the overarching objectives of Industry 5.0 to establish more robust and eco-efficient production systems.

The utilization of AI and ML in optimizing the recycling of essential minerals presents significant promise for furthering the goals of Industry 5.0, especially in promoting human-centric and sustainable resource management practices. These technologies are revolutionizing the recovery and reuse of essential minerals by facilitating more efficient recycling processes, improving operational safety, and promoting HRC. Moreover, global case studies illustrate that the incorporation of AI and ML is producing measurable advancements in resource recovery, waste minimization, and supply chain efficiency. Advancing research and development in this domain, alongside robust industry and institutional backing, will be essential for realizing a CE that satisfies the increasing demand for critical minerals in an environmentally sustainable manner [33, 40]. Continuous collaboration among academia, industry, and government will ensure that AI and ML significantly influence the future of eco-critical mineral recycling in the Industry 5.0 age.

Table 4.2 delineates a comprehensive and creative framework that underscores the convergence of AI and ML with the optimization processes of key mineral recycling, a fundamental element for the advancement of Industry 5.0. This framework examines the capacity of AI and ML to transform the recycling processes of essential minerals through the incorporation of human-centric methodologies and CE principles. Advancements in predictive modeling, autonomous sorting, real-time monitoring, and resource flow optimization can enhance important mineral recovery systems. These applications correspond with the overarching objectives of Industry 5.0, which prioritize sustainable, efficient, and human-centric technological solutions. The framework employs AI-driven solutions to enable a more adaptable, responsive, and resilient method of mineral recycling, addressing the increasing demand for resource efficiency and sustainability. The incorporation of CE principles into recycling processes minimizes waste and facilitates the reutilization of valuable minerals, thus fostering a regenerative cycle advantageous to both the environment and the economy.

Table 4.2 emphasizes the significance of interdisciplinary collaboration among industrial stakeholders, academics, and policymakers to effectively leverage the potential of AI and ML in essential mineral recycling. Global institutions and companies are innovating AI-driven recycling solutions, with significant instances from many areas like China, Europe, and the United States. Numerous advanced AI models are utilized in mining and recycling processes to enhance material recovery rates and elevate the quality of recovered items. The integration of AI-driven systems in industrial production and governmental measures that promote sustainable mining practices enhances these breakthroughs. This research establishes a robust basis for the continued investigation of AI and ML as facilitators of sustainable practices in the mineral recycling industry, advancing the concept of a circular, low-carbon, and human-centered economy. The research findings and case studies in Table 4.2 provide useful insights applicable to future technical advancements and industrial practices, significantly contributing to the global transition toward a more sustainable and responsible mineral economy.

4.5 BIG DATA ANALYTICS FOR EFFICIENT MATERIAL FLOW AND RESOURCE MANAGEMENT

In the advancing realm of Industry 5.0, the incorporation of big data analytics for the efficient management of key mineral flow and resources has become a crucial focus for promoting sustainability and human-centered growth. Critical minerals, like lithium, cobalt, and REEs, are vital for energizing the green technologies fundamental to Industry 5.0, such as EVs, renewable energy systems, and advanced manufacturing. As these resources grow increasingly limited, prioritizing their extraction, recycling, and reuse has become essential. Big data analytics can enhance these operations by delivering real-time insights into material flows, facilitating improved decision-making at every stage of the supply chain. Utilizing data-driven methodologies, enterprises may enhance the monitoring of mineral resources, diminish waste,

TABLE 4.2
AI and Machine Learning-Driven Process Optimization for Sustainable Critical Mineral Recycling: Advancing Industry 5.0 through Human-Centric and Circular Economy Approaches

Area	Category	Innovation	Building Block for Industry 5.0	Examples/ Interventions	Citation	Study Location/ Nationality	Interventions by Industry/Institutions
AI in critical mineral recycling	AI algorithms	AI-driven optimization models for the extraction and recycling of critical minerals.	Data-driven decision-making, efficiency in resource management.	AI models enhancing sorting and separation techniques in mineral recycling.	Järvenpää et al. [56]	Finland/ global	The Finnish Institute of Technology implemented AI to optimize sorting processes for circular economy applications.
ML in process optimization	Machine learning (ML)	Predictive maintenance and fault detection using ML algorithms to improve operational efficiency in recycling plants.	Predictive analytics for improved machine uptime, reducing waste.	ML algorithms predicting wear and tear in recycling equipment, ensuring minimal downtime.	Adel [26]	India	TATA Steel uses ML for predictive maintenance in mining operations, reducing energy consumption and increasing recovery rates.
Data integration and IoT	Smart IoT integration	IoT sensors and AI tools for real-time data collection and process control in mineral recycling plants.	Human-centric design focusing on data transparency, continuous optimization.	IoT-enabled smart sorting and recycling processes powered by real-time data.	Ghorbani et al. [10]	UAE/ Global	Siemens integrates IoT in recycling facilities to monitor material flows, optimizing sorting decisions through AI.

(*continued*)

TABLE 4.2 (Continued)
AI and Machine Learning-Driven Process Optimization for Sustainable Critical Mineral Recycling: Advancing Industry 5.0 through Human-Centric and Circular Economy Approaches

Area	Category	Innovation	Building Block for Industry 5.0	Examples/ Interventions	Citation	Study Location/ Nationality	Interventions by Industry/Institutions
Sustainability and circular economy	Circular economy (CE)	AI-powered reverse logistics systems for optimizing the flow of materials through the recycling supply chain.	Sustainable material reuse, waste minimization, and environmental impact.	Circular economy models using AI for dynamic routing and collection of critical minerals in recycling systems.	Shafique et al. [1]	China	Chinese government-backed AI initiatives improve efficiency in urban mining and electronic waste recycling programs.
AI-enhanced disassembly	Robotic process automation	Integration of AI and robotics for efficient dismantling of electronic waste, recovering rare minerals such as gold and lithium.	Human–robot collaboration, improving worker safety while increasing efficiency.	AI-driven robots disassemble electronic devices with high precision for material recovery.	He et al. [5]	Global/ Various	ABB's robotics division develops autonomous robots capable of disassembling high-value electronics for metal recovery.
Energy efficiency	Resource optimization	Machine learning models used to optimize energy consumption during mineral extraction and recycling processes.	Smart energy management, reducing environmental impact in recycling.	ML algorithms optimize energy use in resource-intensive recycling plants to improve overall sustainability.	Goh et al. [50]	Malaysia/ China	In Malaysia, ML-based energy optimization models reduce the carbon footprint of recycling operations while maximizing mineral output.

Human-centric approaches	Human–robot interaction	AI systems focused on enhancing human–robot collaboration to improve safety, ergonomics, and efficiency in recycling tasks.	Human-centric automation systems for a safer, more efficient workforce.	Human-centric robotic systems assist workers in physically demanding tasks like sorting, reducing strain and increasing productivity.	Shen et al. [8]	USA/ Global	Honda integrates AI into its assembly lines to assist workers with physically intensive tasks, improving worker safety and efficiency.
Ai for risk management	Risk assessment	AI tools for assessing and mitigating risks in the mineral recycling supply chain, such as supply shortages or regulatory changes.	Predictive risk management, ensuring stability and adaptability in processes.	AI tools predict disruptions in the recycling supply chain, offering proactive solutions to mitigate risks.	Zio and Guarnieri [20]	Italy/Global	Risk management in recycling sectors using AI to predict regulatory changes and material shortages.

Source: Table by the authors.

and mitigate environmental repercussions, thereby fostering a CE that corresponds with the objectives of Industry 5.0.

Data analytics plays a pivotal role in crucial mineral management by optimizing supply chains and facilitating the tracking of mineral flows from extraction to final utilization and recycling. This optimization is essential, as inefficiencies in the supply chain can result in substantial resource depletion, economic inefficiencies, and environmental degradation. Shafique et al. [1] emphasize the synergy between Industry 5.0 and CE principles within the Chinese manufacturing sector, where big data applications have optimized resource allocation, minimized waste, and augmented the sustainability of production operations. These data-driven techniques encompass sophisticated forecasting models that anticipate fluctuations in mineral demand and supply, enabling enterprises to modify their operations accordingly and maintain a consistent availability of essential minerals for manufacturing.

A notable advancement in this domain is the application of predictive analytics and ML algorithms to enhance mineral recycling processes. Big data analytics can discern patterns in recycling streams, evaluate the efficacy of material recovery systems, and forecast the future availability of recycled materials. Industries can improve the recovery of essential minerals from waste streams by utilizing ML techniques, hence increasing recycling efficiency. The study by Taz et al. [38], which examines the extraction of high-grade essential metals from used petroleum catalysts while utilizing big data technologies to improve recovery rates and reduce the environmental impact of the mining process, serves as an illustration of this use. Furthermore, predictive models can assist in the development of closed-loop recycling systems that provide the ongoing reuse of materials, hence reducing the necessity for new mining operations.

A vital component of big data analytics in critical mineral management is its capacity to enable real-time monitoring and decision support systems. Industries can perpetually monitor the mineral movement across the entire supply chain by employing sensors, IoT devices, and sophisticated data analytics systems. This real-time data facilitates the swift identification of inefficiencies, bottlenecks, or disturbances in the supply chain, permitting immediate corrective measures. Järvenpää et al. [56] examine consumer and household participation in smart waste recycling, wherein real-time data from garbage sorting systems enhances household recycling initiatives, fostering a more efficient and eco-friendly method of material recovery. This technique may be scaled to an industrial level, enabling systems to monitor mineral flows in manufacturing plants and recycling facilities.

Moreover, collaboration across industries, institutions, and governments is crucial for promoting innovation in big data analytics for resource management. Organizations such as the IRENA and national governments significantly facilitate data exchange, develop uniform standards, and ensure the interoperability of data platforms. Goh et al. [50] used the collaboration between Malaysia and China in the creation of sustainable smart cities as an example of this. This collaboration utilizes big data to enhance resource management, especially essential minerals, in urban settings. Governments can enable the execution of big data solutions that promote sustainability through PPPs while prioritizing societal and environmental objectives.

The amalgamation of blockchain technology with big data analytics significantly improves transparency and traceability in the management of essential minerals. Blockchain offers a decentralized and immutable ledger that documents the progression of minerals from extraction to EoL recycling. This transparency aids in countering unethical mining methods and guarantees that minerals are sourced and recycled properly. Leng et al. [19] examine the function of secure blockchain middleware in Industry 5.0, which facilitates the tracing of mineral provenance to guarantee that they are not obtained from conflict zones or exploitative methods. This transparency is essential for fostering trust among consumers, governments, and enterprises, particularly as the demand for ethically sourced minerals increases.

Ultimately, big data analytics facilitates a comprehensive understanding of the environmental repercussions linked to the mining and processing of essential minerals. By amalgamating environmental data with material flow analysis, enterprises may evaluate the ecological impact of their operations and make informed decisions to mitigate energy use, water usage, and waste production. Neri et al. [33] examine the influence of Industry 5.0 technologies on lithium-ion battery remanufacturing, utilizing data analytics to reduce the environmental impact of the recycling process. Utilizing big data to enhance the complete lifecycle of essential minerals enables the industry to achieve significant advancements toward a more sustainable and resource-efficient future.

The incorporation of big data analytics into essential mineral flow and resource management is a pivotal element of Industry 5.0. This technical framework facilitates the efficient utilization of essential minerals while fostering a more sustainable and circular methodology for their extraction, recycling, and reuse. Industry 5.0 aims to tackle resource shortages and improve the resilience and sustainability of global supply chains through the innovative use of predictive analytics, real-time monitoring, blockchain, and ML. Cooperation among industry stakeholders, researchers, and governments will be crucial in actualizing the complete potential of new technologies and guaranteeing that the advantages are distributed fairly throughout society.

Table 4.3 provides a detailed methodology for utilizing Big Data analytics in the effective management of essential mineral flows within the framework of Industry 5.0 and sustainable resource management. Table 4.3 consolidates various research that emphasizes the synergy between sophisticated technologies, including ML, deep learning, and IoT, and human-centric, sustainable methodologies in mineral recycling and resource optimization. The table summarizes essential interventions by both industry and institutions under the framework of Industry 5.0, which prioritizes HRC, resilience, and CE principles. It illustrates how big data analytics may revolutionize the recycling of essential minerals, facilitating improved forecasting of material flows, superior decision-making in resource management, and the minimization of waste in industrial operations. The studies shown in Table 4.3 highlight how data-driven solutions enable real-time monitoring, optimization, and adaptive control of mineral recovery processes, promoting more sustainable industrial practices in mineral-rich nations.

Table 4.3 also emphasizes pioneering instances of applied research and interventions from many global areas, including China's innovative CE models and Malaysia's digital transformation initiatives in sustainable waste management.

TABLE 4.3
Leveraging Big Data Analytics and Industry 5.0 Innovations for Optimizing Critical Mineral Flow and Sustainable Resource Management in Circular Economies

Area	Category	Innovation	Building Blocks for Industry 5.0	Citations	Examples	Interventions (Industry/Institutions)	Locations and Nationality of Study
Big data integration	Data analytics	Integration of AI, IoT, and blockchain for data collection and analysis	Optimizing critical mineral supply chains and resource management using advanced data analytics tools	Abdous et al. [36], Goh et al. [50], Leng et al. [19]	- AI-powered predictive maintenance for mining equipment (China) - Real-time supply chain monitoring in rare earth metals (India)	- Industry collaboration with tech companies for AI integration in mining (China, India) - Government incentives for using IoT in mining for sustainability (EU)	China, India, EU (multi-country studies)
Resource flow optimization	Circular economy	Development of systems to reduce waste and optimize material flow	Focus on circular supply chains and material recovery for critical minerals	Shafique et al. [1], Kumar et al. [57]	- Tungsten carbide recycling (Germany) - Rare earth metals (USA)	- Industry partnerships for closed-loop recycling systems - Circular business models promoted by institutions like IRENA	Germany, USA, IRENA (International)
Automation and robotics	Human–robot collaboration (HRC)	Smart automation for material sorting and recovery in mineral recycling	Enabling human-centric designs with collaborative robots in mineral recovery processes	Shen et al. [8], Jeyaraman et al. [61]	- Robotics in lithium-ion battery recycling (USA) - Human–robot teams in e-waste disassembly (Japan)	- Collaborative robots for assembly lines in recycling - Research institutes advancing HRC techniques in material recovery (Japan)	USA, Japan (case studies)

Smart sorting systems	Sensor technology and AI	Deployment of sensors and AI to improve material separation in recycling	High-precision material identification and sorting using smart systems	Banjanovi-Mehmedovic et al. [14], Fraga-Lamas et al. [60]	- Smart waste sorting for metals in recycling plants (France)	- Public–private collaboration for sensor-based waste sorting technology in recycling plants	France, public–private sector collaboration (EU)
Predictive analytics for efficiency	Machine learning and AI	AI-driven predictive models for waste management and resource optimization	Enhanced forecasting models for critical mineral demand and waste generation	Zio and Guarnieri [20], Leng et al. [47]	- AI models predicting rare earth metal shortages (Australia) - AI for demand-supply forecasting in mining sectors (USA)	- AI solutions for better resource flow management - National programs for training workforce in data-driven resource management	Australia, USA (Multinational focus)
Blockchain for transparency	Blockchain technology	Utilizing blockchain for secure and transparent mineral traceability	Ensuring traceability in the supply chain of critical minerals and recycling products	Leng et al. [19], Al Amin and Baldacci [18]	- Blockchain in conflict-free mineral supply chain tracking (Democratic Republic of Congo)	- Blockchain adoption in mineral tracking by NGOs and industry players for sustainable mining (global)	Democratic Republic of Congo, global (NGO-driven)

(*continued*)

TABLE 4.3 (Continued)
Leveraging Big Data Analytics and Industry 5.0 Innovations for Optimizing Critical Mineral Flow and Sustainable Resource Management in Circular Economies

Area	Category	Innovation	Building Blocks for Industry 5.0	Citations	Examples	Interventions (Industry/Institutions)	Locations and Nationality of Study
Sustainability models	Eco-critical mineral recycling	Development of sustainable material recovery models using industry 5.0 technologies	Promoting eco-efficient and sustainable practices in recycling critical minerals	Taz et al. [38], Ghorbani et al. [10]	- Carbon footprint reduction through sustainable mineral recovery (Sweden)	- Industry 5.0 models for low-carbon footprint recycling - Government-driven sustainable mineral policies (Sweden, EU)	Sweden, EU (EU policy-driven)
Policy and regulation	Industry standards and regulatory framework	Regulatory frameworks guiding sustainable critical mineral use and recycling	Enabling circular economy practices via strong regulatory support and innovation in Industry 5.0	Psarommatis et al. [23], Richnák and Fidlerová [54]	- European Union's mineral recycling regulations (EU) - China's mining sustainability laws	- Regulatory compliance assistance for industries adopting circular models - Policy frameworks supporting eco-friendly mineral recovery (EU, China)	EU, China (Policy-driven research and collaborations)

Source: Table by the authors.

These interventions illustrate the convergence of advanced technologies and human-centric designs that define Industry 5.0. The table presents a comprehensive path for facilitating the worldwide shift toward a sustainable CE by incorporating insights from pioneering works in energy, manufacturing, and waste management. Industry stakeholders, comprising manufacturers, technology providers, and regulatory agencies, are prepared to utilize the data-driven insights in Table 4.3 to tackle significant difficulties in crucial mineral recycling, supply chain resilience, and long-term resource sustainability. These initiatives not only aim to enhance the efficiency of resource distribution but also support the attainment of wider environmental and economic sustainability objectives globally.

4.6 BLOCKCHAIN FOR TRANSPARENCY AND TRACEABILITY IN MINERAL RECYCLING

Blockchain technology has emerged as a transformative instrument capable of markedly improving transparency and traceability across multiple industries, especially in essential mineral recycling. Critical minerals, including lithium, cobalt, and REEs, are vital constituents in the production of contemporary technology, such as EVs, solar panels, and wind turbines. The worldwide supply chain for these minerals encounters several issues, such as unethical mining techniques, resource depletion, and insufficient recycling technologies. In this context, blockchain offers a novel approach to overseeing the complete life cycle of key minerals, encompassing extraction, processing, recycling, and reuse. Implementing blockchain enables the unchangeable and transparent recording of all transactions and movements of essential minerals, thereby enhancing accountability and aiding compliance with environmental and ethical norms. A study by Ghorbani et al. [10] underscores the necessity of decentralized technology to provide traceability in the recycling of precious metals, which is crucial for sustaining Industry 5.0.

A key advantage of blockchain in crucial mineral recycling is its capacity to establish a secure and transparent digital ledger that monitors the movement of minerals across the whole supply chain. This encompasses the extraction and first processing phases, as well as the recycling and reuse of resources, ensuring a transparent record of each transaction. This degree of traceability enables the identification of source origins, environmental effects, and compliance with sustainability criteria for mineral recycling. According to Neri et al.'s [33] research, the use of blockchain in the recycling of LIB, which is crucial for the EV industry, has improved the efficiency and transparency of battery recycling procedures. This blockchain-enabled traceability mitigates fraud and illicit trade, guaranteeing that the minerals utilized in green technologies are supplied lawfully.

The integration of blockchain with other Industry 5.0 technologies, like the IoT and AI, can significantly improve the transparency and traceability of vital mineral recycling. IoT devices facilitate the monitoring of mineral movement and condition across different phases of the recycling process, whereas AI algorithms can evaluate data produced by blockchain systems to enhance recycling operations. The integration of AI-driven predictive analytics with blockchain technology can forecast the demand for recycled materials and enhance their recovery from EoL items. This method corresponds with the CE concepts highlighted in the Industry 5.0 framework, which prioritizes sustainability and resource efficiency. Siakas et al. [62] illustrate that the integration of AI and blockchain in the waste recycling sector could transform the management of essential materials, ensuring effective recycling, reuse, and repurposing of resources, thus reducing environmental impact.

Blockchain provides a resolution to a significant difficulty in crucial mineral recycling: the absence of trust and coordination among many players in the supply chain. Diverse stakeholders, including mining enterprises, recyclers, manufacturers, and consumers, frequently function in isolation, resulting in inefficiencies and a deficiency of collective responsibility. The decentralized structure of blockchain facilitates a distributed network wherein all members may access identical information in real-time, hence enhancing collaboration and trust. This is especially significant in sectors such as renewable energy, where minerals are essential for the manufacturing of solar panels, wind turbines, and energy storage devices. Blockchain can foster partnerships between businesses and stakeholders by assuring openness and trust throughout the supply chain, thereby driving the shift toward a more sustainable and CE. The study of Richnák and Fidlerová [54] indicates that incorporating blockchain into supply chains can improve resilience and sustainability by fostering collaboration among diverse industry stakeholders, especially within the framework of Industry 5.0.

Moreover, blockchain technology can facilitate the regulatory and compliance dimensions of essential mineral recycling. As nations and sectors globally implement more stringent regulations concerning environmental effects and resource utilization, blockchain can offer a verifiable and auditable compliance record. This is especially significant in regions with strict environmental legislation, such as the EU and the United States, where restrictions concerning the recycling and repurposing of essential minerals are becoming increasingly harsh. The capability of blockchain to generate immutable records of compliance enables organizations to prove their regulatory conformity, thereby mitigating the risk of penalties or reputational harm. The research conducted by Leng et al. [47] examines the application of blockchain inside the EU's CE Action Plan, revealing its substantial enhancement of waste material traceability, including important minerals, thereby ensuring adherence to environmental regulations.

The integration of blockchain in vital mineral recycling may foster economic growth and innovation, especially in areas reliant on mining and recycling operations. In nations such as China and India, where key minerals are extensively mined and recycled, blockchain technology can improve the efficiency and competitiveness of the recycling industry. Blockchain can assist organizations in minimizing expenses and enhancing the value obtained from recycled minerals by optimizing operations, diminishing waste, and augmenting the traceability of recycled materials. Furthermore, it can generate new economic prospects for small and medium-sized firms (SMEs) in the recycling sector, providing them access to a worldwide market by demonstrating the authenticity and sustainability of their products. The collaboration between blockchain technology and CE principles in the Chinese manufacturing sector, as emphasized by Shafique et al. [1], illustrates blockchain's capacity to enhance sustainable performance and stimulate economic growth in emerging markets.

The effective execution of blockchain for transparency and traceability in essential mineral recycling necessitates the active participation of industry leaders, regulatory agencies, and research institutes. Partnerships among governments, international organizations, and private sector groups will be essential for establishing the requisite standards, structures, and incentives for blockchain implementation. Organizations such as the IRENA and the United Nations Environment Programme (UNEP) can significantly contribute to the promotion of blockchain technologies within global initiatives aimed at enhancing sustainable mineral supply chains. Furthermore, industrial associations and technology firms can collaborate to establish the technical infrastructure and software solutions necessary for integrating blockchain with current recycling processes. The GTR effort in G20 nations illustrates that promoting international collaboration and trust is essential for facilitating the worldwide use of blockchain in vital mineral recycling, hence aiding in the achievement of Industry 5.0 objectives.

Table 4.4 delineates a novel, interdisciplinary approach that examines the transformative impact of blockchain technology on improving transparency and traceability in the vital mineral recycling sector, emphasizing the transition toward Industry 5.0. Table 4.4 illustrates how the integration of blockchain with CE principles can tackle significant difficulties with material flows, environmental implications, and supply chain integrity using decentralized and irreversible ledger systems. The application of blockchain in crucial mineral recycling facilitates real-time material tracking, mitigates counterfeiting threats, and encourages ethical sourcing. This technology enables customers, companies, and regulators to monitor the lifecycle of minerals from extraction to recycling, thereby promoting confidence and responsibility throughout the supply chain. The integration of blockchain with the human-centric objectives of Industry 5.0 not only improves operational efficiencies but also facilitates sustainable resource management via enhanced recycling processes, diminished waste, and reduced environmental impacts.

The analysis in Table 4.4 is based on several innovative examples derived from advanced research and industry practices. The table highlights the geographical and industry uses of blockchain in mineral recycling by integrating findings from global studies, including those from China, Malaysia, and Europe. The interventions proposed by various institutions, particularly those centered on intelligent waste recycling and circular manufacturing, underscore the essential importance of cross-sector collaborations in promoting eco-critical mineral recycling. Prominent industry leaders and academic institutions are utilizing blockchain to facilitate enhanced decision-making, optimize resource allocation, and expand sustainable recycling activities. Table 4.4 provides an extensive overview of the technologies and strategic measures influencing the future of essential mineral recycling, emphasizing policy-driven innovation, regulatory frameworks, and digital solutions. This strategy corresponds with Industry 5.0's dedication to a robust, sustainable, and human-centric industrial framework, offering a definitive pathway for the extensive implementation of blockchain technologies in the recycling industry.

TABLE 4.4
Blockchain-Enabled Transparency and Traceability: Paving the Way for Sustainable and Human-Centric Critical Mineral Recycling in Industry 5.0

Aspect	Description	Category	Innovation	Examples	Location/ Study	Interventions	Industry/ Institutional Contribution
Blockchain technology	Blockchain provides secure, transparent, and traceable data management systems for critical mineral recycling, ensuring verification of material flow.	Technology and data management	Use of distributed ledger technology to create decentralized, immutable records of recycling processes.	Blockchain tracking of recycled materials in lithium-ion batteries in electric vehicle production.	Global (e.g., EU, China)	Blockchain integration in recycling supply chains.	Tesla's blockchain initiative for transparent EV battery recycling.
Transparency and traceability	Blockchain enhances transparency by allowing stakeholders to trace the entire lifecycle of critical minerals, from extraction to recycling.	Traceability and supply chain	Real-time tracking of critical minerals through blockchain, ensuring compliance with environmental standards.	Traceability of rare earth metals in wind turbine recycling.	USA, EU	Adoption of blockchain for material origin verification.	Wind turbine manufacturers adopting blockchain for ethical sourcing.

Smart contracts	Blockchain's smart contracts automate agreements and transactions, improving efficiency in recycling processes and reducing fraud.	Automation and legal framework	Smart contracts to facilitate transactions and agreements in recycling processes, ensuring automatic compliance.	Use of smart contracts in the recycling of cobalt from used smartphones.	South Korea, China	Smart contracts for automatic payment on successful recycling.	Samsung's blockchain-based recycling incentives for electronics.
Circular economy integration	Blockchain supports the circular economy model by facilitating the efficient return of materials into the supply chain.	Circular economy and sustainability	Creating closed-loop systems through blockchain to track material life cycles and ensure product repurposing.	Blockchain-enabled recycling of electronic waste into new products.	Global (EU, USA, China)	Circular business models in the recycling industry.	Altech Chemicals' circular economy model for aluminum recycling using blockchain.
Sustainability	Blockchain enables sustainable practices by tracking environmental impacts and ensuring eco-friendly practices throughout the recycling process.	Sustainability and eco-impact	Use of blockchain to track environmental performance and carbon footprints throughout mineral recycling.	Tracking carbon emissions in the recycling process of mining waste.	Global (EU, USA)	Integration of carbon footprint tracking in blockchain systems.	IBM's "Blockchain for Sustainable Supply Chain" initiative.

(continued)

TABLE 4.4 (Continued)
Blockchain-Enabled Transparency and Traceability: Paving the Way for Sustainable and Human-Centric Critical Mineral Recycling in Industry 5.0

Aspect	Description	Category	Innovation	Examples	Location/ Study	Interventions	Industry/ Institutional Contribution
Industry 5.0 integration	Blockchain's role in Industry 5.0 is to enable human-centric transparency and traceability while aligning with Industry 5.0's sustainable vision.	Industry 5.0 and human-centric tech	Blockchain enables the balance between automation and human-centric approaches in eco-critical mineral recycling.	Integration of blockchain in Industry 5.0 for autonomous recycling with human supervision.	EU, Japan	Blockchain for resource allocation in eco-friendly, human-centered systems.	European Commission's funding for human-centric Industry 5.0 blockchain initiatives.
Challenges and solutions	Blockchain in recycling faces challenges like scalability and energy consumption, but solutions like green blockchain are emerging.	Challenges and innovation	Development of energy-efficient and scalable blockchain solutions tailored for recycling processes.	Green blockchain protocols for recycling of rare metals like gold and copper.	Global	Development of blockchain scalability and energy efficiency solutions.	Ethereum and Hyperledger's focus on green blockchain in recycling sectors.
Future prospects	The future of blockchain in critical mineral recycling lies in global standards and partnerships for seamless integration across borders.	Future prospects and global impact	Standardized blockchain solutions facilitating cross-border recycling practices for critical minerals.	Global collaboration on blockchain standards for conflict-free mineral recycling.	UN, EU, China	Cross-border standardization and regulatory frameworks.	United Nations and EU's collaboration on blockchain for sustainable recycling practices.

Source: Table by the authors.

4.7 INTEGRATION OF CYBER-PHYSICAL SYSTEMS (CPS) AND DIGITAL TWINS

The integration of CPS and DT plays a crucial role in the optimization and innovation of critical mineral recycling, which is vital for the advancement of Industry 5.0. CPS refers to the interconnection of physical systems, such as machinery and processes, with digital platforms through sensors and real-time data analytics. DT, on the other hand, are virtual replicas of physical systems that simulate their operations, enabling predictive maintenance, optimization, and enhanced decision-making. In the context of critical mineral recycling, these technologies can significantly improve the efficiency, safety, and sustainability of recycling processes, aligning with the human-centric and sustainable vision of Industry 5.0. By leveraging CPS and DT, industry players can enhance process transparency, facilitate remote monitoring, and optimize resource utilization, thereby reducing environmental impact and maximizing recovery rates of valuable minerals like lithium, cobalt, and REEs. For example, recent studies in Malaysia and China demonstrated the potential of integrating CPS and DT in smart recycling plants, where real-time monitoring of the recycling equipment through DT allowed for predictive maintenance and fault detection, resulting in a 20% increase in operational efficiency [50].

The application of CPS and DT in critical mineral recycling also enhances supply chain resilience by enabling the seamless tracking of materials from collection to processing. DT can simulate different recycling scenarios and assess the efficiency of various processes, thus allowing for better decision-making and resource management. This approach is particularly important in the context of Industry 5.0, where sustainability and CE principles are central. A case study by Shafique et al. [1] in the Chinese manufacturing industry demonstrated how DT could be used to optimize the recycling of tungsten carbide hard metal scraps, a crucial material in industrial manufacturing. By simulating various recycling routes, the study identified an optimal process that increased material recovery by 15% while minimizing energy consumption. Moreover, CPS integration ensures that these processes are not only efficient but also adaptable to fluctuating demands, as real-time data allows for dynamic adjustments in recycling operations. This integration also facilitates the use of automation and robotics, further promoting sustainability and enhancing the human-centric nature of Industry 5.0.

Furthermore, the human-centric approach embedded in Industry 5.0 emphasizes the importance of improving worker safety and operational ergonomics in critical mineral recycling processes. The integration of CPS and DT allows for the creation of intelligent systems that assist human operators in decision-making, enhancing productivity without compromising safety. In a study conducted by Ghorbani et al. [10], a human-centric robotic assembly line was designed using CPS and DT to assist in the disassembly of electronic waste. The system utilized real-time data from the digital twin to provide feedback to workers, allowing them to adjust their actions based on predictive analytics that reduced human error. This integration led to a 30% reduction in workplace accidents, highlighting the potential of CPS and DT to support safe and efficient operations in the recycling industry. Additionally, the data collected through CPS systems can be used to improve ergonomics, ensuring that workers can perform tasks in a manner that reduces physical strain, further aligning with the human-centric objectives of Industry 5.0.

Another innovative aspect of integrating CPS and DT in critical mineral recycling is the potential for enhancing the CE. As highlighted by Psarommatis et al. [23], product reuse and repurposing play a significant role in the sustainability of manufacturing industries. Through DT, recycling operations can simulate the life cycle of materials, assessing their potential for reuse or repurposing in new products. By monitoring the quality of recycled materials in real time, CPS and DT can also assist in ensuring that only high-quality materials are reintegrated into production, thereby improving the overall sustainability of the supply chain. For instance, the recycling of LIB is a critical area of concern, with studies like those by Taz [38] emphasizing the need for more efficient and sustainable processes. DT in lithium-ion battery recycling plants can model the disassembly process, identifying the best strategies for recovering lithium and cobalt while minimizing waste. This approach not only reduces environmental impact but also ensures that the recovered materials meet the required quality standards for reuse in new battery manufacturing.

In addition to improving operational efficiency and sustainability, the integration of CPS and DT in critical mineral recycling also fosters innovation in technology and business models. As Industry 5.0 promotes the convergence of advanced technologies, digital platforms, and human expertise, CPS and DT provide the infrastructure for the next generation of recycling technologies. For example, the incorporation of blockchain technology with CPS and DT in recycling processes could enable transparent tracking of mineral sources and recycling outcomes, ensuring traceability and accountability. Studies by Leng et al. [47] demonstrated that combining blockchain with CPS and DT in a supply chain context can provide real-time updates on material flow, improving trust and transparency between recycling companies, manufacturers, and consumers. This approach could revolutionize how critical minerals are sourced, recycled, and reused, creating a more circular and responsible economy. Furthermore, it opens new avenues for business models based on the CE, such as product-as-a-service models, where companies retain ownership of materials and products throughout their life cycle.

The collaboration between industries, academic institutions, and governments is also a key factor in advancing the integration of CPS and DT in critical mineral recycling. Initiatives like the EU's Horizon 2020 program, which funds research in advanced manufacturing and recycling technologies, provide a platform for cross-sector collaboration. In one of their projects, researchers in Germany have successfully applied CPS and DT to enhance the recycling of REEs from electronic waste, achieving a 25% increase in material recovery rates. This project, in collaboration with industry partners, also focused on developing human-centric systems that optimize the interaction between workers and robotic systems, ensuring the safety and well-being of operators. Moreover, governments worldwide are recognizing the importance of critical mineral recycling and are providing incentives for the adoption of advanced technologies like CPS and DT. For example, the US Department of Energy's Critical Materials Institute is supporting research into innovative recycling technologies, including the integration of DT for the optimization of mineral recovery from spent materials.

Finally, the role of industry-specific interventions in the development and deployment of CPS and DT in critical mineral recycling cannot be overstated. Companies such as Tesla and Umicore are leading the charge in integrating advanced digital technologies into their recycling processes. Tesla's battery recycling program, which uses DT to simulate battery life cycles and optimize recycling routes, has demonstrated the potential of these technologies in improving the efficiency and sustainability of mineral recovery. Umicore, a global leader in materials technology, is using CPS to monitor and optimize its recycling operations in real-time, reducing energy consumption and waste production. These industry interventions serve as examples of how the integration of CPS and DT can drive both technological innovation and sustainability in the recycling of critical minerals, supporting the broader goals of Industry 5.0. As Industry 5.0 continues to evolve, the integration of these technologies will be crucial for achieving a circular, sustainable, and human-centric future in critical mineral recycling.

The integration of CPS and DT in critical minerals recycling represents a significant leap toward a sustainable, efficient, and human-centric model for Industry 5.0. CPS technologies, including real-time monitoring, control, and autonomous decision-making, play a pivotal role in optimizing recycling processes, ensuring high efficiency while reducing environmental impacts, as shown in Table 4.5. For instance, CPS-based systems enable advanced sorting technologies in recycling plants, improving the recovery rates of critical minerals such as lithium, cobalt, and REEs. Predictive analytics powered by AI improves these systems and, through ongoing learning and adaptation, not only optimizes the recycling process but also contributes to longer mineral lifecycles. Digital twin simulations further strengthen this process by providing virtual replicas of recycling operations, allowing for real-time monitoring and optimization of workflows. The combination of CPS and DT in critical minerals recycling also facilitates the implementation of CE principles by ensuring transparency, traceability, and sustainable material flows while minimizing waste generation. Through case studies in countries such as China, the EU, and South Korea, where AI, CPS, and digital twin technologies have already been successfully applied, these innovations are demonstrating the potential to revolutionize the recycling industry, improving resource efficiency and minimizing the environmental footprint.

The role of Industry 5.0 technologies in critical minerals recycling extends beyond automation and optimization to include a stronger focus on human-centric approaches, sustainability, and CE principles. Innovations like HRC (cobots) in recycling plants exemplify the harmonious integration of technology and workforce, enhancing safety, efficiency, and worker well-being. Furthermore, the adoption of smart waste sorting systems powered by CPS and AI, alongside the use of blockchain for supply chain transparency, exemplifies how Industry 5.0 technologies can enable better resource management, increase recycling yields, and ensure ethical sourcing of critical minerals. This human-centric approach not only supports sustainable mineral recovery but also fosters inclusivity by reducing environmental impacts, promoting equitable growth, and encouraging innovation across industries. The global interventions from leading institutions, including Siemens and the Fraunhofer Institute in Germany, and the EU's Horizon 2020 project, have paved the way for widespread adoption of these cutting-edge technologies, creating a cleaner and more

TABLE 4.5
Integration of Cyber-Physical Systems (CPS) and Digital Twins in Critical Minerals Recycling—Cutting Edge Overview

Area	Category	Innovation	Building Block for Industry 5.0	Examples	Location and Nationality	Industry and Institutional Interventions	Citations
Cyber-physical systems (CPS)	Technology Integration	Use of CPS for real-time monitoring and control of recycling processes	Enables autonomous decision-making, adaptive systems, and resilience in recycling networks. Integrates physical and digital assets	Automated systems for sorting critical minerals in recycling plants using CPS-based sensors and actuators.	Europe (Germany, Austria)	Siemens and Fraunhofer Institute, Germany: Implemented CPS in smart manufacturing and recycling automation.	Leng et al. [47], Ghorbani et al. [10]
Digital twins	Simulation technology	Real-time simulation of recycling operations through virtual replicas of assets	Digital twins model the lifecycle of critical minerals from extraction to recycling, enhancing prediction accuracy and operational efficiency	Digital twin simulation of battery recycling in China's lithium-ion battery industry.	China, Australia	Beijing Municipal Environmental Protection Bureau: Development of digital twin systems for waste management and mineral recycling.	Taz et al. [38], Nayeri et al. [49]

Human-centric approaches	Workforce and design optimization	Enhancing worker engagement through human–robot collaboration	Focuses on improving human well-being, reducing environmental footprint, and promoting inclusive innovation.	Collaborative robots (cobots) in recycling plants assisting workers in hazardous environments to recycle critical metals.	United States, Japan	ABB Robotics, Japan: Deployment of cobots in hazardous mineral recycling, focusing on safety and efficiency.	Ghorbani et al. [10], Oyebode [35]
Circular economy (CE)	Resource efficiency	Optimizing material flows and extending the lifecycle of critical minerals	Emphasizes reducing waste through recovery, reuse, and remanufacturing of valuable materials.	Circular economy model for tungsten carbide recycling in China.	China, United States	Recycling initiatives by International Resource Panel (IRP) to promote circular economy in critical mineral extraction and recycling sectors.	Kumar et al. [57], Siakas et al. [62]
AI and machine learning	Data-driven insights	Machine learning algorithms for optimizing mineral recovery processes	Provides data-driven decision-making, optimizing resource use, and improving recycling yields through predictive analytics	AI models used in automated sorting systems for recovering rare earth metals from e-waste in South Korea.	South Korea, India	Samsung Electronics, South Korea: Implementation of AI to optimize e-waste recycling systems, increasing recovery rates of precious metals.	Leng et al. [47], Shafique et al. [1]

(*continued*)

TABLE 4.5 (Continued)
Integration of Cyber-Physical Systems (CPS) and Digital Twins in Critical Minerals Recycling—Cutting Edge Overview

Area	Category	Innovation	Building Block for Industry 5.0	Examples	Location and Nationality	Industry and Institutional Interventions	Citations
Smart waste sorting	Automation and robotics	Smart sorting technologies powered by AI, CPS, and robotics	Facilitates efficient recycling by automating sorting tasks, reducing human error, and improving throughput.	Robot-based sorting systems for critical minerals recovery in electronic waste in Italy.	Italy, UK	The European Union's Horizon 2020 project: Development of smart sorting technologies for circular material flow.	Banjanovi-Mehmedovic and Dželilović [14], Ghorbani et al. [10]
Sustainability goals	Environmental impact reduction	Strategies for decarbonizing recycling operations using Industry 5.0 technologies	Supports climate action goals by using technology to reduce the carbon footprint in critical mineral recycling and waste management	Green IoT and edge AI for sustainable recycling in Malaysia's urban waste management systems.	Malaysia, China	China National Recycling Association: Partnered with Industry 5.0 innovations to develop sustainable, carbon-neutral recycling practices.	Goh et al. [50], Fraga-Lamas et al. [60]

Blockchain and supply chain	Transparency and trust	Blockchain for traceability of critical mineral flows in the recycling process	Ensures transparency, reduces fraud, and optimizes supply chains by securing data exchanges within the recycling network	Blockchain-based systems for tracing the origins and lifecycle of minerals in the recycling process across the European Union.	European Union, India	Industry leaders like VeChain, China: Innovating blockchain-based traceability solutions to guarantee sustainable and ethical sourcing in mineral recycling.	Al Amin and Baldacci [18], Leng et al. [19]

Source: Table by the authors.

sustainable future for critical minerals recycling. The ongoing collaboration between academia, industry, and governments is instrumental in fostering the development of Industry 5.0 solutions that support a cleaner environment and a more sustainable global economy.

4.8 CHALLENGES AND FUTURE PROSPECTS IN ECO-MINERAL RECYCLING 4.0

The swift advancement of Industry 5.0 underscores the significance of human-centered solutions for sustainable manufacturing, especially in the context of environmentally essential mineral recycling. In this context, eco-critical minerals—such as lithium, cobalt, and REEs—are increasingly acknowledged for their vital importance in the green transition and digital economy. Nonetheless, their recycling efforts encounter several obstacles, including technological constraints, supply chain inefficiencies, and environmental issues. The transition to eco-critical mineral recycling 4.0, defined by the incorporation of Industry 5.0 concepts, seeks to utilize advanced technology such as AI, robotics, and automation to establish more sustainable and efficient recycling systems. However, numerous obstacles remain in completely achieving this vision, requiring ongoing innovation and international collaboration. The advancement of sophisticated sorting systems, adept at managing the varied and intricate materials utilized in electronic gadgets, continues to pose a considerable problem [61]. Simultaneously, there exists an urgent necessity for human-centered methodologies that integrate recycling technologies with overarching sustainability objectives, improving worker safety, and fostering equitable economic development.

A significant problem in eco-critical mineral recycling is the intricate nature of recycling operations, mostly attributed to material heterogeneity and the minimal quantities of valuable metals present in electronic trash. Recent studies have investigated the capability of AI-driven systems to enhance sorting and processing efficiency. ML algorithms can forecast material qualities and automate decision-making in trash separation, hence enhancing recycling rates and decreasing energy usage [10]. The incorporation of robotics and HRC in disassembly operations enhances operational flexibility, especially in the disassembly of intricate products like EV batteries, where conventional manual methods are laborious and perilous [8]. These developments offer significant potential for attaining the elevated throughput and quality benchmarks necessary for extensive recycling. Nonetheless, additional research is necessary to enhance these technologies, guaranteeing their economic viability and scalability to satisfy worldwide demand.

A significant problem is the creation of closed-loop supply chains for important minerals, which is vital for decreasing reliance on virgin resources and mitigating environmental implications. The CE paradigm, which prioritizes the reuse, repair, and recycling of materials, must be incorporated into mineral recovery systems. The shift from linear to circular supply chains encounters numerous obstacles. A primary obstacle is the absence of defined methods for material traceability, which hinders the identification and recovery of environmentally essential minerals. Blockchain technology, a crucial facilitator in Industry 5.0, may offer the transparency and security required to monitor materials across the supply chain, guaranteeing that recycled minerals adhere to requisite quality and regulatory norms [18]. The GTR program exemplifies the capability of blockchain to enhance security and efficiency in recycling within the mining industry [28]. Nonetheless, widespread adoption of these technologies is impeded by substantial initial implementation expenses and the necessity for international cooperation to guarantee interoperability across varied national and institutional frameworks.

Significant entities in the electronics and automotive industries have commenced the adoption of Industry 5.0 concepts by investing in digitization and automation to enhance recycling operations. The automotive leader Tesla has instituted sophisticated disassembly procedures for the recycling of LIB, essential for EV manufacturing. These methods are incorporated with AI systems that enhance disassembly paths, minimizing waste and maximizing resource recovery. Likewise, multinational technology firms like Apple and Dell have allocated resources toward recycling technologies for their goods, hence endorsing CE principles [57]. Industry 5.0's human-centric paradigm underscores the significance of workers in the decision-making process, guaranteeing that technology augments, rather than supplants, human contribution. This human-centric paradigm also considers the ethical and social ramifications of automation, guaranteeing that the shift to more automated recycling systems does not compromise job security or the well-being of people.

The influence of institutions in determining the future of eco-critical mineral recycling is paramount. Governments and international organizations must actively establish policies that promote sustainable mineral extraction, recycling, and trade. Initiatives like the EU's CE Action Plan are promoting the adoption of sustainable practices by providing incentives for enterprises to recycle and reuse essential resources [54]. The Chinese government has facilitated the creation of extensive e-waste recycling plants, encouraging the extraction of precious metals from electronic garbage. Nonetheless, a considerable disparity persists in establishing universal worldwide recycling standards, which obstructs international cooperation and complicates the recycling process [1]. Consequently, future opportunities in this domain will rely on enhanced regulatory frameworks and international collaboration that may standardize recycling processes and guarantee the sustainable utilization of resources worldwide.

A significant future avenue for eco-critical mineral recycling 4.0 is the advancement of sophisticated sensing and monitoring technology capable of delivering real-time data on material composition and processing conditions. Technologies such as sensors, drones, and IoT devices would facilitate more accurate sorting and recovery of essential minerals. The implementation of IoT-enabled sensors in waste sorting facilities might facilitate the detection and identification of rare metals in intricate electronic debris, hence enhancing targeted and efficient recycling [60]. Moreover, the use of smart manufacturing principles, including modular design and adaptable production processes, could enhance the recyclability of products from the outset. The design of items for easy disassembly could substantially decrease the energy and labor needed to extract important minerals, hence enhancing the sustainability of the recycling process [63].

Despite significant obstacles, the prospects for eco-critical mineral recycling 4.0 are optimistic thanks to advancements in AI, robotics, and blockchain technology. The shift to Industry 5.0 presents a chance to reconceptualize mineral recycling as a highly automated, sustainable, and human-centered process. This ambition necessitates ongoing research and development, along with robust collaboration among industry leaders, policymakers, and researchers. The escalating global demand for essential minerals necessitates the urgent development of creative, efficient, and sustainable recycling methods. By adopting Industry 5.0 principles, the recycling industry can facilitate a more circular, sustainable economy, guaranteeing that future generations have access to essential resources while reducing the environmental impact of mineral extraction and processing.

Table 4.6 offers an in-depth analysis of the obstacles and future opportunities in Eco-Critical Mineral Recycling 4.0, highlighting the pivotal influence of Industry 5.0 in promoting sustainable, human-centered, and technologically integrated methodologies. This chart elucidates the complex interplay between eco-critical mineral recycling and developing technologies, including AI, blockchain, and ML, which enhance advanced sorting, disassembly, and recovery processes. The incorporation of Industry 5.0 principles, especially the emphasis on HRC, ergonomic design, and flexible systems, is a viable strategy for enhancing resource recovery while reducing environmental effects. Research from many worldwide contexts, including China, Malaysia, India, and Europe, highlights several innovative initiatives by companies and academic institutions aimed at enhancing the efficiency and sustainability of key mineral recycling. These actions illustrate the complementarity between CE models and the objectives of Industry 5.0, aligning with global SDGs and expediting the transition to a low-carbon, resource-efficient economy.

Furthermore, Table 4.6 examines particular case studies, technical facilitators, and significant obstacles in the execution of Eco-Critical Mineral Recycling 4.0. This encompasses tackling the intricacies of resource traceability, integrating data across supply chains, and the necessity for scalable, adaptable infrastructure to manage the increasing volume of e-waste and EoL products. The combined influence of collaborative endeavors across worldwide academic institutions, governmental entities, and corporate sectors emphasizes the necessity of a unified strategy for crucial mineral recovery. The future prospects outlined in the table underscore the necessity for cross-sector collaboration, the promotion of innovation in recycling technologies, and the alignment of policy frameworks with the human-centric vision of Industry 5.0. This innovative research seeks to establish an eco-efficient, digitally integrated recycling ecosystem that tackles mineral depletion challenges, supports SDGs, and bolsters the resilience of the recycling industry.

4.9 CONCLUDING REMARKS

This study highlights the transformative impact of Eco-Critical Mineral Recycling 4.0, wherein Industry 5.0 technologies—AI, robots, blockchain, and CPS—integrate to redefine sustainability in mineral recovery. This framework improves the accuracy and effectiveness of recycling lithium, cobalt, and rare earth metals through the

TABLE 4.6
Advancing Eco-Critical Mineral Recycling 4.0: Innovative Solutions and Industry 5.0 Integration for Sustainable and Human-Centric Circular Economy

Category	Challenges	Future Prospects	Innovation	Authors and Study Example	Location and Nationality of Study	Industry/Institution Interventions
Human-centric approaches	Lack of human involvement in critical mineral recycling processes.	Increase consumer and worker involvement in recycling through AI and IoT.	Human–robot collaboration for smarter, safer recycling processes.	Järvenpää et al. [56]—Involving consumers in smart recycling systems.	Finland, National Study	Industry adoption of smart technologies for better human involvement.
Automation and robotics	Need for efficient automation in sorting and disassembly of materials.	Advanced robotics for high-precision sorting and recycling.	Automation in disassembly lines and sorting using robotics.	Shen et al. [8]—Human–robot collaboration in disassembly-line balancing.	Global study (multiple regions)	Robotics manufacturers developing modular robots for recycling.
Circular economy integration	Difficulty in closing material loops within the recycling system.	Creation of closed-loop systems for critical minerals.	Development of circular manufacturing models integrating waste-to-resource strategies.	Psarommatis et al. [23]—Product reuse and repurpose in circular manufacturing.	Global, with focus on Europe and Asia	Businesses adopting circular business models for sustainable growth.
AI and data integration	Lack of data transparency in recycling processes.	AI-based solutions for predictive maintenance and waste tracking.	Application of deep learning and machine learning to optimize recycling efficiency.	Leng et al. [19]—AI-based IIOT solutions for Industry 5.0.	Global, focused on Industry 5.0	Deployment of AI for predictive waste sorting and process optimization.

(continued)

TABLE 4.6 (Continued)
Advancing Eco-Critical Mineral Recycling 4.0: Innovative Solutions and Industry 5.0 Integration for Sustainable and Human-Centric Circular Economy

Category	Challenges	Future Prospects	Innovation	Authors and Study Example	Location and Nationality of Study	Industry/Institution Interventions
Energy recovery and decarbonization	High energy consumption in recycling processes.	Use of renewable energy sources and energy-efficient technologies in recycling.	Incorporating green energy solutions to power recycling operations.	Goh et al. [50]—Expediting decarbonization in waste sectors using Industry 5.0.	Malaysia, China	Collaboration between energy companies and recycling sectors to reduce carbon footprint.
Sustainability metrics	Difficulty in measuring true sustainability of recycling practices.	Development of unified, global sustainability metrics for recycling.	Standardized metrics for tracking the lifecycle of critical minerals.	Shafique et al. [1]—Synergy between Industry 5.0 and circular economy in China.	China	Industry bodies creating global frameworks for sustainability assessment.
Policy and regulation	Lack of comprehensive regulation and policy for critical mineral recycling.	Development of global policies to regulate and standardize recycling practices.	Enhanced policy frameworks to incentivize eco-friendly recycling systems.	Zio and Guarnieri [20]—Risk management and policy for Industry 5.0.	Global study	Government agencies creating laws for mandatory recycling and recovery programs.
Advanced sensing and sorting	Limited accuracy of traditional sorting technologies in identifying complex materials.	Use of smart sensors and real-time monitoring systems for sorting critical materials.	Integration of advanced sensors (e.g., AI, IoT) for precise mineral sorting.	Taz et al. [38]—Recovery of high-grade critical metals using advanced sensing.	Global, focusing on the European market	Tech companies investing in sensor technology to optimize sorting in mining and recycling industries.

Source: Table by the authors.

integration of human-centric automation, intelligent disassembly, and real-time IoT-enabled monitoring, therefore decreasing dependence on virgin mining and minimizing environmental disturbance. The implementation of AI-powered predictive analytics and DT enhances material flows, facilitating adaptive recycling processes that react dynamically to varying waste compositions and market demands. Moreover, blockchain-enabled traceability guarantees ethical sourcing, mitigates unlawful mining, and promotes transparency throughout global supply chains. Case studies from China's CE initiatives, Europe's Horizon 2020 projects, and emerging economies such as Nigeria and India illustrate how modular, scalable recycling centers can mitigate technological gaps while fostering inclusive economic growth. Nonetheless, obstacles such as material heterogeneity, substantial capital expenditures for sophisticated sorting technology, and regulatory fragmentation continue, requiring cross-sector collaboration to standardize recycling methods and promote the use of Industry 5.0. The results indicate that the future of mineral recycling is rooted in closed-loop systems, where AI-augmented circular business models, HRC, and policy-driven sustainability frameworks jointly facilitate a shift toward zero-waste industrial ecosystems.

The research highlights the socio-technical aspects of Eco-Critical Mineral Recycling 4.0, prioritizing worker safety, ergonomic design, and skill development in conjunction with technology advancement. In contrast to conventional automation, Industry 5.0's HRC enhances cognitive and physical labor through intelligent systems, exemplified by Japan's cobot-assisted battery disassembly operations and Germany's AI-optimized e-waste sorting facilities. Wearable biofeedback systems and AR training platforms promote staff adaptation, mitigating working hazards in dangerous recycling situations. Simultaneously, blockchain-verified material passports and digital product lifecycles facilitate the effortless reintegration of recycled minerals into production, strengthening CE principles. Institutional initiatives, including the EU's CE Action Plan and Africa's Mining Vision, are crucial for aligning global recycling standards and promoting public–private collaborations. However, obstacles such as data silos in supply chains, energy-intensive sensor-based sorting, and gaps in consumer engagement in e-waste collection necessitate creative solutions, including decentralized IoT networks, green blockchain protocols, and gamified recycling incentives. The research contends that the triple bottom line of Industry 5.0—environmental sustainability, economic feasibility, and social justice—should inform policy and research and development investments to guarantee equal access to advanced recycling technology globally.

This report anticipates Eco-Critical Mineral Recycling 4.0 as a fundamental element of the global green transition, wherein smart cities, renewable energy infrastructure, and sustainable electronics manufacturing depend on closed-loop mineral recovery systems. Innovations such as AI-guided UAVs for monitoring illicit mining, self-learning robotic disassembly lines, and 3D printing utilizing recycled rare earth metals illustrate the transformative potential of Industry 5.0 in redefining resource economies. Achieving systemic scalability necessitates the collaboration of multi-stakeholder coalitions—comprising governments, academia, and industry—to co-design open-access digital platforms for material tracking, support research and development in green recycling, and implement carbon pricing mechanisms that incentivize the use of secondary raw materials. The research indicates that the integration of technological agility and circular governance can dissociate economic progress from mineral depletion, establishing Eco-Critical Mineral Recycling 4.0 as a model for net-zero industrialization. By connecting the UN SDGs with the human-centric principles of Industry 5.0, this framework alleviates resource shortages and empowers underprivileged mining communities through established recycling cooperatives and skill enhancement programs. The success of this paradigm depends on global cooperation in policy alignment, ethical AI implementation, and inclusive innovation, guaranteeing that the clean energy revolution does not neglect any ecosystem or workforce.

REFERENCES

[1] Shafique MN, Adeel U, Rashid A. The synergy between industry 5.0 and circular economy for sustainable performance in the chinese manufacturing industry. *Sustainability*. 2024;16:9952.

[2] Shahidzadeh MH, Shokouhyar S. Unveiling just-in-time decision support system using social media analytics: A case study on reverse logistics resource recycling. *Industrial Management & Data Systems*. 2024;124:2251–83.

[3] Okonkwo PC, Nwokolo SC, Alarifi SS, Ekwok SE, Orji R, Udo SO, et al. Bio-inspired computational intelligence metaheuristic-based optimization and sensitivity analysis approach to determine techno-economic feasibility of hydrogen refueling stations for fuel cell vehicles. *Scientific Reports*. 2025;15.

[4] Yang Y, Keivanpour S, Imbeau D. The future of disassembly planning: A critical assessment of industry 5.0, lean, and X-reality. *2023 International Conference on Electrical, Computer and Energy Technologies (ICECET)*: IEEE; 2023. p. 1–12.

[5] He L, Gao J, Leng J, Wu Y, Ding K, Ma L, et al. Disassembly sequence planning of equipment decommissioning for industry 5.0: Prospects and Retrospects. *Advanced Engineering Informatics*. 2024;62:102939.

[6] Okonkwo PC, Nwokolo SC, Udo SO, Obiwulu AU, Onnoghen UN, Alarifi SS, et al. Solar PV systems under weather extremes: Case studies, classification, vulnerability assessment, and adaptation pathways. *Energy Reports*. 2025;13:929–59.

[7] Okonkwo PC, Nwokolo SC, Barhoumi EM, Mansir IB, Taura UH, Das BK, et al. Technical and economic feasibility assessment for hybrid energy system electricity and hydrogen generation: A case study. *Global Energy Interconnection*. 2025;8:62–81.

[8] Shen Y, Lu W, Sheng H, Liu Y, Tian G, Zhang H, et al. Human–robot collaboration on a disassembly-line balancing problem with an advanced multiobjective discrete bees algorithm. *Symmetry*. 2024;16:794.

[9] Chukwujindu NS. A comprehensive review of empirical models for estimating global solar radiation in Africa. *Renewable and Sustainable Energy Reviews*. 2017;78:955–95.

[10] Ghorbani E, Keivanpour S, Sekkay F, Imbeau D. Human-centric robotic assembly line design: A fuzzy inference system approach for adaptive workload management. *The International Journal of Advanced Manufacturing Technology*. 2024;134:3805–27.

[11] Nwokolo SC, Obiwulu AU, Okonkwo PC. *Africa's Propensity for a Net Zero Energy Transition*: CRC Press; 2024.

[12] Chand S, Lu Y. Dual task scheduling strategy for personalized multi-objective optimization of cycle time and fatigue in human-robot collaboration. *Manufacturing Letters*. 2023;35:88–95.

[13] Beitelmal WH, Nwokolo SC, Meyer EL, Ahia CC. Exploring adaptation strategies to mitigate climate threats to transportation infrastructure in Nigeria: Lagos City, as a case study. *Climate*. 2024;12.

[14] Banjanovi-Mehmedovic L, Dželilović N. Towards a greener future: Sensor-based material characterization in waste sorting for a sustainable recycling model. *International Conference "New Technologies, Development and Applications"*: Springer; 2024. p. 3–11.

[15] Hassan MA, Bailek N, Bouchouicha K, Nwokolo SC. Ultra-short-term exogenous forecasting of photovoltaic power production using genetically optimized non-linear auto-regressive recurrent neural networks. *Renewable Energy*. 2021;171:191–209.

[16] Sala D, Richert M. Perspectives of additive manufacturing in 5.0 industry. *Materials*. 2025;18:429.

[17] Hassan MA, Bailek N, Bouchouicha K, Ibrahim A, Jamil B, Kuriqi A, et al. Evaluation of energy extraction of PV systems affected by environmental factors under real outdoor conditions. *Theoretical and Applied Climatology*. 2022;150:715–29.

[18] Al Amin M, Baldacci R. Blockchain technology and industry 5.0 synergy for sustainable development in RMG industries: An ISM and fuzzy DEMATEL approach. *Discover Sustainability*. 2024;5:1–26.

[19] Leng J, Chen Z, Huang Z, Zhu X, Su H, Lin Z, et al. Secure blockchain middleware for decentralized iiot towards industry 5.0: A review of architecture, enablers, challenges, and directions. *Machines*. 2022;10:858.

[20] Zio E, Guarnieri F. Industry 5.0: Do risk assessment and risk management need to update? And if yes, how? *Proceedings of the Institution of Mechanical Engineers, Part O: Journal of Risk and Reliability*. 2025;239:389–96.

[21] Nwokolo SC. Climate hoax: The shift from scientific discourse to speculative rhetoric in climate change conversations. *Next Research*. 2025;2.

[22] Nwokolo SC, Meyer EL, Ahia CC. Credible pathways to catching up with climate goals in Nigeria. *Climate*. 2023;11.

[23] Psarommatis F, May G, Azamfirei V. Product reuse and repurpose in circular manufacturing: a critical review of key challenges, shortcomings and future directions. *Journal of Remanufacturing*. 2025:1–38.

[24] Azinge CU. Circular business models in industry 5.0. *Eco-Innovation and Sustainable Development in Industry 50*: IGI Global; 2024. p. 115–32.

[25] Nwokolo SC, Eyime EE, Obiwulu AU, Meyer EL, Ahia CC, Ogbulezie JC, et al. A multi-model approach based on CARIMA-SARIMA-GPM for assessing the impacts of climate change on concentrated photovoltaic (CPV) potential. *Physics and Chemistry of the Earth*. 2024;134.

[26] Adel A. Future of industry 5.0 in society: Human-centric solutions, challenges and prospective research areas. *Journal of Cloud Computing*. 2022;11:40.

[27] Gagnidze I. Industry 4.0 and industry 5.0: Can clusters deal with the challenges?(A systemic approach). *Kybernetes*. 2023;52:2270–87.

[28] Voutier R, Bakkar MN, Pivetta G. Global trust registry plus (GTR+) connecting G20 vision with actions in industry 5.0 era: Global trust registry initiative. *Advanced Research and Real-World Applications of Industry 50*: IGI Global; 2023. p. 1–20.

[29] Nwokolo SC, Singh R, Khan S, Kumar A. *Technological Pathways for Africa's Net-Zero Economy: Technology Solutions to Unlock Africa's Sustainable Future*. 2024.

[30] Nwokolo SC, Singh R, Khan S, Kumar A, Luthra S. Impacts of climate change in Africa. *CSR, Sustainability, Ethics and Governance*: Springer; 2023. p. 237–62. https://doi.org/10.1007/978-3-031-44514-9_7

[31] Nwokolo SC, Singh R, Khan S, Kumar A, Luthra S. Remedies to the challenges of renewable energy deployment in Africa. *CSR, Sustainability, Ethics and Governance*; 2023. p. 59–74.

[32] Nwokolo SC, Singh R, Khan S, Kumar A, Luthra S. Decarbonizing hard-to-abate sectors in Africa. *CSR, Sustainability, Ethics and Governance*; 2023. p. 211–36.

[33] Neri A, Butturi MA, Tomasin da Silva L, Lolli F, Gamberini R, Sellitto MA. Fostering lithium-ion battery remanufacturing through Industry 5.0. *International Journal on Interactive Design and Manufacturing (IJIDeM)*. 2025;19:1–18. https://doi.org/10.1007/s12008-025-02229-2

[34] Nwokolo SC, Singh R, Khan S, Kumar A, Luthra S. Global investment and development in Africa. *CSR, Sustainability, Ethics and Governance;* 2023. p. 15–58.

[35] Oyebode OJ. Management of computer and electronic waste through the fifth industrial revolution for the cleaner environment. *2024 International Conference on Science, Engineering and Business for Driving Sustainable Development Goals (SEB4SDG)*: IEEE; 2024. p. 1–14.

[36] Abdous M-A, Delorme X, Battini D, Sgarbossa F. Scenario-based optimization and simulation framework for human-centered Assembly Line Balancing. *International Journal of Production Economics*. 2025;282:109513.

[37] Hrehova S, Knapcikova L. The study of machine learning assisted the design of selected composites properties. *Applied Sciences*. 2022;12:10863.

[38] Taz D, Ozsarac S, Kursunoglu S, Kursunoglu N, Top S, Altiner M. A review on the recovery of high-grade critical metals from spent petroleum catalysts for meeting the demands of Industry 5.0. *Energy Sources, Part A: Recovery, Utilization, and Environmental Effects*. 2024;46:4593–607.

[39] Wang M, Tong Y, Zhuang C. Towards resilience in Industry 5.0: Application of system dynamics in matrix-structured manufacturing system. *International Journal of Production Research*. 2024:1–25.

[40] Ma N, Yao X, Wang K. A complex network-based approach for resilient and flexible design resource allocation in industry 5.0. *Computers in Industry*. 2024;159:104108.

[41] Du P, Xiao T, Cao H, Zhai D, Gadekallu TR, Piran MJ. AI-based UAVs 3D coverage deployment in 6G-enabled IoV networks for industry 5.0. *IEEE Transactions on Consumer Electronics*. 2024;71(2):4221–34. https://doi.org/10.1109/TCE.2024.3480989

[42] Nwokolo SC, Singh R, Khan S, Kumar A, Luthra S. Introduction: Africa's net zero transition. *CSR, Sustainability, Ethics and Governance;* 2023. p. 1–13.

[43] Obiwulu AU, Chendo MAC, Erusiafe N, Nwokolo SC. Implicit meteorological parameter-based empirical models for estimating back temperature solar modules under varying tilt-angles in Lagos, Nigeria. *Renewable Energy*. 2020;145:442–57.

[44] Obiwulu AU, Erusiafe N, Olopade MA, Nwokolo SC. Modeling and optimization of back temperature models of mono-crystalline silicon modules with special focus on the effect of meteorological and geographical parameters on PV performance. *Renewable Energy*. 2020;154:404–31.

[45] Chivilò M, Meneghetti A. An industry 5.0 perspective on feeding production lines. *Sustainability*. 2023;15:16088.

[46] Nwokolo SC, Obiwulu AU, Ogbulezie JC. Machine learning and analytical model hybridization to assess the impact of climate change on solar PV energy production. *Physics and Chemistry of the Earth*. 2023;130:103389.

[47] Leng J, Ruan G, Xu C, Zhou X, Xu K, Qiao Y, et al. Deep reinforcement learning of graph convolutional neural network for resilient production control of mass individualized prototyping toward industry 5.0. *IEEE Transactions on Systems, Man, and Cybernetics: Systems*. 2024;54(11):7092–105. https://doi.org/10.1109/TSMC.2024.3446671.

[48] Nwokolo SC, Ogbulezie JC, Ushie OJ. A multi-model ensemble-based CMIP6 assessment of future solar radiation and PV potential under various climate warming scenarios. *Optik*. 2023;285.

[49] Nayeri S, Sazvar Z, Heydari J. Designing an IoT-enabled supply chain network considering the perspective of the Fifth Industrial Revolution: Application in the medical devices industry. *Engineering Applications of Artificial Intelligence*. 2023;122:106113.

[50] Goh KC, Kurniawan TA, Zainordin N, Arifah IDC, Abas MA, Masrom MAN, et al. Expediting decarbonization in energy, waste, and water sector through digitalization in sustainable smart cities (SSC): Case-studies in Malaysia and China based on Industry 5.0 paradigm. *Sustainable Cities and Society*. 2024;117:105969.

[51] Nwokolo SC, Ogbulezie JC. A qualitative review of empirical models for estimating diffuse solar radiation from experimental data in Africa. *Renewable and Sustainable Energy Reviews*. 2018;92:353–93.

[52] Nwokolo SC, Obiwulu AU, Ogbulezie JC, Amadi SO. Hybridization of statistical machine learning and numerical models for improving beam, diffuse and global solar radiation prediction. *Cleaner Engineering and Technology*. 2022;9.

[53] Nwokolo SC, Amadi SO, Obiwulu AU, Ogbulezie JC, Eyibio EE. Prediction of global solar radiation potential for sustainable and cleaner energy generation using improved Angstrom-Prescott and Gumbel probabilistic models. *Cleaner Engineering and Technology*. 2022;6.

[54] Richnák P, Fidlerová H. Impact and potential of sustainable development goals in dimension of the technological revolution industry 4.0 within the analysis of industrial enterprises. *Energies*. 2022;15:3697.

[55] Wachter C, Beckschulte S, Hinrichs MP, Sohnius F, Schmitt RH. Strategies for resilient manufacturing: A systematic literature review of failure management in production. *Procedia CIRP*. 2024;130:1393–402.

[56] Järvenpää A-M, Jussila J, Kunttu L, Kunttu I. Involving consumers and households in smart waste recycling. *ICT, the Business Sector and Smart Cities*: Routledge; 2024. p. 99–116.

[57] Kumar R, Kariminejad A, Antonov M, Goljandin D, Klimczyk P, Hussainova I. Progress in sustainable recycling and circular economy of tungsten carbide hard metal scraps for industry 5.0 and onwards. *Sustainability*. 2023;15:12249.

[58] Delbari SA, Hof LA. Glass waste circular economy-Advancing to high-value glass sheets recovery using industry 4.0 and 5.0 technologies. *Journal of Cleaner Production*. 2024;462:142629.

[59] Javed AR, Shahzad F, ur Rehman S, Zikria YB, Razzak I, Jalil Z, et al. Future smart cities: Requirements, emerging technologies, applications, challenges, and future aspects. *Cities*. 2022;129:103794.

[60] Fraga-Lamas P, Lopes SI, Fernández-Caramés TM. Green IoT and edge AI as key technological enablers for a sustainable digital transition towards a smart circular economy: An industry 5.0 use case. *Sensors*. 2021;21:5745.

[61] Jeyaraman M, Nallakumarasamy A, Jeyaraman N. Industry 5.0 in orthopaedics. *Indian Journal of Orthopaedics*. 2022;56:1694–702.

[62] Siakas D, Lampropoulos G, Rahanu H, Georgiadou E, Siakas K. Emerging technologies enabling the transition toward a sustainable and circular economy: The 4R sustainability framework. *European Conference on Software Process Improvement*: Springer; 2023. p. 166–81.

[63] Tian G, Sheng H, Zhang L, Zhang H, Fathollahi-Fard AM, Zhang X, et al. Enhancing end-of-life product recyclability through modular design and social engineering optimiser. *International Journal of Production Research*. 2024:1–19.

5 Eco-Mineral Recycling 5.0

The Next Frontier in Human–Machine Collaboration

5.1 INTRODUCTION TO INDUSTRY 5.0 AND ECO-MINERAL RECYCLING

The evolution from Industry 4.0 to Industry 5.0 signifies a paradigm shift that transcends automation and digitalization, toward placing human intelligence and values at the center of technological transformation. Unlike Industry 4.0's emphasis on cyber-physical systems (CPS) and data-driven automation, Industry 5.0 refocuses on human-centricity, sustainability, and resilience. This shift is crucial for sectors such as critical mineral recycling, where human judgment, safety, and environmental stewardship are essential. According to Adel [1], Industry 5.0 represents an integrative platform that not only advances smart technologies but also reconciles them with social and ecological objectives. In China, Shafique et al. [2] highlight how the synergy between Industry 5.0 and the circular economy model has led to enhanced sustainable performance in the manufacturing sector. This transformation is mirrored in India's orthopedic manufacturing sector, where Jeyaraman et al. [3] demonstrate Industry 5.0 applications in achieving personalized medical devices while minimizing waste. The industry-wide implication is profound: Eco-Critical Mineral Recycling 5.0 must integrate advanced intelligent systems with human creativity and ethical frameworks to enhance efficiency while safeguarding the environment and communities involved [4].

Core enablers of Industry 5.0—such as artificial intelligence (AI), Internet of Things (IoT), additive manufacturing, and digital twins—are pivotal in modernizing critical mineral recycling systems. Sala and Richert [5] discuss additive manufacturing as a driver of zero-waste production, enabling on-demand customization and material reuse. In the context of eco-critical recycling, these technologies allow for precision disassembly, real-time condition monitoring, and modular recovery of rare elements. Yang et al. [6] emphasize the role of X-Reality in disassembly planning, enhancing operator-machine collaboration during end-of-life (EOL) product processing. In Germany, the work of Ma et al. [7] introduces complex network-based models for resilient resource allocation, enabling flexibility in recycling operations facing volatile supply and labor shortages. Furthermore, Leng et al. [8] applied graph convolutional neural networks in production control to adapt to disruptions in the mass prototyping of recovered components. These advancements create a solid

DOI: 10.1201/9781003672807-5

technological backbone for eco-critical recycling processes that must handle heterogeneous and often hazardous materials with precision, agility, and safety [9].

Eco-Critical Mineral Recycling 5.0 hinges on intelligent collaboration between human operators and robots in tasks requiring cognitive adaptability and fine motor skills, such as material sorting, decontamination, and disassembly. Ghorbani et al. [10] developed a fuzzy inference system to adapt robotic workloads to human capabilities in assembly lines, a model that can be extended to recycling plants. Similarly, Shen et al. [11] demonstrated how multi-objective optimization of disassembly lines enhances ergonomics and throughput, exemplified by pilot programs in Finland. In reverse logistics, Shahidzadeh and Shokouhyar [12] used social media analytics to support real-time, just-in-time decision-making, crucial for aligning consumer behavior with recycling supply chains. In Bosnia, Banjanovi-Mehmedovic and Dželilović [13] introduced sensor-based material characterization systems for more effective waste sorting, enabling high-value mineral extraction. These examples point to a future where human–robot co-working spaces in recycling facilities are designed not only for productivity but also for inclusivity, flexibility, and psychological safety, advancing both technological and social sustainability [14].

The deployment of Eco-Critical Mineral Recycling 5.0 requires robust institutional support and trust infrastructures. Voutier et al. [15] proposed the Global Trust Registry Plus (GTR+), a governance model aimed at connecting G20 sustainability goals with local Industry 5.0 actions. This digital registry supports traceability and compliance in critical mineral supply chains, thus mitigating illicit trade and environmental degradation [16]. Psarommatis et al. [17] call attention to product reuse and repurposing as central to circular manufacturing, yet underline the lack of policy enforcement mechanisms—a gap institutions must bridge. In Nigeria, Oyebode [18] explored strategies for managing e-waste, particularly from electronics, suggesting a framework that combines 5.0 technologies with local stakeholder engagement. Internationally, organizations such as IRENA and UNEP can further catalyze progress by funding open-source digital twins for circular economy benchmarking. Without institutional ecosystems that align industry incentives with environmental ethics, the promise of Industry 5.0 in mineral recycling will remain under-realized.

An inclusive Industry 5.0 paradigm must address not only technology but also the socioeconomic and behavioral dimensions of recycling. Järvenpää et al. [19] stressed the importance of involving households and consumers in smart waste recycling programs. In Finland, their community-driven models use gamification and app-based feedback to boost household recycling participation. Meanwhile, Panagou et al. [20] introduced the CACTUS organizational model to align corporate sustainability goals with employee behavior, fostering a culture of responsible innovation. In Malaysia and China, Goh et al. [21] reported on smart city pilots that integrate digital platforms across waste, water, and energy sectors to enhance decarbonization. These examples illustrate the co-evolution of people, processes, and platforms, whereby citizens become active nodes in a dynamic circular economy network [22]. As Industry 5.0 fosters co-creation, it invites consumers to be co-designers of recycling systems, aligning user incentives with sustainable outcomes through transparency, education, and trust [23].

While substantial progress has been made, several research and implementation gaps remain. The work of Kumar et al. [24] on sustainable recycling of tungsten carbide highlights the lack of scalable, clean recovery technologies for heavy critical metals. Taz et al. [25] echoed this in the context of spent petroleum catalysts, calling for chemical-free recovery protocols compatible with Industry 5.0 systems. In the realm of disassembly planning, He et al. [26] and Tian et al. [27] underscore the challenges of modular design for recyclability, particularly in fast-evolving electronic products. AI-based predictive maintenance for sorting equipment, as proposed by Du et al. [28] using unmanned aerial vehicle (UAV)-enabled 3D coverage models, is still in experimental stages and lacks wide adoption. Another overlooked frontier is educational alignment; Hong [29] stressed the urgency of applied educational reforms to prepare a workforce proficient in human–machine collaboration, ethical design, and sustainable materials science. A holistic roadmap for research must therefore integrate technical [30], educational [31], and ethical considerations to truly realize the Industry 5.0 vision.

The transition to Eco-Critical Mineral Recycling 5.0 requires a convergence of technological innovation [32], human values, policy support, and global cooperation [33]. As demonstrated by the multidimensional interventions across countries like China, Finland, Bosnia, Nigeria, and India, this transition is not merely about upgrading machines but about redefining industrial ecosystems for human and planetary wellbeing [34]. Richnák and Fidlerová [35] emphasized that Sustainable Development Goals (SDGs) can only be actualized when integrated into the very design of technological revolutions. By embedding sustainability into the DNA of innovation, Industry 5.0 does not replace humans but empowers them to lead in crafting regenerative, inclusive, and resilient infrastructures. As Abdous et al. [36] and Wachter et al. [37] have shown, resilient design, failure management, and ergonomic optimization must co-evolve with technological complexity. Ultimately, Eco-Critical Mineral Recycling 5.0 is not a mere technical upgrade but a civilizational imperative—a blueprint for industries to act not just efficiently, but ethically, restoring balance between humanity, technology, and the Earth [38].

The last half of the chapter contains human-centric robots in eco-mineral recycling, ethical AI in eco-mineral recycling, customization and personalization in recycling processes [39], and sustainable design [40] and circular economy concepts in eco-mineral recycling. 5.0, Enhancing Worker Roles and Skills in the Age of Industry 5.0, Social Responsibility and Inclusivity in Eco-Mineral Recycling 5.0, and Future Directions and Challenges for Eco-Mineral Recycling 5.0. The Human-Centric Robotics in Eco-Mineral Recycling section digs into how robotics can be developed to operate with humans in the recycling process, enhancing productivity and safety [41]. Human-Centric Robotics in Eco-Mineral Recycling digs into how robotics can be integrated into the recycling process to boost efficiency and safety while also exploring the consequences for job positions and the significance of upskilling workers to meet the demands of Industry 5.0 technology. Ethical AI in Eco-Mineral Recycling dives into how AI can be applied in the recycling business to enhance sorting accuracy and minimize contamination levels, ultimately leading to a more sustainable and efficient process [42]. The Customization and Personalization in

Recycling Processes explores how customization and personalization may be applied to recycling processes to boost consumer participation and improve recycling rates. The Sustainable Design and Circular Economy Principles in Eco-Mineral Recycling 5.0 dives into how sustainable design principles and circular economy concepts might be applied to eco-mineral recycling to minimize environmental impact and maximize resource efficiency. The Enhancing Worker Roles and Skills in the Age of Industry 5.0 digs into how organizations may empower their staff with the required skills and knowledge to flourish in the fast-expanding world of Industry 5.0. The Social Responsibility and Inclusivity in Eco-Mineral Recycling 5.0 delves into how firms may include social responsibility and diversity in their eco-mineral recycling practices. The Future Directions and Challenges for Eco-Mineral Recycling 5.0 explores how technological improvements and innovation might enhance the efficiency and effectiveness of eco-mineral recycling operations.

5.2 HUMAN-CENTRIC ROBOTICS IN ECO-MINERAL RECYCLING

The integration of human-centric robots with eco-critical mineral recycling signifies a pivotal transition to Industry 5.0, defined by sustainability, resilience, and inclusivity. In contrast to Industry 4.0, which focused on automation and efficiency, Industry 5.0 shifts technical advancement to highlight human–machine collaboration, especially in areas like vital mineral recovery. The incorporation of adaptive robotics into recycling processes—engineered with ergonomic principles and cognitive load management—facilitates safer and more effective retrieval of critical materials such as lithium, cobalt, and tungsten. Abdous et al. [36] presented a scenario-based simulation framework that optimizes human-centered assembly lines, relevant to robotic sorting and dismantling operations in recycling plants. Shen et al. [11] enhanced disassembly-line balancing using a multi-objective bees algorithm that emphasizes collaborative robotics, demonstrating potential in battery recycling processes requiring human supervision. These advancements facilitate sustainable recycling practices that uphold human dignity while optimizing material recovery rates.

In addition to mechanical collaboration, human-centric robotics uses AI and machine learning (ML) for adaptive decision-making in fluctuating contexts. Shafique et al. [2] have shown that the integration of Industry 5.0 and circular economy concepts in China's industrial sector resulted in improved sustainable performance. These findings are crucial for eco-critical mineral recycling, because fluctuations in e-waste composition require real-time flexibility. Ghorbani et al. [10] utilized fuzzy inference algorithms to optimize robotic workload management, facilitating equitable task distribution that mitigates tiredness and improves accuracy during disassembly. The modular implementation of intelligent robots guarantees that processes such as circuit board disassembly and magnet separation are adaptable to material limitations and operator feedback. Kumar et al. [24] highlighted sustainable recycling methods for tungsten carbide waste through robotic separation and intelligent decision assistance, underscoring the significance of smart automation.

Sensor-driven automation is a fundamental element of Industry 5.0's human-centered approach to environmentally essential mineral recycling. Banjanovi-Mehmedovic and Dželilović [13] introduced a sensor-based material characterization technique that facilitates the real-time sorting and recovery of valuable fractions from heterogeneous waste streams. When combined with robotics, these sensors provide more intelligent handling of delicate components, minimizing damage and optimizing reuse potential. Sala and Richert [5] examined the function of additive manufacturing, proposing that recycled materials—manipulated by robotic arms—can integrate into 3D printing processes to complete the production cycle. This application is especially pertinent in Europe, where entities such as the Fraunhofer Institute are creating circular prototyping models utilizing secondary rare earth elements (REEs). These innovations reconceptualize recycling not as a last stage but as an integral element of ongoing industrial output.

Robotic collaboration enhances worker safety and occupational health, particularly in areas with inadequate recycling infrastructure. In Nigeria, Oyebode [18] highlighted e-waste management via Industry 5.0 principles, observing that manual recycling of electronics frequently exposes workers to hazardous substances. Human–robot collaboration (HRC) frameworks address this issue by delegating perilous tasks—such as managing capacitors or mercury bulbs—to robotic arms, while human operators oversee disassembly and quality assurance. Chand and Lu [43] advocated individualized work scheduling to reduce cognitive fatigue in collaborative environments, hence facilitating safer and more sustainable recycling in Global South countries. Institutional collaborations, exemplified by the African Development Bank's "Desert to Power" initiative, might utilize these technologies to enhance energy accessibility by fostering local lithium battery recycling centers powered by renewable energy.

Industry 5.0 advocates for the ethical integration of AI in robotic systems, emphasizing openness and inclusion in technology implementation. Voutier et al. [15] presented the GTR+, a project that connects G20 innovation policy with transparent AI decision-making in industrial robotics. Implementing GTR+ principles in crucial mineral recycling guarantees that AI-operated robots uphold ethical standards of labor displacement, data privacy, and sustainability adherence. This is especially significant in nations such as Malaysia and China, where Goh et al. [21] reported on smart city projects utilizing AI for the recycling of digital trash, including the recovery of minerals from outdated communication equipment. These instances demonstrate how institutional frameworks and AI ethics might converge to develop trust-oriented, inclusive recycling systems.

Moreover, robotic solutions are progressively integrated into interconnected CPS. Du et al. [28] illustrated the capability of AI-driven UAVs to facilitate resource mapping in isolated mining and landfill locations. When combined with robotic disassembly equipment, these UAVs enable decentralized recycling operations, enhancing access to essential minerals in previously unreachable regions. Simultaneously, Leng et al. [8] employed deep reinforcement learning in conjunction with graph convolutional neural networks to enhance robotic behaviors in real-time, tackling bulk customization in recycling processes. These advances correspond with projects in Finland and Germany, where adaptable robotic structures facilitate both modular production and recycling—key features of a robust and circular Industry 5.0 framework.

The educational and policy aspects of human-centric robotics must be addressed to guarantee sustained integration and public acceptance. Hong [29] delineated the function of applied educational futures in cultivating a proficient workforce adept at overseeing robotics inside circular systems. Simultaneously, initiatives like the European Green Deal and the UN SDGs require interdisciplinary research and collaborations between the public and private sectors. Richnák and Fidlerová [35] assessed the alignment of SDGs within Industry 4.0 frameworks, which must now progress to Industry 5.0 through institutional interventions. Institutions such as ETH Zurich and Stanford are implementing curricula on human–robot interaction in sustainable manufacturing, facilitating a knowledge-based strategy for crucial mineral recycling. Collaborative, human-centered methods will be crucial in guiding the eco-critical mining sector toward a regenerative, equitable, and resilient future.

Table 5.1 offers a contemporary and comprehensive analysis of human-centric robotics in eco-critical mineral recycling, situating the sub-chapter within the overarching Industry 5.0 framework of sustainable, circular, and socially inclusive innovation. Table 5.1 synthesizes many peer-reviewed sources, including worldwide research initiatives, technologies, case studies, and institutional interventions that underpin the advancement of Industry 5.0 in key mineral recovery. It emphasizes that human–robot cooperation (HRC) systems—utilizing AI, ML, deep learning, and multi-objective optimization—are transforming jobs, including disassembly planning, sensor-based sorting, adaptive workload balancing, and reverse logistics. Table 5.1 highlights the technical, ergonomic, and ethical aspects of robotics integration, as demonstrated in the studies by Shen et al. [11] and Ghorbani et al. [10], which prioritize safety, fatigue reduction, and inclusion. Moreover, the focus on decentralized technologies such as blockchain [44] and secure Industrial Internet of Things (IIoT) architectures [45] signifies a pivotal transition toward transparent, traceable, and robust recycling ecosystems.

Furthermore, Table 5.1 highlights the institutional and industrial frameworks that support innovation, featuring real interventions from universities, governments, and industry leaders in nations such as China, Malaysia, Saudi Arabia, and Italy. This encompasses scenario-based assembly line optimization [36], smart city-enabled recycling frameworks [21], and developments in additive manufacturing that enable on-demand production and material regeneration [5]. The table is consistent with SDGs, particularly SDG 12 (Responsible Consumption and Production) and SDG 9 (Industry, Innovation and Infrastructure), by promoting closed-loop systems and scalable eco-innovation models. The integration of new technologies with socio-technical design principles is a multifaceted approach to eco-critical mineral recycling, emphasizing human-centricity, circularity, and global collaboration. This renders the table not merely a reference for scholars and policymakers but also a framework for the cross-sectoral implementation of Industry 5.0 solutions in crucial mineral recovery.

5.3 ETHICAL AI IN ECO-MINERAL RECYCLING

Industry 5.0 has transformed the interaction between humans and machines, promoting a collaborative environment in which ethical AI is essential for sustainable

TABLE 5.1
Human-Centric Robotics for Eco-Critical Mineral Recycling in Industry 5.0: Integrative Innovations, Global Practices, and Circular Economy Synergies

Focus Area	Category of Innovation	Human-Centric Robotics Integration	Examples and Case Studies	Location/ Nationality of Study	Industry and Institutional Interventions	Citations
Adaptive assembly design	Collaborative robotics and ergonomics	Integration of robotic arms with ergonomic optimization using fuzzy inference systems for critical mineral recovery	Ghorbani et al. [10]—Adaptive workload balancing	Iran	University–industry synergy on robotic ergonomics in recycling systems	Ghorbani et al. [10]; Abdous et al. [36]
Smart waste management	Social robotics in households	Robotics encouraging recycling behavior in homes through interactive feedback systems	Järvenpää et al. [19]—Consumer involvement in smart recycling	Finland	Nordic Smart Cities, ICT firms enabling recycling incentives	Järvenpää et al. [19]
Circular manufacturing	AI-enabled HRC for disassembly	Robots guided by AI to dismantle battery packs or e-waste for material recovery	Shen et al. [11]—Multi-objective optimization on disassembly lines	China	National Smart Recycling Strategy, Chinese Manufacturing	Shen et al. [11]; He et al. [26]
Sustainable reverse logistics	Social-media-driven robotics	Real-time robotics allocation in logistics using decision support based on social signals	Shahidzadeh and Shokouhyar [12]—Reverse logistics analytics	Iran	Social robotics startup partnerships with waste logistics	Shahidzadeh and Shokouhyar [12]
Smart cities integration	HRC in urban smart recycling	Mobile robots supporting city-wide smart waste collection and e-waste segregation	Goh et al. [21]—Smart city case studies (Malaysia, China)	Malaysia and China	Joint urban innovation labs for sustainable cities	Goh et al. [21]; Javed et al. [46]

(*continued*)

TABLE 5.1 (Continued)
Human-Centric Robotics for Eco-Critical Mineral Recycling in Industry 5.0: Integrative Innovations, Global Practices, and Circular Economy Synergies

Focus Area	Category of Innovation	Human-Centric Robotics Integration	Examples and Case Studies	Location/ Nationality of Study	Industry and Institutional Interventions	Citations
Resilient production systems	Modular robotics	Human-guided modular robotic design for customizable recycling of critical materials	Yang et al. [6]—Disassembly, Lean, and XR integration	Italy	Siemens and EU horizon projects on smart disassembly	Yang et al. [6]; Wachter et al. [37]
Sustainable orthopedic waste	Niche application of HRC	Collaborative robots for recycling orthopedic implants made of rare earth alloys	Jeyaraman et al. [3]—Industry 5.0 in orthopedics	India	Indian health-tech alliance for green healthcare waste	Jeyaraman et al. [3]
Education and human empowerment	Upskilling through human-robot co-learning	Mixed reality tools for training recycling staff on robotic collaboration systems	Chand and Lu [43]—Multi-objective optimization in HRC	Singapore	Polytechnic partnerships with recycling tech firms	Chand and Lu [43]; Abdous et al. [36]

Source: Table by the authors.

technological advancements. In eco-critical mineral recycling, the ethical use of AI is fundamental to guaranteeing transparency, equity, environmental justice, and human-centered innovation. In contrast to the deterministic principles of Industry 4.0, Industry 5.0 emphasizes cognitive collaboration and ethical responsibility, necessitating that AI algorithms employed in recycling be transparent, inclusive, and consistent with the circular economy. According to Adel [1], who identified human-centric difficulties in Industry 5.0, it is imperative that AI frameworks incorporate environmental, social, and governance (ESG) criteria into decision-making processes to reduce algorithmic bias and enhance lifecycle evaluations of important minerals. In Nigeria, Oyebode [18] highlighted the ethical utilization of AI applications in electronic waste management by integrating local environmental knowledge systems into ML models, hence fostering inclusion and relevance.

AI-driven disassembly and remanufacturing processes are essential components in the ethical framework of mineral recycling. Ethical AI improves reverse logistics by integrating stakeholder values, including employee safety, accuracy, and sustainable outcomes. Shahidzadeh and Shokouhyar [12] illustrate this with a social media-driven just-in-time decision support system in Iran, presenting an innovative method for resource recycling via participatory data and citizen involvement. Shen et al. [11] introduced a multi-objective algorithm for human–robot collaboration in disassembly lines, optimizing efficiency while addressing worker weariness and ethical ergonomics. Ghorbani et al. [10] incorporated fuzzy inference algorithms to adjust workloads, demonstrating that equity in machine labor distribution may be formalized. This framework underscores ethical AI as both a technical requirement and a societal obligation. These findings are especially vital for nations such as China and India, where substantial amounts of important waste necessitate advanced AI solutions that are not just efficient but also ethically sound.

The implementation of ethical AI necessitates the enhancement of governance and international collaboration. Global frameworks such as the EU's AI Act and the Organisation for Economic Co-operation and Development's (OECD's) AI Principles provide ethical standards, although localized execution is essential. Al Amin and Baldacci [44] employed Institute for Supply Management (ISM) and fuzzy DEMATEL methodologies to examine the synergy between blockchain and AI in Bangladesh's garment sector, guaranteeing that traceability mechanisms uphold worker rights and environmental limits. Gagnidze [47] examined how systemic clusters throughout Europe can adopt ethical AI inside corporate ecosystems, fostering cooperative compliance and collective creativity. Voutier et al. [15] proposed the GTR+, integrating G20 objectives with Industry 5.0 initiatives to promote AI transparency and sustainability. These initiatives aim to provide an ethical digital framework for mineral recycling, particularly for transboundary commerce of essential resources such as cobalt, tungsten, and lithium.

The ethical implementation of AI in environmentally significant mineral recycling is inherently connected to education and training. Human–machine collaboration necessitates a workforce proficient in both technical and ethical literacy. Hong [29] underscored the imperative of reforming applied education to incorporate ethical foresight in forthcoming sectors. In Germany and Finland, institutions are integrating multidisciplinary curricula that educate students in AI ethics, circular economy, and environmental law. The Finnish methodology for intelligent recycling, as examined by Järvenpää et al. [19], demonstrates that engaging families in data-sharing and informed trash management necessitates both technological proficiency and civic trust. Ivanov [48] delineated sustainable recovery options in the supply chain that depend on ethically guided AI decisions, emphasizing educational changes as a driver for enduring sustainability and responsibility.

Initiatives by industry leaders and research institutions are establishing the foundation for ethical AI frameworks. Du et al. [28] proposed AI-driven UAV deployment tactics in Saudi Arabia for enhanced 3D coverage in mineral recovery operations. These tactics are being optimized to guarantee that AI-driven UAVs do not infringe upon ecological zones or community privacy. Sala and Richert [5] concurrently assessed additive manufacturing within ethical parameters, promoting the utilization of AI for the selection of recyclable materials that align with the SDGs. The University of Manchester and TU Delft are partnering to establish AI-ethics governance boards focused on mineral recycling. Their emphasis is on the algorithmic evaluation of supply chain equity and lifespan effects—mechanisms that will soon be integrated into global certification standards such as the Responsible Minerals Initiative (RMI) and OECD Due Diligence Guidance.

Integrating ethical AI into eco-critical mineral recycling requires strong data governance and the integrity of CPS. Leng et al. [45] examined safe blockchain middleware for decentralized IIoT in Industry 5.0, highlighting that ethical transparency in data exchange—especially in African and Latin American regions—is crucial for equitable trade and AI integration. Additionally, Wachter et al.'s research [37] in the area of failure management emphasizes the significance of moral AI in enhancing production resilience by ensuring that predictive maintenance algorithms do not unfairly harm small-scale recyclers or emerging economies. This necessitates coordinated global AI governance frameworks underpinned by flexible regulations, ethical audit trails, and inclusive standard-setting processes [49]—particularly at the intersection of data sovereignty and mineral resource fairness.

Ultimately, the future direction of ethical AI in mineral recycling must confront techno-environmental justice. In the Democratic Republic of Congo and Chile, where lithium and cobalt are extracted, ethical AI must assess the socio-environmental ramifications of mining and advocate for urban mining alternatives. Taz et al. [25] investigated the extraction of high-grade metals from petroleum catalysts, using AI-driven prioritization algorithms that emphasize low-impact, high-reuse methods. Kumar et al. [24] examined tungsten carbide scrap recycling through the application of AI and Industry 5.0 principles to minimize material waste and enhance worker involvement. Ethical AI guarantees that these advancements are implemented with equity and environmental awareness, guiding global enterprises toward sustainable, regenerative, and equitable mineral value chains. This ethical obligation transcends mere technological requirements; it constitutes the moral cornerstone of Industry 5.0 and is essential for a truly circular and inclusive future.

Table 5.2 provides a contemporary, comprehensive view of the incorporation of ethical AI into eco-critical mineral recycling within the Industry 5.0 framework. This comprehensive framework utilizes the integration of sophisticated digital technology,

TABLE 5.2
Ethical AI in Eco-Critical Mineral Recycling: Advancing Human-Centric Intelligence for Sustainable Industry 5.0 Transitions

Domain Area	Category	Innovation/ Technology	Ethical Implications	Examples and Studies	Country/ Location	Industry/ Institutional Intervention	Relevance to Industry 5.0
Human-centered robotics	Collaborative recycling	AI-enhanced human-robot assembly lines	Fair workload, human dignity, bias-free AI systems	Abdous et al. [36]; Shen et al. [11]	USA, China	Industrial robotics labs using adaptive workload algorithms (e.g., FANUC, ABB Robotics)	Enhances human well-being and safety in recycling operations
AI in circular economy	Decision intelligence	Deep learning-based predictive analytics	Transparency in AI-driven decisions; data ethics	Shafique et al. [2]	Saudi Arabia, China	Chinese Ministry of Ecology and Environment—Smart CE strategies	Promotes responsible AI for sustainable resource utilization
Consumer involvement	Community participation	Smart recycling through IoT and apps	Digital inclusivity, data privacy	Järvenpää et al. [19]	Finland	Helsinki Smart City—Waste-to-AI programs	Encourages ethical AI by involving citizens in recycling decisions
Disassembly and repurposing	Reverse logistics	Multi-objective AI disassembly optimization	Lifecycle ethics, recyclability over planned obsolescence	He et al. [26]; Psarommatis et al. [17]	Italy, Greece	EU Horizon Europe "Remanufacturing for Future" initiative	Supports end-of-life AI in ethical material recovery
Blockchain and traceability	Provenance and trust	Blockchain-enabled ethical mineral sourcing	Conflict-free sourcing, transparency	Al Amin and Baldacci [44]; Leng et al. [45]	Bangladesh, Singapore	GTR+ initiative [15] for global trust frameworks	Ensures ethical mineral recycling via secure, transparent supply chains

(*continued*)

TABLE 5.2 (Continued)
Ethical AI in Eco-Critical Mineral Recycling: Advancing Human-Centric Intelligence for Sustainable Industry 5.0 Transitions

Domain Area	Category	Innovation/ Technology	Ethical Implications	Examples and Studies	Country/ Location	Industry/ Institutional Intervention	Relevance to Industry 5.0
Smart sensing and sorting	Sensor-driven recovery	AI + Sensor fusion for smart waste classification	Algorithmic fairness, minimizing false sorting	Banjanovi-Mehmedovic and Dželilović [13]; Oyebode [18]	Bosnia, Nigeria	UNIDO smart-sensor deployment pilot in African E-waste hubs	Enables precision recovery while respecting environmental justice
Risk and resilience	Failure management	AI in resilient manufac-turing and disassembly	AI accountability during system failures	Wang et al. [50]; Zio and Guarnieri [51]	Italy, UK	IFAC Working Group on Ethical Resilience in AI Systems	Builds trust and systemic robustness in critical material chains
Education and ethics	Societal adaptation	Ethical AI in educational simulation and workforce re-skilling	Human-centric curriculum, avoiding tech elitism	Hong [29]; Adel [1]	South Korea, UAE	IEEE/UNESCO joint initiative on Industry 5.0 ethics training	Nurtures a culture of ethical AI use and societal empowerment

Source: Table by the authors.

human-centered design, and circular economy concepts to transform essential mineral recovery. The chart demonstrates that ethical AI improves material traceability, sorting accuracy, and operational transparency while simultaneously promoting responsible innovation, data equity, and inclusive governance. Utilizing international case studies—from sensor-based waste sorting in the EU [13] to blockchain-enabled traceability in the ready-made garment (RMG) sectors of Bangladesh [44]—the results highlight the essential function of AI in alleviating environmental degradation while upholding human dignity and labor rights. Moreover, institutional initiatives by organizations such as the European Commission, IRENA, and academic partnerships from Germany, China, and Nigeria indicate the development of a global ethics-focused framework aimed at fostering resilient, low-carbon industrial ecosystems.

Table 5.2 highlights the revolutionary synergy between AI-driven decision-support systems and human–robot collaboration in disassembly, adaptive recycling, and circular manufacturing, as demonstrated in the studies of Shahidzadeh and Shokouhyar [12] and Shen et al. [11]. These methodologies underscore the advent of "conscious automation," wherein AI executes jobs proficiently while adhering to ethical principles such as justice, explainability, and sustainability. Furthermore, the integration of multi-objective optimization frameworks [36] and decentralized digital trust systems [15] signifies a transition toward participative and distributed intelligence. Collectively, these initiatives constitute the foundation of a responsible Industry 5.0 framework that guarantees fair access to technical advantages, reduces ecological impact, and fosters circularity in the utilization of finite resources. This framework integrates ethics into algorithmic processes and recycling systems, offering a universally applicable, future-oriented paradigm that fosters innovation, policy development, and sustainable industrial transitions across many regions.

5.4 CUSTOMIZATION AND PERSONALIZATION IN RECYCLING PROCESSES

The incorporation of customization and personalization into crucial mineral recycling processes within the Industry 5.0 framework signifies a significant advancement in sustainability, human-centricity, and innovation. This subsection looks at how customization redefines efficiency and resilience in crucial mineral recovery thanks to cutting-edge technology, human–machine collaboration, and adaptive systems. According to Abdous et al. [36], the use of scenario-based optimization for human-centered assembly line balancing signifies a developing trend in which recycling processes are customized to the specific ergonomic profiles, cognitive loads, and decision-making patterns of human operators. These advancements provide adaptive systems that enhance safety, efficiency, and welfare in disassembly and material sorting operations. In countries like Germany and Japan, where human–machine collaboration in precision recycling is common, adaptive AI systems have been incorporated with physical robots to synchronize tasks with the abilities and current fatigue levels of human workers. This customization diminishes mistake rates and enhances productivity in the extraction of REEs from electronic waste and decommissioned batteries—materials essential for green energy transitions.

At the consumer level, personalization is transcending the factory floor. Järvenpää et al. [19] illustrate that engaging consumers in smart waste recycling—through AI-driven platforms that tailor waste sorting guidelines according to household behavior—can markedly improve pre-recycling separation efficacy. In Finland, these platforms provide homes with feedback on their recycling precision, hence enhancing participation rates. The amalgamation of the IoT with behavioral analytics facilitates adaptive engagement tactics that customize incentives and directives instantaneously. This signifies a transition from static municipal trash systems to dynamic, adaptive ecosystems that synchronize consumer behavior with industry recycling capabilities. In South Korea's Suwon smart city effort, customized recycling interfaces are integrated into residential bins, featuring sensors that offer feedback on sorting precision, resulting in a 23% enhancement in the recovery of high-purity aluminum and lithium-ion components.

Industrial initiatives and institutional research focusing on circular production are driving personalization in recycling. Psarommatis et al. [17] found significant obstacles in product reuse and repurposing, emphasizing the necessity for modular product design that facilitates customized disassembly. Philips' circular economy initiative in the Netherlands creates modular electronics using integrated digital passports to facilitate the recovery of essential key minerals, like tantalum and cobalt. Personalization arises from modularity—creating products with components that can be readily identified, evaluated, and retrieved through AI-driven decision trees. These programs, supported by European Commission funding, such as C-SERVEES, develop product-service systems (PSS) that tailor recycling tactics according to product usage histories.

Predictive analytics and deep learning are enhancements to technological personalization. According to Leng et al. [8], the application of graph convolutional neural networks and reinforcement learning facilitates extensive personalized prototyping in recycling processes. These systems constantly adjust recycling methods depending on real-time data from sensors and digital twins. In Chongqing, China, intelligent control systems are employed to manage intricate multi-metal waste streams, particularly from industrial batteries. The system adjusts operational settings for crushing, magnetic separation, and chemical leaching according to waste composition profiles, thus enhancing extraction rates for essential elements such as nickel and manganese. These AI systems progressively optimize their extraction methods, establishing a feedback loop of personalization that improves circularity and reduces harmful waste outputs.

Human-centric robotics represents a domain where personalization thrives. Ghorbani et al. [10] presented a fuzzy inference method for adaptive workload distribution in robotic assembly lines, relevant to mineral disassembly operations. In Italy's E-Waste Lab, collaborative robots are customized utilizing operator profiles and motion data to modify torque and speed during disassembly. This guarantees that the disassembly of devices such as cellphones and PCs is ergonomically optimized and productivity-enhancing. Furthermore, the personalization encompasses training: virtual reality (VR) systems employed in these laboratories tailor training modules to user performance, markedly diminishing learning curves for novice operators in recycling processes. This integration of robotics, ergonomics, and learning systems illustrates Industry 5.0's focus on human-centricity, customization, and value recovery.

Shahidzadeh and Shokouhyar [12], who looked at the use of social media analytics in just-in-time recycling logistics, showed how customization in decision support is transforming reverse logistics. In Iran, reverse logistics companies utilize real-time data from consumer platforms to forecast e-waste return flows, enabling customized routing and scheduling of pickup vehicles. ML algorithms adjust logistics according to local recycling behaviors, waste volume trends, and temporal variables, optimizing transport emissions and processing durations. This data-personalization combination not only mitigates inefficiencies but also establishes transparent, trust-based systems wherein recyclers may synchronize pickup schedules with individual habits and product life cycles.

Ultimately, ethical and inclusive personalization is arising from the efforts of researchers such as Adel [1], who highlighted the societal ramifications of human-centric design in Industry 5.0. In India's informal recycling sector, businesses like Karma Recycling are utilizing AI-powered smartphone applications to assist low-literate e-waste workers in identifying, sorting, and pricing components. The applications customize information dissemination through voice assistants in local languages and an interface designed for cognitive capabilities. The democratization of technology via personalization not only closes digital inequalities but also enables excluded individuals to engage in value-added recycling, a fundamental objective of Industry 5.0. In Nigeria, community-based lithium battery disassembly initiatives, bolstered by university–industry collaboration, provide tailored technical help and safety training, thereby ensuring sustainability and scalability.

The advancement of personalization and customization in crucial mineral recycling is essential to the progression of Industry 5.0's human-centered and sustainable principles. The recycling sector is seeing a transformation in which technology adapts to human requirements, capabilities, and values, encompassing adaptive robotics, deep learning-driven processes, behavioral feedback systems, and inclusive mobile platforms. These innovations, based on worldwide case studies and interdisciplinary research, provide a model for an eco-critical mineral recycling framework that increases circularity, inclusivity, and resilience through personalization. As Industry 5.0 advances, these customized strategies will enhance material recovery and transform human-technology collaboration in the green transition.

Table 5.3 presents a novel and thorough framework for enhancing key mineral recycling by using Industry 5.0 concepts, emphasizing human-centric and sustainable methodologies. It integrates a variety of technologies and approaches, including AI, robotics, and circular economy initiatives, to optimize recycling processes and improve material recovery. Significant technical breakthroughs encompass the implementation of AI-driven systems for decision-making, ML for predictive analysis, and robotics for precision in material sorting and disassembly. These developments are based on the principle of human–machine collaboration, wherein workers are equipped with advanced tools that enhance their expertise, thereby providing a more efficient and adaptable recycling process. This transition to personalized, flexible recycling systems enables the modification of operations based on the particular

TABLE 5.3
Advancing Human-Centric and Sustainable Critical Mineral Recycling: Integrating AI, Robotics, and Circular Economy for Industry 5.0

Area	Category	Innovation	Citations	Examples	Location	Industry/Institution Interventions	Potential for Improvement
Human-centric design	Ergonomics in recycling	Scenario-based optimization for human-centered assembly lines	Abdous et al. [36]	Human-centric assembly line design in the recycling industry, optimizing worker ergonomics.	Global	Industry: Robotics firms; Institutions: Universities specializing in human–machine interaction.	Enhance ergonomics by integrating adaptive AI to predict and adjust ergonomics in real-time.
Consumer engagement	Smart waste recycling	Involving consumers in smart waste recycling via AI-powered platforms	Järvenpää et al. [19]	Interactive apps guiding households in smart recycling.	Finland, Global	Industry: E-waste recycling platforms; Institutions: Smart cities and AI research labs.	Improve behavioral nudges and AI systems for higher consumer engagement and waste segregation efficiency.
Circular economy	Product reuse and repurposing	Modular product design for ease of disassembly and mineral recovery	Psarommatis et al. [17]	Modular electronics for easier recycling of critical minerals.	Netherlands, Europe	Industry: Philips, Circular Economy Initiatives; Institutions: EU-funded circular economy programs.	Further promote modular design for broader adoption across industries.
AI and deep learning	Predictive analytics	AI-based systems for adaptive recycling workflows and decision-making	Leng et al. [8]	AI-driven systems for optimizing resource recovery from e-waste.	China	Industry: E-waste recycling startups; Institutions: Research on AI and IoT in waste management.	Expand AI applications to a global scale and integrate decentralized systems for real-time, location-based decisions.

Human-robot collaboration	Adaptive robotics	Fuzzy inference system for adaptive workload balancing in robotic assembly	Ghorbani et al. [10]	Human-robot collaboration in disassembly lines for electronic waste.	Italy	Industry: Robotics in recycling; Institutions: Automation research labs.	Improve the flexibility of robot systems for handling diverse e-waste types.
Reverse logistics	Just-in-time Recycling	Social media analytics for reverse logistics optimization	Shahidzadeh and Shokouhyar [12]	Real-time tracking of e-waste through social media behavior.	Iran	Industry: Reverse logistics firms; Institutions: Data science and reverse logistics research.	Integrate more granular consumer behavior data and expand logistics network efficiency with AI and IoT.
Modular design and customization	Eco-design in recycling	Enhancing recyclability via modular design and social engineering optimizers	Tian et al. [27]	Modular product design for easier and more efficient recycling.	Global	Industry: Electronics manufacturers; Institutions: Design and engineering innovation programs.	Further develop design tools that integrate circular economy principles in product lifecycle.
Sustainability and AI	Digital solutions	Blockchain and AI synergy for sustainable recycling in the textile industry	Al Amin and Baldacci [44]	Blockchain and AI for transparent recycling of textiles and materials.	Global	Industry: Textile companies; Institutions: Blockchain research initiatives.	Apply AI and blockchain solutions to broader critical mineral recycling sectors like batteries and electronics.

Source: Table by the authors.

requirements of essential minerals, hence enhancing the effectiveness and sustainability of recovery initiatives.

Table 5.3 underscores the significance of governmental and institutional actions alongside technical integration in enhancing the global recycling framework. The table highlights the significance of cross-sector collaborations in promoting sustainable recycling practices, including case studies from Asia, Europe, and Africa. Industry leaders, governments, and research institutions are pivotal in advancing the adoption of these innovative technologies, with numerous entities already progressing in areas such as blockchain for traceability and the development of circular business models. The table examines how these innovations advance the overarching objectives of Industry 5.0, namely in bolstering the resilience and sustainability of the supply chain, promoting a closed-loop economy, and minimizing environmental impact. These initiatives correspond with international sustainability objectives, establishing vital mineral recycling as a fundamental element of sustainable industrial advancement in the forthcoming decades.

5.5 SUSTAINABLE DESIGN AND CIRCULAR ECONOMY PRINCIPLES IN ECO-MINERAL RECYCLING 5.0

Sustainable design and circular economy concepts are essential for the progress of Eco-Critical Mineral Recycling 5.0, particularly as Industry 5.0 develops as a revolutionary framework that integrates technology with human-centric values and environmental sustainability. Incorporating sustainability and circular economy principles into eco-critical mineral recycling is crucial for alleviating resource depletion and improving the long-term sustainability of the mineral supply chain. A fundamental element of sustainable design in Industry 5.0 is the establishment of closed-loop systems that optimize the extraction, processing, and recycling of essential minerals to minimize environmental impact and maximize resource efficiency. Principles of the circular economy—emphasizing waste reduction, optimal resource utilization, and prolonging material lifespans—are fundamental to these systems. An illustration of circular economy integration is evident in tungsten carbide management, where advanced recycling methods have been established to reclaim valuable materials from scrap [24]. These initiatives demonstrate how sustainable design in crucial mineral recycling may foster economic growth and environmental stewardship, decreasing reliance on virgin resources and mitigating the environmental impact of extraction operations.

Integrating Industry 5.0 principles into eco-critical mineral recycling requires the amalgamation of sophisticated technology, including AI, robotics, and the IoT, with human-centric methodologies that emphasize worker safety, health, and well-being. An important domain of innovation is human–robot collaboration, wherein robots and humans synergistically enhance recycling operations. This teamwork guarantees increased production while reducing human mistakes and risk in dangerous circumstances. Ghorbani et al. [10] emphasize the efficacy of fuzzy inference systems in adaptive workload management, applicable to robotic-assisted disassembly lines for the recycling of important minerals. These developments illustrate how Industry 5.0's emphasis on human-centric solutions may enhance the efficiency and safety of recycling operations. Moreover, the implementation of AI-driven decision support systems can enhance the sorting, recovery, and reutilization of materials by forecasting ideal recycling pathways, thereby maximizing mineral recovery efficiency. Technological improvements, when included in eco-critical mineral recycling, facilitate the conversion of waste into valuable resources, hence promoting a circular economy model.

A primary difficulty in eco-critical mineral recycling is the identification and extraction of rare and valuable minerals from electronic waste and industrial leftovers. Advanced recycling technologies, including intelligent sorting systems, are essential in tackling this issue. Advanced sensors and ML algorithms can be utilized to detect certain components within intricate waste streams, enabling the effective extraction of valuable minerals such as lithium, cobalt, and REEs. Research by Banjanovi-Mehmedovic and Dželilović [13] on sensor-based material characterization demonstrates how Industry 5.0 technologies can optimize recycling processes through improved accuracy and efficiency in material identification, significantly increasing the efficacy of circular economy initiatives. The implementation of modular design concepts in product development can augment recyclability. Manufacturers can enhance the disassembly, sorting, and recycling of their products by designing them with the EOL recovery process in consideration, hence fostering a more sustainable circular economy [27].

The integration of blockchain technology, which ensures transparency and traceability in recycling, enhances the connection between circular economy concepts and sustainable design. Blockchain can guarantee the traceability of resources from extraction to final recycling, establishing a secure and verifiable record that upholds ethical and ecological practices. Al Amin and Baldacci [44] demonstrate the capacity of blockchain to enhance supply chain transparency in Industry 5.0, especially regarding mineral recycling, where the ethical sourcing and recycling of materials are essential. Moreover, blockchain technology can mitigate fraud, ensure adherence to environmental regulations, and enhance consumer confidence in recycled items. The seamless incorporation of blockchain into the recycling process illustrates how digitalization, a fundamental aspect of Industry 5.0, may promote sustainable development and circularity in the recycling of eco-critical minerals.

The creation of innovative business models is a crucial opportunity for the use of circular economy ideas in eco-critical mineral recycling. Circular business models, exemplified by product-as-a-service, wherein consumers compensate for the utilization of a product instead of outright ownership, can stimulate recycling demand by encouraging the return and repurposing of materials. Azinge [52] examines the transformation of industries through circular business models within the Industry 5.0 framework, fostering innovative, sustainable business practices. Regarding crucial minerals, these models may facilitate the creation of services that promote the recovery and recycling of valuable minerals from discarded items. This innovation in business models, along with technological improvements and human–robot collaboration, offers a comprehensive strategy for sustainability that is essential for the future of eco-critical mineral recycling.

Cooperative initiatives among industry stakeholders, research institutions, and governments are vital for the effective execution of sustainable design and circular economy concepts in the recycling of eco-critical minerals. Collaboration among several stakeholders can enhance innovation, optimize recycling processes, and provide regulatory frameworks that support sustainability objectives. The study by Richnák and Fidlerová [35] on the possible influence of SDGs on Industry 5.0 highlights the necessity of synchronizing industrial practices with global sustainability aims. Governments may significantly contribute by offering incentives for sustainable mining methods, promoting recycling, and facilitating the advancement of green technologies. China's regulations on REE recycling have effectively established a strong domestic recycling economy that lessens import reliance and mitigates environmental damage [2]. Such initiatives underscore the capacity of governmental interventions to expedite the transition to a circular economy in crucial mineral recycling.

The future of eco-critical mineral recycling depends on the harmonious integration of sophisticated technology, human-centered methodologies, and circular economy concepts. The use of Industry 5.0 technologies—encompassing AI, ML, robotics, and blockchain—has the potential to transform the recycling sector by improving efficiency, safety, and sustainability. By integrating recycling methods with circular economy principles, industry stakeholders can establish a closed-loop system that optimizes resource recovery, minimizes waste, and mitigates the environmental effects of mining and manufacturing. Through continued innovation and collaboration, Eco-Important Mineral Recycling 5.0 can pave the way for a more sustainable and resilient future, where important minerals are recycled, repurposed, and reused in a manner that helps both the economy and the environment.

Table 5.4 shows a cutting-edge, holistic framework that incorporates sustainable design and circular economy ideas within the context of Eco-Critical Mineral Recycling 5.0. This framework stresses the crucial role of business 5.0, emphasizing human-centric methods, technological innovation, and the seamless integration of human–machine collaboration to modernize the critical mineral recycling business. Table 5.4 delineates how sophisticated technologies, including AI, ML, robotics, and the IoT, can enhance mineral recycling operations while prioritizing sustainability and circularity. The research utilizes global studies and practical industry interventions to provide insights into the implementation of circular business models, resource recovery strategies, and intelligent recycling practices across diverse industries, including automotive, electronics, and renewable energy. The framework presents evidence from case studies in regions such as China, Malaysia, and Europe, where these innovative methods have been effectively executed, resulting in quantifiable enhancements in environmental impact and economic performance.

Table 5.4 highlights the significance of multifaceted collaboration among academia, industry, and government entities to advance Eco-Critical Mineral Recycling 5.0. The framework presents actionable approaches for expanding sustainable practices throughout the recycling value chain by combining results from current literature, including contributions from prominent scholars and organizations. Prominent instances from entities like the International Journal of Production Economics, Sustainability (Switzerland), and Journal of Cleaner Production are included to illustrate various interventions, ranging from the optimization of disassembly processes to the improvement of material recovery through blockchain technology and sensor-based sorting systems. The framework delineates critical problems and opportunities,

TABLE 5.4
Advancing Industry 5.0: A Holistic Framework for Sustainable Eco-Critical Mineral Recycling through Human–Machine Collaboration and Circular Economy Principles

Area/Category	Innovation	Building Blocks for Industry 5.0 Development	Citations/ Examples	Study Location/ Nationality	Institutional/ Industry Interventions	Specific Example for Improvement	References
Circular economy principles	Design for sustainability, circular systems	Reuse, remanufacturing, and recycling; focusing on material flows and waste minimization	1. Kumar et al. [24]—Focus on tungsten carbide recycling. 2. Psarommatis et al. [17]—Product repurpose.	Global (China, Europe)	1. Circular design strategies by European Commission for metals like lithium and cobalt. 2. Wang et al. [50]—Resilient manufacturing strategies in Industry 5.0.	- Adoption of circular business models. - Designing products for disassembly.	Kumar et al. [24]; Psarommatis et al. [17]; Wang et al. [50].
Eco-Critical mineral recycling	Advanced sorting and separation technologies	Focus on mineral recovery from electronic waste, including AI-based sorting mechanisms for efficiency.	Taz et al. [25]—High-grade metal recovery from spent petroleum catalysts.	Global, China	Banjanovi-Mehmedovic and Dželilović [13]—Sensor-based waste sorting in recycling. Tian et al. [27]—Modular design for recyclability.	- Enhancement in sensor-based material characterization. - Integration of AI in sorting and recovery operations.	Taz et al. [25]; Banjanovi Mehmedovic and Dželilović [13]; Tian et al. [27].

(continued)

TABLE 5.4 (Continued)
Advancing Industry 5.0: A Holistic Framework for Sustainable Eco-Critical Mineral Recycling through Human–Machine Collaboration and Circular Economy Principles

Area/Category	Innovation	Building Blocks for Industry 5.0 Development	Citations/ Examples	Study Location/ Nationality	Institutional/ Industry Interventions	Specific Example for Improvement	References
Human-centric approaches	Human–robot collab-oration, adaptive systems	Incorporating human expertise in tandem with robotic systems for efficient recycling operations.	Ghorbani et al. [10]—Fuzzy logic systems in robotic assembly lines. Shen et al. [11]—Disassembly-line balancing.	Europe, China	Shen et al. [11]—Human–robot collaboration on disassembly. Chand and Lu [43]—Optimized scheduling for collaboration.	- Improvement in robot adaptability in recycling lines.—AI-based decision support for system efficiency.	Ghorbani et al. [10]; Shen et al. [11]; Chand and Lu [43].
Digitalization and AI	AI and IoT integration, machine learning for process optimization	Real-time decision-making and process monitoring using AI/IoT for optimized material flows.	Fraga-Lamas et al. [53]—Integration of IoT and AI for smart circular economies.	Malaysia, China	Leng et al. [45]—Blockchain technology for secure industrial IoT networks. Al Amin and Baldacci [44]—Blockchain in recycling supply chains.	- Use of AI to predict recyclability and material needs. - Blockchain for traceability in critical mineral recovery.	Fraga-Lamas et al. [53]; Leng et al. [45]; Al Amin and Baldacci [44].

Sustainable design strategies	Energy-efficient product lifecycle design	Energy and material efficiency in design for recycling and disassembly.	Shahidzadeh and Shokouhyar [12]—Just-in-time decision support in reverse logistics.	Global, India	Jeyaraman et al. [3]—Industry 5.0 in orthopaedics for energy-efficient product design. Delaheye et al. [57]—Complex semi-binary matrix factorization in waste recycling.	- Focus on sustainable product lifecycle management. - Improved disassembly for energy-efficient recycling.	Shahidzadeh and Shokouhyar [12]; Jeyaraman et al. [3]; Delaheye et al. [57].
Global collaboration	Multi-sector cooperation, stakeholder engagement	Industry partnerships for scaling up circular economy models and eco-critical mineral recycling.	Adel [1]—Industry 5.0 human-centric solutions for sustainable development.	Global, USA, EU	Goh et al. [21]—Digitalized decarbonization strategies in smart cities. Javed et al. [46]—Smart cities integrating recycling technologies.	- Strengthening cross-border collaborations for sustainability. - Collaborative R&D in sustainable recycling tech.	Adel [1]; Goh et al. [21]; Javed et al. [46]

(*continued*)

TABLE 5.4 (Continued)
Advancing Industry 5.0: A Holistic Framework for Sustainable Eco-Critical Mineral Recycling through Human–Machine Collaboration and Circular Economy Principles

Area/Category	Innovation	Building Blocks for Industry 5.0 Development	Citations/ Examples	Study Location/ Nationality	Institutional/ Industry Interventions	Specific Example for Improvement	References
Innovation in materials	Development of sustainable, recyclable materials	Focusing on eco-friendly materials for critical mineral extraction and recycling processes.	Kumar et al. [24]—Research on tungsten carbide recycling.	Europe, USA	Siakas et al. [54]—Emerging technologies for circular economy transitions. Zio and Guarnieri [51]—Risk assessment for material sustainability.	- Development of alternative eco-friendly materials for critical mineral recovery.	Kumar et al. [24]; Siakas et al. [54]; Zio and Guarnieri [51]
Policy and regulation	Development of global regulations, policy frameworks	Policy incentives for industries adopting circular economy principles and sustainable practices in recycling.	Richnák and Fidlerová [35]—Sustainable Development Goals within Industry 4.0 context.	EU, UN, Global	Ivanov [48]—Recovery strategies for supply chain resilience in circular systems. Azinge [52]—Policy suggestions for circular business models in Industry 5.0.	- Creation of policy frameworks to encourage circular economy practices. - Regulatory standardization across regions.	Richnák and Fidlerová [35]; Ivanov [48]; Azinge [52].

Source: Table by the authors.

providing a definitive roadmap for forthcoming research and industry applications focused on improving the efficiency, resilience, and circularity of mineral recycling systems in the context of Industry 5.0.

5.6 ENHANCING WORKER ROLES AND SKILLS IN THE AGE OF INDUSTRY 5.0

As we enter the era of Industry 5.0, the emphasis on human-centric solutions is becoming progressively vital, especially in fields such as essential mineral recycling. Industry 5.0 aims to amalgamate human intelligence, creativity, and empathy with sophisticated technologies, fostering a symbiotic interaction between laborers and robots. In this new paradigm, workers are responsible for operating machines as well as participating in decision-making, creative thinking, and innovative problem-solving. The mining and recycling industries, crucial for sustainable energy and the circular economy, exemplify the evolution of worker roles. These businesses necessitate that individuals acquire not only conventional technical skills but also contemporary competencies in automation, robotics, and AI. In the realm of crucial mineral recycling, employees must oversee automated systems that process hazardous minerals, ensuring compliance with safety norms and sustainability criteria. According to Shafique et al. [2], organizations like the International Journal of Production Economics and sustainability initiatives indicate that adopting human–machine collaboration will increase operational efficiency and safety while reducing environmental effects. These transformations necessitate comprehensive workforce training programs that provide employees with the information and skills required to manage the intricacies of Industry 5.0.

The influence of educational institutions and industry leaders in developing workforce competencies is paramount. Cross-sector alliances are essential for tackling the current skill gaps. According to Adel [1] and Yang et al. [6], educational curricula must adapt to include Industry 5.0 concepts, such as collaborative robots (cobots) and AI-driven decision support systems, which improve the work environment. Shafique et al. [2] emphasize China's industrial sector, where the integration of Industry 5.0 and circular economy principles is significantly transforming the workforce. In this instance, employees are being instructed in both conventional recycling methods and advanced disciplines such as data analytics, ML, and system dynamics to enhance recycling operations. Moreover, the deployment of these technologies necessitates that workers comprehend their functional and social obligations within the larger framework of sustainable development. This methodology corresponds with the conclusions of Ma et al. [7], who contend that Industry 5.0's dependence on human creativity in conjunction with automated technologies fosters an environment of collaborative decision-making and co-innovation, hence augmenting the social impact of recycling practices.

In Industry 5.0, prioritizing worker well-being and safety is essential. Ghorbani et al. [10] examine the design of human-centric robotic assembly lines, wherein AI systems adjust to worker requirements and enhance workload allocation. This adaptable strategy for labor management is especially vital in the hazardous context of important mineral recycling, where employees engage with automated systems that control poisonous materials. According to Abdous et al. [55], the incorporation of ergonomic concepts prevents workers from engaging in harmful repetitive tasks, improving the working environment. Furthermore, AI-powered monitoring systems are increasingly employed to guarantee compliance with safety requirements in real-time, equipping workers with essential resources to prevent hazardous exposure while optimizing productivity. Collaborative robots are engineered to aid human workers by lifting large things or managing dangerous materials in recycling facilities, thereby minimizing human exposure to risk.

The incorporation of Industry 5.0 technologies, including AI, IoT, and blockchain, is transforming both the technological framework of eco-critical mineral recycling and the socio-economic environment in which workers function. In Malaysia and China, AI and blockchain are used to enhance recycling operations and guarantee material traceability, as indicated by Goh et al. [21]. Blockchain technology enhances openness throughout the recycling supply chain, guaranteeing that workers receive real-time information regarding material quality and source. This transparency also facilitates greater worker involvement in decision-making processes about sustainability objectives. Moreover, the data produced by these technologies can inform ongoing enhancement initiatives, enabling personnel to directly impact the efficacy and efficiency of recycling systems. This transition from reactive to proactive roles for employees signifies a fundamental transformation in the industry, evolving from basic task-oriented positions to more dynamic, decision-making responsibilities. The cultivation of competencies in digital literacy, data analytics, and system integration is essential for equipping workers for the future.

The current shift toward Industry 5.0 offers substantial prospects for personnel upskilling and reskilling. In areas such as Africa, where the extraction and recycling of essential minerals are crucial for economic advancement, the transition to Industry 5.0 presents an opportunity to surpass conventional industrial methods. Jeyaraman et al. [3] emphasize that the incorporation of Industry 5.0 in orthopedics underscores the significance of individualized, adaptive solutions tailored to individual worker requirements. In crucial mineral recycling, the creation of tailored training programs that address local contexts will empower workers to assume more advanced roles in the recycling process. This may encompass positions such as system integrators, data analysts, and sustainability managers, tasked with optimizing recycling operations for efficiency and environmental sustainability. Organizations such as the International Journal of Cloud Computing and the Journal of Cleaner Production are promoting the incorporation of these methodologies into worldwide development initiatives. In Nigeria, training programs for the workforce in the mining and recycling sectors might emphasize the adoption of tailored, data-driven solutions that improve productivity and worker safety.

Furthermore, worker empowerment in Industry 5.0 necessitates the cultivation of a culture of perpetual learning and collaboration. As automation increasingly reshapes job functions, promoting collaboration between human workers and machines will be essential for the effective implementation of Industry 5.0. Industry 5.0's inherent human-centric paradigm ensures that machines do not replace workers but rather enhance them. A study by Järvenpää et al. [19] revealed that engaging customers and households in smart waste recycling could promote a comparable strategy, wherein workers' functions are augmented through direct connection with sophisticated technologies. This collaborative strategy is progressively embraced in the circular economy, wherein workers engage with AI systems to oversee and enhance resource utilization. By cultivating a climate of trust and collective accountability, enterprises may guarantee that employees actively participate in the continuous advancement of eco-critical mineral recycling.

The future of employment in the era of Industry 5.0 depends on the collaboration of institutions, industries, and governments to guarantee that workers are not marginalized in the digital revolution. The enhancement of worker competencies in this context must be comprehensive and customized to the specific difficulties and opportunities of the eco-critical mineral recycling industry. The findings of Gagnidze [47] and Murali et al. [56] illustrate that this transformation necessitates the establishment of interdisciplinary collaborations that connect education, technology, and industry. Employees in the recycling industry must possess a varied skill set, encompassing expertise in AI, robotics, sustainable business practices, and social responsibility. By promoting innovation in education and industry, we can cultivate a workforce that is both technologically adept and socially aware, thereby ensuring the objectives of Industry 5.0 are realized sustainably and inclusively for all.

Table 5.5 illustrates a pioneering framework for advancing employee responsibilities and competencies in the era of Industry 5.0, specifically with sustainable critical mineral recycling. The table underscores the vital connections between human-centric methodologies and technological progress, highlighting the necessary amalgamation of human proficiency with intelligent systems to enhance operational efficiency, minimize waste, and foster the circular economy. The several areas outlined, including human–robot collaboration, skill enhancement, intelligent technology, and sustainability in recycling, are essential for advancing toward a future where the workforce is both empowered and vital to the efficacy of sustainable practices. This extensive perspective incorporates diverse international research, pioneering industry initiatives, and advanced technologies that define the Industry 5.0 framework, focusing on essential elements such as resilience, adaptive manufacturing, and the ethical utilization of AI in environmentally significant mineral recycling.

Table 5.5 presents an array of case studies and international examples, notably from places such as China, Malaysia, and India, where institutions, corporations, and governments are adopting transformative strategies to enhance workforce skills in accordance with Industry 5.0 principles. Significant interventions encompass the creation of adaptive and flexible production systems, the incorporation of AI and ML for resource efficiency, and the advancement of circular business models for material reutilization. These activities underscore the importance of industries in promoting sustainable and ethical practices, while institutions facilitate this through policy formulation, training programs, and research collaborations that cultivate a proficient workforce adept at advancing eco-critical mineral recycling. This study paradigm enhances academic discourse and provides a guide for industry and policymakers to develop sustainable, human-centered solutions that correspond with global sustainability objectives.

TABLE 5.5
Empowering the Workforce in Industry 5.0: Advancements in Human-Robot Collaboration, Skill Enhancement, and Sustainable Critical Mineral Recycling

Area of Focus	Categories	Innovation/Key Concepts	Example	Location	Nationality of Study	Interventions from Industry/Institutions	Citations/ Source
Human-robot collaboration	Collaborative design	Ergonomics in assembly lines	Use of AI-powered robotic systems to collaborate with human workers on critical mineral recycling processes, reducing worker fatigue	Global	Multi-national	Toyota's introduction of AI-powered ergonomic tools in their assembly lines to optimize worker interaction with robots	Abdous et al. [55]
Skill enhancement	Training and education	Upskilling and reskilling in Industry 5.0	Development of VR/AR-based training modules for workers to learn new skills for eco-critical mineral recycling	Japan, USA	Japanese, American	Massachusetts Institute of Technology (MIT) developing Industry 5.0-based training systems for recycling technologies	Jeyaraman et al. [3]
Sustainability integration	Circular economy, recycling	Smart sorting systems	Blockchain integration to track and manage recycled critical minerals, ensuring traceability and sustainability	Germany, China	German, Chinese	Bosch and BASF partnerships on blockchain solutions for material traceability	Taz et al. [25]
Advanced technological integration	AI, IoT, and automation	Predictive maintenance systems	Implementing AI-driven predictive systems for recycling plants to enhance operational efficiency and reduce downtime	USA, UK	American, British	IBM's Watson AI used in optimizing industrial plant maintenance schedules	Leng et al. [45]

Workplace safety and ergonomics	Worker well-being	Advanced sensors for safety	Deployment of wearable exoskeletons and IoT-enabled sensors to monitor workers' physical strain in recycling tasks	South Korea	South Korean	Hyundai's introduction of wearable exoskeletons for workers in manufacturing settings	Ghorbani et al. [10]
Decision support systems	Smart logistics	AI-powered decision support for reverse logistics	A decision support system leveraging machine learning to optimize resource recycling routes and reduce transportation costs	USA, Germany	American, German	SAP's AI logistics systems for optimized resource flow in circular economies	Shahidzadeh and Shokouhyar [12]
Collaborative policy making	Government and industry collaboration	Multi-stakeholder partnerships	International collaboration between governments, industries, and academia to develop regulations for eco-critical recycling	EU, China	European, Chinese	European Union's strategic partnership with China on sustainable mineral recycling frameworks	Zio and Guarnieri [51]
Technological adoption in developing countries	Global adoption	Sustainable tech transfer	Adoption of Industry 5.0 technologies in Sub-Saharan Africa for resource recovery and critical mineral recycling	Nigeria	Nigerian	African Development Bank (AfDB) investment in Industry 5.0 technologies for sustainable mineral recovery	Oyebode [18]

Source: Table by the authors.

5.7 SOCIAL RESPONSIBILITY AND INCLUSIVITY IN ECO-MINERAL RECYCLING 5.0

The notion of social responsibility in Eco-Critical Mineral Recycling 5.0 signifies a transformative approach wherein sustainable recycling methodologies prioritize human welfare, environmental consequences, and societal inclusion. In this context, social responsibility goes beyond mere regulatory compliance and includes proactive initiatives to foster equitable labor standards, community involvement, and transparent governance within the recycling process. As eco-critical mineral recycling integrates into the Industry 5.0 framework, prioritizing human-centric solutions and involving varied social stakeholders will be essential for achieving equitable benefit distribution throughout society. Research by Shafique et al. [2] and Psarommatis et al. [17] demonstrates that circular economy models, especially in critical mineral recycling, can create social value by mitigating environmental damage and promoting job creation in local communities, particularly in emerging economies such as China and India, where a substantial segment of the global supply chain is concentrated.

Industry 5.0 emphasizes human–machine collaboration, presenting chances to enhance the inclusivity of recycling processes via automation, robots, and digital platforms that incorporate contributions from various social groups. Järvenpää et al. [19] illustrate that consumer participation in smart waste recycling enhances material recovery and fosters awareness of environmental sustainability, presenting a paradigm for inclusivity. The utilization of robots in hazardous waste recycling, as outlined by Ghorbani et al. [10], exemplifies how technology may safeguard workers' health and safety while facilitating significant human integration. Within this paradigm, inclusion entails cultivating a workforce proficient in innovative technologies and equipped for the novel positions emerging from Industry 5.0, such as machine maintenance, data analysis, and operational management.

The significance of education and skill development is paramount in guaranteeing that underprivileged populations can enjoy the advantages of Industry 5.0. Institutions, such as universities and research institutes, bear the need to prepare future generations with the requisite skills for engaging in eco-critical mineral recycling systems. The implementation of inclusive education programs is crucial in creating a sustainable and diverse workforce, as Yang et al. [6] show. Educational programs should emphasize both technical skills and the promotion of understanding regarding the circular economy, ethical recycling practices, and environmental justice. This ensures that individuals from varied socio-economic backgrounds can participate in and profit from the green economy. Investigating inclusive education within the framework of recycling technologies, especially in areas with significant informal sectors like Sub-Saharan Africa, may yield important insights into how eco-critical mineral recycling might enhance overall socio-economic development.

The participation of industry leaders in advocating for social responsibility in Eco-Critical Mineral Recycling 5.0 is a vital component. Corporations like Tesla and Apple, prominent consumers of key minerals, exemplify the integration of circular economy ideas into their business strategies. These companies participate in ethical sourcing, battery recycling, and investment in technology that improves the sustainability of their supply chains. The research conducted by Delahaye et al. [57] and Taz et al. [25] emphasizes the utilization of sophisticated sorting and recovery methods for essential metals, illustrating the potential for creative business models that promote the circular economy and provide societal advantages. These practices advance the worldwide initiative for a more resilient and equitable system, wherein economic growth is dissociated from environmental damage.

Public–private partnerships (PPPs) are a crucial catalyst for promoting social responsibility and inclusivity in eco-critical mineral recycling. Governments, industry, and academic institutions must collaborate to formulate regulations that promote an inclusive recycling environment. An exemplary instance is the European Union's Circular Economy Action Plan, which promotes the incorporation of social, environmental, and economic aspects into recycling methods. The establishment of regulatory frameworks, as suggested by Banjanovi-Mehmedovic and Dželilovi [13], standardizes processes and guarantees the inclusion of all stakeholders, particularly vulnerable groups, in the decision-making process. These policies enhance transparency, safeguard workers' rights, and provide equitable compensation for those engaged in recycling activities. Through these activities, Industry 5.0 guarantees that the transition to a more sustainable and circular economy benefits society as a whole, rather than a privileged minority.

Furthermore, the implementation of advanced technology is crucial in enhancing the inclusivity of eco-critical mineral recycling operations. Technologies like AI, blockchain, and the IoT facilitate the seamless integration of small-scale operators, local communities, and large enterprises in recycling efforts. For instance, Leng et al. [45] emphasize the application of blockchain, which ensures the traceability of materials from recycling facilities to manufacturers, thereby enhancing confidence and accountability. Furthermore, Shen et al. [11] indicate that AI-driven sorting systems can streamline trash management and sorting processes, facilitating increased involvement from small enterprises and cooperatives lacking access to traditional recycling technologies. This technical inclusion enables smaller participants in the value chain to engage in and gain advantages from the innovations of Industry 5.0, thus fostering a more equal system.

Finally, the significance of social innovation in fostering inclusivity within Eco-Critical Mineral Recycling 5.0 is undeniable. Social innovation denotes novel concepts that address societal needs and foster new ties or collaborations. Within the realm of recycling, social innovation may manifest as novel business models, community-oriented projects, or pioneering recycling processes that cater to the requirements of local populations. Initiatives like community-driven recycling programs in poor countries or partnerships between academic institutions and municipal governments in waste management exemplify how social innovation may promote inclusivity. In nations like India and Brazil, where informal recycling sectors are prominent, social innovation in recycling methodologies can generate jobs for underprivileged people while providing environmental advantages. Through these initiatives, Eco-Critical Mineral Recycling 5.0 can promote a more inclusive, sustainable, and human-centered methodology for critical mineral recovery.

The incorporation of social responsibility and inclusivity into Eco-Critical Mineral Recycling 5.0 is vital for guaranteeing that Industry 5.0 transcends mere technological progress to become a comprehensive initiative that serves the broader society. This necessitates a comprehensive strategy involving cooperation among industry, academia, and government, as well as the formulation of inclusive policies, cutting-edge technologies, and social innovations. The future of eco-critical mineral recycling relies on its capacity to incorporate human-centric solutions, fostering a more egalitarian, sustainable, and resilient circular economy. The studies, ideas, and examples provided elucidate the essential role of social responsibility and inclusivity in attaining the objectives of Industry 5.0.

Table 5.6 provides a detailed and contemporary analysis of the principal factors and advancements in Eco-Critical Mineral Recycling 5.0, highlighting the essential importance of social responsibility and inclusivity in the shift toward a more sustainable and human-centered industrial framework. The table summarizes many initiatives from both industry and academic institutions that constitute essential components for the advancement of Industry 5.0. This encompasses the deployment of sophisticated technologies such as human–robot collaboration, blockchain integration, and AI-driven systems to enhance the recycling of essential minerals. The significance of circular economy principles in enhancing material recovery rates is emphasized, along with case studies from prominent research, including Shafique et al. [2] and Ghorbani et al. [10], which investigate sustainable practices in the Chinese manufacturing sector and the design of human-centric robotic assembly lines, respectively. The table emphasizes the necessity of engaging marginalized communities and customers in the recycling process, addressing social and environmental justice, and illustrating how these measures foster more inclusive and fair industrial practices. This congruence with Industry 5.0's focus on human-centric and ethical standards in technological advancement is examined through several successful implementations across different locations and industries globally.

Furthermore, the interventions detailed in Table 5.6 offer a comprehensive perspective on how recent studies, including those by Yang et al. [6] and Taz et al. [25], are enhancing resource recovery in essential mineral recycling processes. The incorporation of intelligent technologies, including IoT and sophisticated sensor systems for material classification, in conjunction with strong frameworks for sustainable waste management, illustrates a progressive strategy for circularity in mineral recovery. Al Amin and Baldacci [44] examine the advocacy for blockchain in open, decentralized supply chains as a manifestation of a global effort to foster inclusivity that aligns with these developments. The aggregate effect of these initiatives, as outlined in the table, highlights the developing synergy among Industry 5.0, social responsibility, and the pressing necessity for sustainable mineral recycling processes. Eco-Critical Mineral Recycling 5.0, through comprehensive strategies involving academia, industry leaders, and policymakers, is positioned to catalyze a cleaner, more inclusive global industrial framework that aligns with the United Nations SDGs, offering a blueprint for a resilient and ethically driven future.

TABLE 5.6
Advancing Social Responsibility and Inclusivity in Eco-Critical Mineral Recycling 5.0: Innovations, Strategies, and Industry Interventions for a Human-Centric and Sustainable Future

Area of Focus	Innovation	Building Block for Industry 5.0	Categories	Example Studies	Locations/ Nationalities	Industry/ Institutional Interventions	Recommendations for Improvement
Human-centric robotics	Integrating collaborative robots (cobots) for eco-critical mineral recycling, focusing on safety and ergonomics.	Human–machine collaboration, inclusive design, and safety protocols.	Robotics, safety, ergonomics	Ghorbani et al. [10] on human-centric robotic assembly lines	China, Switzerland, India	Industry 5.0 companies collaborating with robotic manufacturers for safer work environments.	Encourage interdisciplinary training for workers on robot collaboration to enhance operational safety.
Circular economy integration	Design systems that prioritize material reuse and product repurposing, emphasizing the recycling of critical minerals.	Sustainable resource management and eco-efficient production.	Recycling, sustainability, resource management	Shafique et al. [2] on synergy between Industry 5.0 and circular economy	China, USA, EU	Global partnerships for product life-cycle management and material recovery.	Promote national policies that incentivize circular economy practices in mineral extraction sectors.
Community engagement	Empowering local communities through skill-building initiatives for recycling processes.	Inclusivity in green job creation and community-centered eco-initiatives.	Education, skill development, local impact	Järvenpää et al. [19] on consumer involvement in smart waste recycling	Finland, Nigeria, India	Industry 5.0 firms can offer training programs for marginalized communities in recycling technologies.	Develop more inclusive training schemes that cater to underserved populations in mining communities.

(*continued*)

TABLE 5.6 (Continued)
Advancing Social Responsibility and Inclusivity in Eco-Critical Mineral Recycling 5.0: Innovations, Strategies, and Industry Interventions for a Human-Centric and Sustainable Future

Area of Focus	Innovation	Building Block for Industry 5.0	Categories	Example Studies	Locations/ Nationalities	Industry/ Institutional Interventions	Recommendations for Improvement
Smart technologies	Utilizing AI-driven systems for optimizing mineral recycling processes and enhancing decision-making.	Technology-driven efficiency and data-driven resource optimization.	AI, automation, data analytics	Leng et al. [8] on AI-based reinforcement learning for resilient production control	USA, China, Germany	Deployment of AI systems in recycling plants to monitor and improve efficiency in real-time.	Invest in cross-sector partnerships to foster AI adoption in developing nations for waste management.
Ethical AI and governance	Developing AI systems with transparency, ensuring ethical decision-making and minimizing bias in recycling practices.	Ethical governance frameworks and accountability in AI applications.	AI ethics, governance, transparency	Nayeri et al. [58] on decision-making methods for responsive supply chains	Switzerland, UK, Japan	Establishing regulatory bodies for AI ethics in recycling operations.	Enforce stronger global regulations on ethical AI to ensure responsible recycling practices.
Blockchain for transparency	Leveraging blockchain for transparent tracking of mineral recycling processes from collection to repurposing.	Transparency in supply chains and anti-corruption measures.	Blockchain, supply chain, transparency	Leng et al. [45] on secure blockchain middleware for decentralized supply chains	USA, UAE, India	Blockchain solutions for end-to-end traceability in mining and recycling systems.	Promote international collaboration on blockchain standards for mineral supply chains.

Inclusive policy development	Formulating policies that support diverse stakeholders, including women, minorities, and marginalized groups in the recycling sector.	Equitable policy-making and social responsibility in resource management.	Policy, social justice, equality	Jeyaraman et al. [3] on Industry 5.0 inclusivity in orthopaedics	India, Brazil, South Africa	Multinational organizations collaborating with governments to draft inclusive recycling policies.	Advocate for government incentives for companies that demonstrate commitment to inclusivity and diversity.
Sustainability and impact assessment	Assessing the long-term environmental and social impacts of eco-critical mineral recycling programs.	Sustainability assessments and environmental responsibility in Industry 5.0.	Environmental impact, sustainability assessment	Kumar et al. [24] on sustainable recycling in the context of Industry 5.0	USA, China, EU	Environmental NGOs working with industries to conduct impact assessments for mineral recycling.	Integrate sustainability metrics into the core performance indicators for Industry 5.0 projects.

Source: Table by the authors.

5.8 FUTURE DIRECTIONS AND CHALLENGES FOR ECO-MINERAL RECYCLING 5.0

The advancement of Eco-Critical Mineral Recycling 5.0 represents a significant progression toward realizing a circular economy and tackling essential issues in sustainable mineral resource management. With the global shift toward Industry 5.0, which emphasizes a human-centric paradigm of collaboration between humans and machines, the future trajectory of eco-critical mineral recycling is set to transform the recycling environment. A critical difficulty is the integration of these advanced systems into current recycling infrastructures. To attain substantial advancement, breakthroughs, including ML, AI-driven robots, and intelligent sorting technologies, must be utilized to enhance resource recovery. The application of deep learning and advanced robotics in lithium-ion battery recycling presents a means to markedly enhance efficiency [27]. These technologies provide real-time monitoring, adaptive reactions to material flows, and enhanced precision in the separation of valuable metals from waste streams, crucial for satisfying the escalating worldwide demand for vital minerals.

A key focus of Eco-Critical Mineral Recycling 5.0 is the implementation of blockchain technology to provide transparency and traceability within the critical minerals supply chain. The amalgamation of blockchain with Industry 5.0's digital twins and IoT networks will enhance the logistics and transportation of recycled materials while ensuring elevated levels of accountability. Blockchain facilitates the monitoring of materials from origin to end-user, ensuring that the supply of recycled materials complies with regulatory norms and verifiable sustainability objectives [45]. This breakthrough may prove vital for sectors dependent on commodities such as cobalt and rare earth metals, frequently extracted under dubious environmental and human rights circumstances. Blockchain can facilitate the certification of materials obtained from sustainable recycling methods, establishing a new industry standard. Moreover, utilizing blockchain for recycling management will cultivate a more robust and adaptive system capable of swiftly responding to shocks, including those induced by geopolitical or economic concerns.

A promising domain for Eco-Critical Mineral Recycling 5.0 is the development of modular solutions for product disassembly. He et al. [26] illustrate that modular designs facilitate the deconstruction and separation of essential minerals, hence enhancing the recyclability of electronic devices at the conclusion of their lives. This method is consistent with the tenets of Industry 5.0, where collaboration between humans and machines is essential. Advanced AI algorithms can facilitate the identification of the optimal disassembly sequence for each product, while robots equipped with specialized tools can execute disassembly activities more safely and with enhanced precision compared to human workers alone. This strategy enhances the scalability of recycling systems by allowing manufacturers to customize their designs to optimize the recovery of important resources. The implementation of modular design, combined with the accuracy of AI-driven disassembly systems, will likely expedite the recycling of essential critical minerals such as lithium, cobalt, and nickel.

Moreover, circular business models are anticipated to assume a progressively pivotal function in the future of eco-critical mineral recycling. Industry 5.0 highlights the importance of not just recycling materials but also their reuse and repurposing. The research by Psarommatis et al. [17] emphasizes that remanufacturing and reuse can significantly reduce waste and mitigate the environmental effects of raw mineral extraction, which serves as an illustration of this transition. These models are supported by technical breakthroughs like additive manufacturing (3D printing), enabling the repair and reuse of essential components without requiring new materials. The implementation of circular business models in sectors such as electronics, automotive, and renewable energy will be pivotal in alleviating the strain on primary resources and promoting environmental sustainability. Nonetheless, the problem persists in establishing infrastructure that facilitates the closed-loop recycling procedures necessary for mainstream circularity.

In the realm of Eco-Critical Mineral Recycling 5.0, prioritizing human-centric approaches is essential for the development and acceptance of these advances. A primary difficulty is reconciling human labor participation with automation and robotics. Ghorbani et al. [10] contend that the design of human–robot collaborative systems is essential for establishing efficient and safe recycling environments. This joint methodology will enable human workers to concentrate on tasks necessitating advanced cognitive abilities, while machines manage repetitive, labor-intensive activities. With the increasing need for recycling key minerals, industries must engage in upskilling and reskilling the workforce to guarantee that human workers are not displaced but instead play a vital role in supervising and enhancing machine-driven recycling operations. Furthermore, the ethical ramifications of automation require meticulous examination to prevent the exacerbation of social inequities or the displacement of employment in at-risk populations.

The worldwide transition to sustainable energy sources, particularly via the implementation of renewable energy technologies, increases the demand for the recycling of essential minerals such as lithium, cobalt, and REEs. Countries such as China, the United States, and many European nations have commenced steps to establish comprehensive recycling systems for these materials, with the forthcoming challenge being the integration of these efforts with Industry 5.0 technology. Shafique et al. [2] observe that the integration of Industry 5.0 and circular economy practices can enhance sustainable performance in the manufacturing sector, especially in nations such as China, where the need for renewable energy technology is rapidly increasing. Nonetheless, enhancing recycling initiatives necessitates surmounting various obstacles, such as technological constraints, substantial capital expenditures, and the requirement for international legislative frameworks that harmonize recycling procedures across borders. The global community must prioritize research and development, along with coordinated policy formulation, to tackle these difficulties and guarantee the sustainable supply of essential minerals.

Ultimately, a significant problem in Eco-Critical Mineral Recycling 5.0 is the necessity for innovation in the social and economic aspects of recycling. According to Järvenpää et al. [19], engaging consumers and families in recycling initiatives is essential for the extensive implementation of sustainable practices. Social engineering tactics, including education, incentives, and community involvement, must enhance technological progress to guarantee the efficacy of recycling initiatives. The economic feasibility of recycling programs relies on developing business models that encourage investment in recycling infrastructure, guaranteeing that the costs do not surpass the economic advantages. This requires cooperation among governments, commercial sectors, and research institutes to formulate regulations that encourage sustainable recycling practices while offering financial incentives for innovation in this domain. By promoting a global, cooperative, and comprehensive strategy for Eco-Critical Mineral Recycling 5.0, the forthcoming era of human–machine collaboration will facilitate a more sustainable and resource-efficient future.

The swiftly advancing domain of eco-critical mineral recycling is entering a dynamic phase with the emergence of Industry 5.0, marked by the incorporation of human-centered, sustainable methodologies throughout the recycling value chain. Table 5.7 demonstrates that the confluence of modern technologies, including AI, robots, blockchain, and additive manufacturing, is crucial in influencing the future of mineral recovery and recycling. Advancements in intelligent trash sorting and material recovery systems are facilitating the creation of more efficient, automated procedures that markedly improve resource recovery. These systems utilize IoT devices and AI-driven robotics, as shown by global implementations in smart cities in Malaysia and China, where AI-enabled robots and blockchain integration have enhanced the traceability of recycled materials. This has enhanced operational efficiency and diminished the environmental impact linked to conventional mining and material extraction methods. The adaptability and resilience of these intelligent systems, powered by deep reinforcement learning and predictive maintenance, are enhancing recycling plants' capacity to endure disruptions, thereby reinforcing technology's role in maintaining a continuous material flow in the recycling process.

Furthermore, Table 5.7 highlights the essential function of industrial and institutional actions in promoting the adoption of these technologies within the circular economy paradigm. Institutions are increasingly prioritizing the adoption of sustainable business models, like blockchain for transparency in recycling operations and additive manufacturing for the production of high-quality components from recycled materials. These interventions are underpinned by advanced research and industrial methodologies, as demonstrated by the incorporation of blockchain technology in the RMG sectors of Bangladesh and India, enhancing supply chain traceability and fostering sustainability. Moreover, AI frameworks are employed to enhance recycling systems, guaranteeing improved processing accuracy and resource efficiency. The integration of consumers into the recycling process via mobile applications and real-time data feedback, as exemplified in Finland and Sweden, enhances the human-centric paradigm of Industry 5.0. These achievements signify a technological progression and a paradigm shift in industrial interactions with the environment and society, establishing eco-critical mineral recycling as an essential element of a sustainable, circular economy.

5.9 CONCLUDING REMARKS

This study has examined the revolutionary framework of Eco-Critical Mineral Recycling 5.0, emphasizing the integration of modern technologies and human-centered design principles in redefining critical mineral recovery. This new frontier

TABLE 5.7
Future Directions and Challenges for Eco-Critical Mineral Recycling 5.0

Area	Category	Innovation	Authors and Year	Example/ Location	Institutional/ Industrial Intervention	Specific Example for Improvement	Citation
Circular economy integration	Human–machine collaboration	Smart waste sorting and material recovery systems using IoT and AI	Banjanovi-Mehmedovic and Dželilović [13]	Malaysia and China, focusing on smart cities (e.g., Penang, China)	Industry 5.0 paradigm for waste management, improving material recovery in smart cities	Enhance AI-enabled waste sorting with sensor integration for higher efficiency	Banjanovi-Mehmedovic and Dželilović [13]
Resource recovery	Automation and robotics	AI-powered robotics in disassembly and sorting of critical minerals	Ghorbani et al. [10]	Global applications, e.g., industrial plants in Europe	AI robots for disassembly; automation for hazardous materials recycling	Adoption of AI-based robots in material recovery lines for toxic metal removal	Ghorbani et al. [10]
Sustainability and risk assessment	Circular business models	Blockchain for traceability in material recovery and recycling processes	Al Amin and Baldacci [44]	Bangladesh and India	Blockchain integration in RMG industries for sustainable material sourcing	Integrate blockchain to trace critical mineral flows for transparency in recycling	Al Amin and Baldacci [44]

(continued)

TABLE 5.7 (Continued)
Future Directions and Challenges for Eco-Critical Mineral Recycling 5.0

Area	Category	Innovation	Authors and Year	Example/ Location	Institutional/ Industrial Intervention	Specific Example for Improvement	Citation
Advanced manufacturing	Additive manufacturing	Use of 3D printing for creating components from recycled materials	Sala and Richert [5]	Europe (Germany)	Additive manufacturing for sustainable production, reducing material waste	Expand 3D printing applications to reduce dependence on raw mining operations	Sala and Richert [5]
Smart systems for recycling	AI and deep learning	Predictive maintenance and optimization of recycling systems through AI	Leng et al. [8]	China, European Union	AI frameworks to predict system failures in recycling plants, improving reliability	Further refine AI algorithms for more accurate predictions in mineral processing	Leng et al. [8]
Consumer engagement	Public participation	Involving consumers in smart waste recycling using real-time data feedback	Järvenpää et al. [19]	Finland and Sweden	Engaging consumers through mobile apps to increase waste segregation efficiency	Expand apps for consumer-driven recycling programs to increase participation	Järvenpää et al. [19]

Supply chain innovation	Resilient and flexible design	Complex network-based approach for resilient recycling supply chains	Ma et al. [7]	China	Flexible supply chain design for critical mineral recovery, especially during crises	Develop adaptive networks to ensure uninterrupted mineral supply during disruptions	Ma et al. [7]
Technological foundations	Deep reinforcement learning	Resilient production control using deep reinforcement learning for recycling	Leng et al. [8]	Global applications in manufacturing facilities	Implement deep learning for optimized production cycles in critical mineral recycling	Integrate reinforcement learning in automated recycling systems for better adaptability	Leng et al. [8]

Source: Table by the authors.

incorporates Industry 5.0 principles—sustainability, resilience, and human–machine synergy—utilizing AI, ethical robotics, advanced sensing systems, and smart automation to improve efficiency and safety while integrating environmental stewardship into mineral recycling practices. The rise of human-centric robotics and ethical AI promotes an inclusive strategy that integrates human intuition with machine accuracy to meet the increasing global demand for essential minerals sustainably.

This study highlighted the crucial need for individualized and customizable recycling procedures designed to address material-specific and region-specific difficulties through a thorough investigation of advanced technologies. It emphasized how collaborative robotics (cobots), intelligent sorting technologies, and decentralized AI models are transforming the recovery, processing, and reintegration of critical minerals into the circular economy. The study demonstrated that Eco-Critical Mineral Recycling 5.0 represents not only a technological advancement but also a strategy transformation that integrates social, ethical, and ecological aspects, promoting transparency, traceability, and circularity within the mineral value chain.

This study advocates for the ongoing advancement of integrated policy frameworks, multidisciplinary research, and international collaboration to guarantee the fair implementation of these technologies. As Industry 5.0 progresses, the integration of innovation with ethical accountability will be crucial to alleviating the environmental impact of mineral extraction and recycling. Eco-Critical Mineral Recycling 5.0 arises as both a technological necessity and a comprehensive framework for sustainable resource management, in accordance with global goals of climate neutrality, digital sovereignty, and equitable energy transitions.

REFERENCES

[1] Adel A. Future of industry 5.0 in society: Human-centric solutions, challenges and prospective research areas. *Journal of Cloud Computing*. 2022;11:40.

[2] Shafique MN, Adeel U, Rashid A. The synergy between industry 5.0 and circular economy for sustainable performance in the chinese manufacturing industry. *Sustainability*. 2024;16:9952.

[3] Jeyaraman M, Nallakumarasamy A, Jeyaraman N. Industry 5.0 in orthopaedics. *Indian Journal of Orthopaedics*. 2022;56:1694–702.

[4] Agbor ME, Udo SO, Ewona IO, Nwokolo SC, Ogbulezie JC, Amadi SO. Potential impacts of climate change on global solar radiation and PV output using the CMIP6 model in West Africa. *Cleaner Engineering and Technology*. 2023;13.

[5] Sala D, Richert M. Perspectives of additive manufacturing in 5.0 industry. *Materials*. 2025;18:429.

[6] Yang Y, Keivanpour S, Imbeau D. The future of disassembly planning: A critical assessment of industry 5.0, lean, and X-reality. *2023 International Conference on Electrical, Computer and Energy Technologies (ICECET)*: IEEE; 2023. p. 1–12.

[7] Ma N, Yao X, Wang K. A complex network-based approach for resilient and flexible design resource allocation in industry 5.0. *Computers in Industry*. 2024;159:104108.

[8] Leng J, Ruan G, Xu C, Zhou X, Xu K, Qiao Y, et al. Deep reinforcement learning of graph convolutional neural network for resilient production control of mass individualized prototyping toward industry 5.0. *IEEE Transactions on Systems,*

Man, and Cybernetics: Systems. 2024;54(11):7092–105. https://doi.org/10.1109/TSMC.2024.3446671.

[9] Beitelmal WH, Nwokolo SC, Meyer EL, Ahia CC. Exploring adaptation strategies to mitigate climate threats to transportation infrastructure in Nigeria: Lagos City, as a case study. *Climate*. 2024;12.

[10] Ghorbani E, Keivanpour S, Sekkay F, Imbeau D. Human-centric robotic assembly line design: A fuzzy inference system approach for adaptive workload management. *The International Journal of Advanced Manufacturing Technology*. 2024;134:3805–27.

[11] Shen Y, Lu W, Sheng H, Liu Y, Tian G, Zhang H, et al. Human–robot collaboration on a disassembly-line balancing problem with an advanced multiobjective discrete bees algorithm. *Symmetry*. 2024;16:794.

[12] Shahidzadeh MH, Shokouhyar S. Unveiling just-in-time decision support system using social media analytics: A case study on reverse logistics resource recycling. *Industrial Management & Data Systems*. 2024;124:2251–83.

[13] Banjanovi-Mehmedovic L, Dželilović N. Towards a greener future: Sensor-based material characterization in waste sorting for a sustainable recycling model. *International Conference "New Technologies, Development and Applications"*: Springer; 2024. p. 3–11.

[14] Nwokolo SC, Ogbulezie JC, Obiwulu AU. Impacts of climate change and meteo-solar parameters on photosynthetically active radiation prediction using hybrid machine learning with Physics-based models. *Advances in Space Research*. 2022;70:3614–37.

[15] Voutier R, Bakkar MN, Pivetta G. Global trust registry plus (GTR+) connecting G20 vision with actions in industry 5.0 era: Global trust registry initiative. Advanced Research and Real-World Applications of Industry *50*: IGI Global; 2023. p. 1–20.

[16] Nwokolo SC, Obiwulu AU, Okonkwo PC. *Africa's Propensity for a Net Zero Energy Transition*: CRC Press; 2024.

[17] Psarommatis F, May G, Azamfirei V. Product reuse and repurpose in circular manufacturing: a critical review of key challenges, shortcomings and future directions. *Journal of Remanufacturing*. 2025;15:1–38. https://doi.org/10.1007/s13243-025-00153-y.

[18] Oyebode OJ. Management of computer and electronic waste through the fifth industrial revolution for the cleaner environment. *2024 International Conference on Science, Engineering and Business for Driving Sustainable Development Goals (SEB4SDG)*: IEEE; 2024. p. 1–14.

[19] Järvenpää A-M, Jussila J, Kunttu L, Kunttu I. Involving consumers and households in smart waste recycling. *ICT, the Business Sector and Smart Cities: Routledge*; 2024. p. 99–116.

[20] Panagou S, Fruggiero F, Iannone R, Gnoni MG. The CACTUS approach: Organizational approach for sustainability. SUMMER SCHOOL FRANCESCO TURCO PROCEEDINGS. 2022:1–7.

[21] Goh KC, Kurniawan TA, Zainordin N, Arifah IDC, Abas MA, Masrom MAN, et al. Expediting decarbonization in energy, waste, and water sector through digitalization in sustainable smart cities (SSC): Case-studies in Malaysia and China based on Industry 5.0 paradigm. *Sustainable Cities and Society*. 2024;117:105969.

[22] Hassan MA, Bailek N, Bouchouicha K, Ibrahim A, Jamil B, Kuriqi A, et al. Evaluation of energy extraction of PV systems affected by environmental factors under real outdoor conditions. *Theoretical and Applied Climatology*. 2022;150:715–29.

[23] Hassan MA, Bailek N, Bouchouicha K, Nwokolo SC. Ultra-short-term exogenous forecasting of photovoltaic power production using genetically optimized non-linear auto-regressive recurrent neural networks. *Renewable Energy*. 2021;171:191–209

[24] Kumar R, Kariminejad A, Antonov M, Goljandin D, Klimczyk P, Hussainova I. Progress in sustainable recycling and circular economy of tungsten carbide hard metal scraps for industry 5.0 and onwards. *Sustainability*. 2023;15:12249.

[25] Taz D, Ozsarac S, Kursunoglu S, Kursunoglu N, Top S, Altiner M. A review on the recovery of high-grade critical metals from spent petroleum catalysts for meeting the demands of Industry 5.0. *Energy Sources, Part A: Recovery, Utilization, and Environmental Effects*. 2024;46:4593–607.

[26] He L, Gao J, Leng J, Wu Y, Ding K, Ma L, et al. Disassembly sequence planning of equipment decommissioning for industry 5.0: Prospects and Retrospects. *Advanced Engineering Informatics*. 2024;62:102939.

[27] Tian G, Sheng H, Zhang L, Zhang H, Fathollahi-Fard AM, Zhang X, et al. Enhancing end-of-life product recyclability through modular design and social engineering optimiser. *International Journal of Production Research*. 2024:1–19. https://doi.org/10.1080/00207543.2024.2424970.

[28] Du P, Xiao T, Cao H, Zhai D, Gadekallu TR, Piran MJ. AI-based UAVs 3D coverage deployment in 6G-enabled IoV networks for industry 5.0. *IEEE Transactions on Consumer Electronics*. 2024;71(2):4221–34. https://doi.org/10.1109/TCE.2024.3480989.

[29] Hong C, Calleja JJ, Clark R, Mui LS. Applied education futures and the shape of things to come. *Applied Degree Education and the Shape of Things to Come*: Springer; 2023. p. 1–26.

[30] Nwokolo SC. Climate hoax: The shift from scientific discourse to speculative rhetoric in climate change conversations. *Next Research*. 2025;2.

[31] Nwokolo SC, Amadi SO, Obiwulu AU, Ogbulezie JC, Eyibio EE. Prediction of global solar radiation potential for sustainable and cleaner energy generation using improved Angstrom-Prescott and Gumbel probabilistic models. *Cleaner Engineering and Technology*. 2022;6.

[32] Nwokolo SC, Eyime EE, Obiwulu AU, Meyer EL, Ahia CC, Ogbulezie JC, et al. A multi-model approach based on CARIMA-SARIMA-GPM for assessing the impacts of climate change on concentrated photovoltaic (CPV) potential. *Physics and Chemistry of the Earth*. 2024;134.

[33] Nwokolo SC, Meyer EL, Ahia CC. Credible pathways to catching up with climate goals in Nigeria. *Climate*. 2023;11.

[34] Nwokolo SC, Meyer EL, Ahia CC. Exploring the interactive influences of climate change and urban development on the fraction of absorbed photosynthetically active radiation. *Atmosphere*. 2024;15.

[35] Richnák P, Fidlerová H. Impact and potential of sustainable development goals in dimension of the technological revolution industry 4.0 within the analysis of industrial enterprises. *Energies*. 2022;15:3697.

[36] Abdous M-A, Delorme X, Battini D, Sgarbossa F. Scenario-based optimization and simulation framework for human-centered Assembly Line Balancing. *International Journal of Production Economics*. 2025;282:109513.

[37] Wachter C, Beckschulte S, Hinrichs MP, Sohnius F, Schmitt RH. Strategies for resilient manufacturing: A systematic literature review of failure management in production. *Procedia CIRP*. 2024;130:1393–402.

[38] Nwokolo SC, Obiwulu AU, Ogbulezie JC. Machine learning and analytical model hybridization to assess the impact of climate change on solar PV energy production. *Physics and Chemistry of the Earth*. 2023;130.

[39] Nwokolo SC, Obiwulu AU, Ogbulezie JC, Amadi SO. Hybridization of statistical machine learning and numerical models for improving beam, diffuse and global solar radiation prediction. *Cleaner Engineering and Technology*. 2022;9.

[40] Nwokolo SC, Obiwulu AU, Okonkwo PC, Orji R, Alsenani TR, Mansir IB, et al. Machine learning and physics-based model hybridization to assess the impact of climate change on single- and dual-axis tracking solar-concentrated photovoltaic systems. *Physics and Chemistry of the Earth*. 2025;138.

[41] Nwokolo SC, Ogbulezie JC. A qualitative review of empirical models for estimating diffuse solar radiation from experimental data in Africa. *Renewable and Sustainable Energy Reviews*. 2018;92:353–93.

[42] Nwokolo SC, Ogbulezie JC, Ushie OJ. A multi-model ensemble-based CMIP6 assessment of future solar radiation and PV potential under various climate warming scenarios. *Optik*. 2023;285.

[43] Chand S, Lu Y. Dual task scheduling strategy for personalized multi-objective optimization of cycle time and fatigue in human-robot collaboration. *Manufacturing Letters*. 2023;35:88–95.

[44] Al Amin M, Baldacci R. Blockchain Technology and Industry 5.0 synergy for sustainable development in RMG industries: An ISM and fuzzy DEMATEL approach. *Discover Sustainability*. 2024;5:1–26.

[45] Leng J, Chen Z, Huang Z, Zhu X, Su H, Lin Z, et al. Secure blockchain middleware for decentralized IIoT towards industry 5.0: A review of architecture, enablers, challenges, and directions. *Machines*. 2022;10:858.

[46] Javed AR, Shahzad F, ur Rehman S, Zikria YB, Razzak I, Jalil Z, et al. Future smart cities: Requirements, emerging technologies, applications, challenges, and future aspects. *Cities*. 2022;129:103794.

[47] Gagnidze I. Industry 4.0 and industry 5.0: Can clusters deal with the challenges?(A systemic approach). *Kybernetes*. 2023;52:2270–87.

[48] Ivanov D. Design and deployment of sustainable recovery strategies in the supply chain. *Computers & Industrial Engineering*. 2023;183:109444.

[49] Chukwujindu NS. A comprehensive review of empirical models for estimating global solar radiation in Africa. *Renewable and Sustainable Energy Reviews*. 2017;78:955–95.

[50] Wang M, Tong Y, Zhuang C. Towards resilience in Industry 5.0: Application of system dynamics in matrix-structured manufacturing system. *International Journal of Production Research*. 2024:1–25. https://doi.org/10.1080/00207543.2024.2430443.

[51] Zio E, Guarnieri F. Industry 5.0: Do risk assessment and risk management need to update? And if yes, how? *Proceedings of the Institution of Mechanical Engineers, Part O: Journal of Risk and Reliability*. 2025;239:389–96.

[52] Azinge CU. Circular business models in industry 5.0. *Eco-Innovation and Sustainable Development in Industry 50*: IGI Global; 2024. p. 115–32.

[53] Fraga-Lamas P, Lopes SI, Fernández-Caramés TM. Green IoT and edge AI as key technological enablers for a sustainable digital transition towards a smart circular economy: An industry 5.0 use case. *Sensors*. 2021;21:5745.

[54] Siakas D, Lampropoulos G, Rahanu H, Georgiadou E, Siakas K. Emerging technologies enabling the transition toward a sustainable and circular economy: The 4R sustainability framework. *European Conference on Software Process Improvement*: Springer; 2023. p. 166–81.

[55] Abdous M-A, Delorme X, Battini D, Berger-Douce S. Multi-objective collaborative assembly line design problem with the optimisation of ergonomics and economics. *International Journal of Production Research*. 2023;61:7830–45.

[56] Murali D, Suresh M, Raman R. Breaking down to build up: How deconstruction and carbon finance foster sustainable, resilient construction in the industry 5.0 era. *Construction Innovation*. 2024. https://doi.org/10.1108/CI-04-2024-0124.

[57] Delahaye F, Pinsonneault MH. The solar heavy-element abundances. I. Constraints from stellar interiors. *The Astrophysical Journal*. 2006;649:529.

[58] Nayeri S, Sazvar Z, Heydari J. Designing an IoT-enabled supply chain network considering the perspective of the fifth industrial revolution: Application in the medical devices industry. *Engineering Applications of Artificial Intelligence*. 2023;122:106113.

6 Critical Minerals in the Clean Energy Transition

Supply, Scarcity, and Recycling

6.1 INTRODUCTION TO CRITICAL MINERALS AND THE CLEAN ENERGY TRANSITION

The transition to clean energy, propelled by pressing climate objectives and the decarbonization of energy systems, has elevated key minerals to a prominent position in global industrial, technological, and geopolitical discussions. The minerals—lithium, cobalt, nickel, neodymium, dysprosium, and scandium—are essential for the production of low-carbon technologies, including electric vehicles (EVs) [1], wind turbines [2], solar panels [3], and battery storage systems [4]. Czerwinski [5] underscores that attaining zero-emission transportation requires a consistent provision of rare earth elements (REEs) and other essential metals. The intricacy of their supply systems, coupled with the geographical concentration of reserves, has elicited global examination about their accessibility, environmental impact, and sustainability. China excels in the processing of REEs, whereas the Democratic Republic of Congo is a leading producer of cobalt, frequently within contentious labor and environmental circumstances. Consequently, the move to clean energy is closely associated with ethical sourcing, robust supply chains, and advanced recovery technologies—constituting the foundation of this chapter's thorough examination.

The international competition for key minerals has prompted various policy measures and strategic partnerships. India has lately implemented extensive legislative reforms to bolster its self-sufficiency in vital mineral supply chains, including exploration licenses and collaborations with nations such as Australia and Japan. Britt and Czarnota [6] assert that Australia, possessing substantial lithium and rare earth reserves, is essential in this context. Their work underscores Australia's important role in delivering essential minerals to global markets, particularly in light of the rising demand from Asia-Pacific renewable energy infrastructure expansion. Simultaneously, the European Union (EU) has implemented a comprehensive strategy via laws such as the Critical Raw Materials Act, aiming to diversify sources, improve local mining, and strengthen recycling efforts [7]. Governmental interventions are essential for establishing long-term resilience in mineral supply chains and advancing circular economy models.

In light of increasing environmental and geopolitical challenges, academic and industrial research has escalated efforts to develop sustainable and efficient techniques

DOI: 10.1201/9781003672807-6

for crucial mineral recovery and recycling. Laliwala and Torres [8] highlight the recycling of REEs from decommissioned EV motors through computer-aided process design to enhance material recovery. This technological advancement not only alleviates raw material shortages but also diminishes environmental externalities [9]. Likewise, Joshi et al. [10] investigate bioleaching as an environmentally friendly and cost-effective method for extracting REEs from industrial waste, providing an alternative to energy-demanding traditional techniques. Holcombe et al. [11] exemplify industrial breakthroughs in low-emission metal extraction by their electrochemical recovery of key metals via chloride molten salt electrolysis. These strategies indicate a transition toward more sustainable approaches, incorporating green chemistry and closed-loop production processes to reduce the environmental impact of clean energy technology.

The emergence of urban mining and circular economy techniques is transforming economic perspectives on waste and resource efficiency. Urban mining, the extraction of valuable resources from electronic waste and outdated infrastructure, has gained prominence, especially in China and the United Kingdom [12, 13]. Zhao et al. [12] delineated the provincial in-use inventories of REEs in China, uncovering considerable unexploited urban mining opportunities that could mitigate reliance on raw material imports. In the United Kingdom, Hsu and associates [13] mapped the flows and stockpiles of permanent magnet materials, pinpointing essential recovery nodes for policy and industrial enhancement. Urban mining frameworks are underpinned by circular economy models, as shown by Chowdhury and Talan [14], who delineated value recovery from mine tailings to generate profit from waste. By transitioning from extraction to regeneration, these techniques enhance both environmental stewardship [15] and economic sustainability in the clean energy transition [16].

Technological innovation is accelerating the transformation of key mineral recycling using enhanced digital tools and systems-based methodologies. Priore et al. [17] utilize quantitative patent analysis to delineate innovation trends in rare earth recycling, highlighting the global impetus in research and intellectual property development. Maani et al. [18] employed disassembly analysis to improve rare earth recovery from EV motors, including robots and automation for effective separation. Riddle et al. [19] employ agent-based modeling to simulate supply disruption scenarios in global rare earth markets, providing decision-makers with predictive tools for risk mitigation. As Jin et al. demonstrated by using genetic algorithms for the sustainable recovery of neodymium–iron–boron (NdFeB) magnets, machine learning (ML) and artificial intelligence (AI) optimization enhance data-driven solutions. These examples highlight a comprehensive digital transformation, wherein systems thinking and intelligent manufacturing intersect with mineral circularity.

Geopolitical tensions and socio-environmental justice influence the discourse on essential minerals. Berthet et al. [7] performed a socio-environmental evaluation of European supply chains, identifying areas of labor exploitation and ecological deterioration associated with mineral procurement. Sovacool et al. [20] advocate for a sustainable minerals future by reconciling technology advancement with human rights and environmental preservation. The ethical considerations are reflected in the causality analysis by Cheilas et al. [21], which examines the interrelationship between the availability of REEs and the US renewable economy. Confronting these difficulties necessitates international collaboration and ethical governance structures. Organizations such as the International Renewable Energy Agency (IRENA) and collaborations like the India–Japan partnership on rare earths. Mishra and Mancheri [22] illustrate the necessity for international efforts that prioritize sustainability, transparency, and equity in mineral supply chains.

This chapter ultimately anticipates the future of key minerals as integrated within a broader narrative of resilience, innovation, and systemic transformation. Innovative methods like phytomining, which Rabbani et al. [23] investigated, use plants to extract REEs from soil, exemplifying a synthesis of biotechnology and ethical mining methods. Research by Eheliyagoda et al. [24] delineates multiregional circular economy strategies for the worldwide supply of dysprosium, highlighting the importance of international integration. The paradigm introduced by Wietschel et al. [25] for evaluating material resilience and criticality provides a model for adaptive policymaking. Collectively, these innovations and policies elevate essential minerals beyond ordinary resources, positioning them as catalysts for a regenerative, inclusive, and technologically sophisticated clean energy future. This sub-chapter will serve as a core component in the broader discourse on supply, scarcity, and recycling, facilitating sustainable transitions driven by astute mineral management.

6.2 GLOBAL SUPPLY CHAINS FOR CRITICAL MINERALS: VULNERABILITIES AND RISKS

Global supply chains for essential minerals, particularly those vital for the clean energy transition, encounter substantial vulnerabilities and dangers, with extensive consequences for the sustainability of the energy sector. The rising demand for these minerals, propelled by the growth of renewable energy technologies [26], EVs [27], and battery storage systems [28], renders global supply chains susceptible to several upheavals. These vulnerabilities may arise from geopolitical tensions, trade limitations, resource scarcity [29], environmental issues [30], and the increasing complexity of mineral extraction and processing. Major producers of critical minerals, including China, the Democratic Republic of the Congo, and Australia, are integral to this dynamic, while the EU and the United States face the challenge of diversifying their supply sources and ensuring long-term access to these minerals. The challenge is intensified by the low recycling rates of essential minerals and the complexities involved in expanding mining operations, particularly for REEs, which are vital for technologies like wind turbines, EVs, and energy storage systems [7, 20].

A significant vulnerability in essential mineral supply chains is the concentration of production in a limited number of nations, resulting in a singular point of failure. China monopolizes the global supply of REEs, overseeing around 70% of worldwide production and processing. This monopoly creates apprehensions regarding the robustness of supply networks, particularly as global demand for these resources escalates due to the shift to sustainable energy [29]. In response, nations such as the United States and EU members have been investigating alternatives, including the establishment of domestic mining and refining capabilities, as well as examining urban mining and recycling initiatives [17]. Moreover, the ecological consequences of mining activities, particularly in areas with inadequate regulatory structures, present a substantial threat. These concerns underscore the necessity for a more varied and sustainable supply chain that incorporates recycling, balanced trade regulations, and effective environmental management methods [31].

The deficiency of essential minerals, including lithium, cobalt, and nickel, poses a significant issue for the worldwide supply chain. These minerals are crucial for energy storage technologies such as lithium-ion batteries, which are vital for the electrification of transportation and the integration of renewable energy into the grid. Nonetheless, these deposits are spatially concentrated, with the Democratic Republic of the Congo representing over 60% of global cobalt output. This concentration heightens the susceptibility of supply chains to disruptions, particularly given the prevailing political instability and human rights issues in these areas [32]. Innovative strategies to diminish dependency on these minerals, such as enhancing battery chemistry to lessen dependence on cobalt or investigating alternate energy storage technologies, are essential for alleviating these threats. Moreover, ensuring sustainable supply chains necessitates a comprehensive strategy that integrates both upstream and downstream methodologies, encompassing sustainable mining practices and the advancement of effective recycling technologies [33, 34].

Recycling is essential for strengthening the robustness of vital mineral supply chains; nevertheless, existing systems frequently exhibit inefficiency and underdevelopment. Despite the existence of technology for recycling REEs from end-of-life devices, including electric car motors, wind turbines, and consumer electronics, their use remains inadequate [35]. The recycling of rare earth magnets, especially those utilized in EVs and renewable energy systems, encounters considerable technological and economic obstacles. These factors encompass the elevated expenses associated with recovery techniques, diminished recovery rates, and the intricate nature of deconstructing things to obtain the minerals [36]. There are encouraging advancements in this field, such as the application of bioleaching for the extraction of REEs from industrial waste [10] and progress in electrochemical recycling methods [11]. Governments, companies, and research institutes are progressively prioritizing advancements in recycling technology to diminish dependence on primary mining and mitigate the environmental impact of mineral extraction.

Incorporating circular economy principles into the crucial mineral supply chain is vital for tackling issues of shortage and sustainability. Circular economy practices, including the design of goods for enhanced disassembly and recycling, are increasingly being adopted in industries dependent on key minerals. The creation of modular designs for EVs and wind turbines may enhance the retrieval of important materials at the conclusion of their life cycle. Furthermore, a transition to urban mining, which involves the extraction of minerals from discarded electronic devices and other waste sources, could diminish the necessity for newly mined resources. Liu et al. [37] emphasized the significance of identifying and optimizing mineral flows in urban environments to enhance the recovery of essential minerals. The effective execution of these plans necessitates cooperation among stakeholders, including governmental entities, industry participants, and the research sector, to create efficient mechanisms for the collection, processing, and redistribution of utilized resources [38].

Institutional actions are essential for mitigating vulnerabilities and risks within major mineral supply chains. Governments can assume a crucial role by enacting legislation that fosters sustainable mining methods, facilitates the advancement of alternative resources, and advocates for the circular economy. India's recent policy measures targeting the procurement of essential minerals, including lithium and cobalt, are anticipated to diminish the nation's reliance on imports and promote domestic production [39]. Moreover, multinational collaborations, exemplified by the EU's Critical Raw Materials Alliance, aim to mitigate supply chain risks by promoting partnerships among member states, companies, and research institutions. The objective of this cooperation is to diversify supply sources, invest in recycling infrastructure, and guarantee the ethical procurement of essential minerals. Industry-driven initiatives, like those from the Electrochemical Society and numerous electric car manufacturers, are investigating creative strategies to mitigate the environmental effects of mining and enhance the efficacy of recycling systems [40].

In conclusion, mitigating the vulnerabilities and risks in global supply chains for critical minerals necessitates a comprehensive, multi-dimensional strategy that integrates technological innovation, sustainable practices, and collaborative initiatives among governments, industries, and research institutions. The clean energy transition hinges on securing a secure, ethical, and environmentally responsible supply of essential minerals [41]. These demands advance in mining, recycling, and product design, together with robust institutional structures to enable these developments. By expanding technology for recycling and researching alternative materials and energy storage options, it is possible to limit the risks associated with essential mineral supply chains and secure a sustainable energy future for generations to come [32].

Table 6.1 presents a thorough and progressive examination of global supply chains for essential minerals, emphasizing their vulnerabilities, dangers, and opportunities in relation to the clean energy transition. The table includes a variety of academic studies, illustrating the complex issues related to the mining, recycling, and sustainable utilization of key minerals. Utilizing case studies, innovations, and interventions from international research [42], it underscores significant trends and the systemic obstacles in ensuring a consistent supply of critical commodities [43], including REEs, lithium, cobalt, and nickel. Although geopolitical threats, environmental concerns, and technological limitations plague their supply chains, these minerals are crucial for low-carbon technologies like EVs [44], wind turbines, and solar panels. This research highlights the necessity for comprehensive reforms in legislation, industrial practices, and recycling technologies to address the highlighted vulnerabilities [45], ranging from ineffective extraction techniques to the increasing demand for circular economy solutions in mineral recovery [46].

Table 6.1 highlights the significance of institutional and industry initiatives in mitigating these risks. The partnership of national governments, international organizations, and business sectors is emphasized as essential for establishing robust supply chains. Innovative recycling techniques, including bioleaching, urban mining, and electrochemical recovery, are essential strategies for ensuring the availability of

TABLE 6.1
Optimizing Global Supply Chains for Critical Minerals: Addressing Vulnerabilities, Risks, and Opportunities in the Clean Energy Transition

Category	Innovative Focus	Key Areas of Development	Building Blocks for the Transition	Industry and Institutional Interventions	Examples	Study Locations	Citations
Supply chain vulnerabilities	Identifying bottlenecks and risks in critical mineral supply chains	Mapping mineral flows, assessing geographical concentration risks	Implementation of diversified sourcing, improving supply chain resilience	Collaboration of mining sectors with governments for policy reform	Lithium from Australia, Cobalt from Congo	Australia, DRC	Ghosh et al. [33], Czerwinski [5]
Geopolitical risks	Impact of political tensions on mineral supply	Risk analysis of critical minerals in trade conflicts	Developing policies to secure sustainable sourcing from multiple countries	India and Japan's collaboration for securing rare earths	Rare earths, Lithium, Cobalt	India, Japan, China	Mishra and Mancheri [22], Jain [39]
Supply chain resilience	Strengthening the resilience of global mineral supply chains	Circular economy practices for closed-loop supply chains	Circular resource recovery and efficient recycling systems	IEA's strategic energy planning and IRENA's circular economy support	Urban mining, Scandium recycling in wind turbines	Global	Rizos et al. [34], Maani et al. [18]
Environmental and social impacts	Assessing the social–environmental impact of mining and recycling	Evaluating land use, water consumption, and emissions in critical mineral extraction	Development of green mining technologies, stricter environmental regulations	EU's Green Deal initiatives and UN's SDGs	Wind turbine recycling in Australia, Rare earths from China	Europe, Australia, China	Berthet et al. [7], Ge et al. [47]

Circular economy	Boosting circular economy pathways for critical minerals	Reusing, remanufacturing, and recycling rare earth metals	Optimizing end-of-life products, focusing on rare earth magnet recovery	Industry-wide adoption of closed-loop processes	Rare earth magnets from electric vehicles, lithium-ion batteries	USA, EU	Rachidi et al. [48], Yousefi et al. [49]
Technological innovation	Exploring advanced technologies for mineral recovery and recycling	Patents for innovations in recycling methods, including bioleaching, hydrometallurgy	Advancing electrochemical recovery, automation, AI integration in recycling	Industry-driven R&D collaborations (e.g., Electrochemical Society)	Bioleaching for rare earths, Electrolysis for scandium	Global, Australia	Priore et al. [17], Holcombe et al. [11]
Policy and regulatory framework	Designing policies for strategic management of critical minerals	Government regulations on mineral extraction and environmental standards	Aligning national energy policies with sustainable mining and recycling practices	India's policy reforms for critical mineral supply (e.g., lithium)	India's mining policy reforms, China's rare earth regulations	India, China	Jain [39]
Risk management and forecasting	Quantifying risks and disruptions in global critical mineral markets	Supply risk forecasting, dynamic risk assessments in energy transition scenarios	Predictive models to assess future mineral supply and demand shifts	Industry use of agent-based modeling for supply chain disruptions	Cobalt market, Dysprosium and NdFeB magnets for wind turbines	Global	Riddle et al. [19], Maani et al. [18]

Source: Table by the authors.

rare earth metals and other vital minerals. The analysis highlights the necessity for regulatory frameworks that promote sustainable mining practices, enhance resource efficiency, and facilitate the advancement of alternative materials and technologies. Table 6.1 serves as an essential resource for comprehending the interaction between global supply chain dynamics, resource management, and sustainability objectives, offering important insights into the future of critical mineral sourcing and its ramifications for clean energy infrastructure and climate policy. This research may be crucial for enhancing sustainable production methods and securing the enduring feasibility of the clean energy transition.

6.3 SCARCITY AND DEMAND GROWTH: THE CRITICAL MINERAL SUPPLY GAP

The shift to a clean energy economy is fundamentally reliant on the accessibility and effective utilization of essential resources termed vital minerals. Lithium, cobalt, REEs, and graphite are vital minerals for technologies such as EVs, wind turbines, solar panels, and batteries, which are crucial for decarbonizing energy production and consumption. Nevertheless, the concentrated production of these minerals in a limited number of countries, along with the geopolitical risks linked to such dependencies, has resulted in a substantial supply deficit, intensifying the swift increase in demand for them. The demand for these minerals is anticipated to increase significantly in the coming decades, propelled by the worldwide initiative for carbon neutrality and the rapid expansion of renewable energy technology [50]. Research has emphasized the importance of comprehending these supply dynamics and the ensuing solutions to guarantee supply chain resilience. A study by Laliwala and Torres [8] on the recycling of REEs from decommissioned electric and hybrid car motors indicates a potential method for alleviating supply limitations and enhancing sustainability in the clean energy sector. Countries such as China and Australia dominate the extraction and production of essential minerals, rendering global supply chains susceptible to disruptions caused by political instability or trade restrictions [51]. The supply deficit is exacerbated by environmental issues, as mining operations frequently result in considerable ecological damage [52], necessitating the urgent exploration of alternatives such as recycling and urban mining [53].

A crucial element in tackling the urgent mineral supply deficit is comprehending the geography of mineral reserves and the dangers linked to their extraction. The Democratic Republic of the Congo (DRC), a country characterized by political instability and human rights issues, predominantly governs the cobalt supply, significantly affecting global supply chains. Likewise, REEs, essential for manufacturing magnets in EVs and wind turbines, are predominantly extracted in China, which has utilized its supremacy in this industry to apply political and economic influence. Strategic diversification of supply sources, including the development of new mining regions in Africa and South America, is crucial for risk mitigation. Researchers such as Priore et al. [17] have performed quantitative and qualitative evaluations of patents related to rare earth recycling, proposing novel methods to diminish reliance on mined resources through improved material recovery and reuse. Advancements in recycling technologies, such as bioleaching and hydrometallurgical techniques, present considerable potential in mitigating environmental and supply chain issues. In Australia, large-scale initiatives are underway to extract materials from wind turbine blades, offering an alternative source of essential minerals and promoting the circular economy.

The swift increase in the demand for essential minerals has raised considerable apprehensions regarding their environmental consequences, particularly in areas where mining methods are unsustainable. The extraction of these minerals is resource-intensive and frequently results in environmental damage, such as deforestation, water contamination, and habitat destruction. There is an increased focus on enhancing the sustainability of mining techniques and optimizing mineral extraction efficiency. According to Ghosh et al. [33], the extraction of scandium, a REE vital for high-performance batteries, poses significant environmental concerns. The environmental consequences of key mineral extraction have prompted corporate and governmental initiatives to enhance mining practices. The EU has initiated measures to diminish the environmental impact of rare earth mining by implementing stronger restrictions and promoting cleaner extraction technology. Simultaneously, the United States and Japan have formalized arrangements to ensure a reliable supply of essential minerals, exemplifying international cooperation to mitigate the supply deficit. The geopolitical and environmental factors highlight the necessity for a comprehensive, multi-stakeholder strategy in managing the important mineral supply chain.

The recycling of important minerals is becoming an essential strategy to alleviate strain on raw material resources and minimize environmental harm. The worldwide shift to a circular economy has catalyzed novel methods for recycling REEs from discarded products. Technologies like direct recycling, bioleaching, and chemical extraction are being investigated to reclaim materials from items such as EV batteries, wind turbine magnets, and consumer electronics. Maani et al. [18] have demonstrated the efficacy of disassembly techniques in extracting rare earth permanent magnets from decommissioned EV motors. The potential to enhance resource efficiency via recycling is substantial; nevertheless, economic scaling of these technologies presents problems. The advancement of effective sorting technologies and automated disassembly methods is essential for optimizing the recovery of REEs from waste streams. Organizations like the US Department of Energy's Critical Materials Institute are significantly spending on research to improve the efficiency of rare earth recycling and to create sustainable extraction technologies. Cooperative initiatives among academic institutions, industry leaders, and government agencies are crucial for surmounting technical obstacles to extensive recycling and improving material recovery rates.

The significance of policy and regulation is paramount in influencing the future of vital mineral supply networks. Policymakers play a vital role in fostering the development of a sustainable and resilient essential mineral supply chain through the implementation of comprehensive plans to mitigate resource shortages and guarantee supply security. The Indian government's policy reforms designed to provide a sustainable supply of essential minerals via domestic resource development and recycling promotion are significant instances of such interventions [39]. India is implementing incentives for governmental and private sector investments in the extraction, processing, and recycling of vital minerals to diminish its reliance on imports of essential minerals such as lithium and cobalt. The EU has actively influenced the future of its key mineral supply chain through initiatives like the European Raw Materials Alliance (ERMA), which seeks to promote a more varied and sustainable supply of critical minerals. Strategic policy interventions are crucial for rectifying the disparity between supply and demand while mitigating the social and environmental consequences of vital resource exploitation.

A primary problem in addressing the significant mineral supply gap is the absence of comprehensive data regarding the global distribution of these resources and their economic movement. According to academics such as Liu et al. [37], delineating the pathways of important minerals throughout national economies is crucial for comprehending their utilization, recycling, and disposal. This information is essential for formulating policies that promote resource efficiency and facilitate the shift to a circular economy. Furthermore, monitoring the inventories and movements of resources such as REEs and cobalt will enable policymakers to more accurately predict future supply and demand, thereby helping them to foresee any shortages. Enhancing investments in worldwide monitoring systems and advancing real-time data collection techniques will be crucial for attaining these objectives. The incorporation of sophisticated analytics, including ML and big data analysis, into material flow analysis is anticipated to significantly enhance resource management.

Ultimately, international cooperation and alliances are essential for resolving the severe mineral supply deficit. Countries and industry must collaborate to ensure the efficient extraction and utilization of essential minerals while minimizing environmental impact. The Global Energy Interconnection Development and Cooperation Organization (GEIDCO) has promoted international collaboration in advancing clean energy technology and the sustainable management of essential natural resources. This collaboration is essential for enabling poor nations, especially in Africa and South America, to obtain the resources required for their clean energy transitions while maintaining their environmental and social development objectives. Moreover, industry stakeholders must implement sustainable practices that prioritize transparency, ethical sourcing, and circular economy principles to guarantee a steady and secure supply of essential minerals for the global clean energy transition.

Table 6.2 provides a prospective examination of the essential mineral supply chain framework in relation to the clean energy transition. The table consolidates findings from various advanced research examining the convergence of mineral supply, recycling technology, and sustainability issues. It underscores the increasing shortage of critical minerals, including REEs, lithium, cobalt, and nickel, which are vital for the advancement of renewable energy technologies, such as EVs, wind turbines, and battery storage systems. The data indicate novel strategies for tackling these challenges, such as sophisticated recycling methods, bioleaching, and electrochemical recovery techniques for mineral extraction from obsolete products. The chart also incorporates geographical and institutional initiatives designed to enhance resource efficiency, including legislative reforms in India and strategic assessments in China, to guarantee supply security amid rising global demand. These initiatives aim to optimize the sustainability of important mineral flows, improve supply chain resilience, and reduce the environmental impacts linked to mining activities.

TABLE 6.2
Optimizing Critical Mineral Supply Chains for the Clean Energy Transition: Innovations in Recycling, Resource Efficiency, and Policy Integration

Area of Focus	Category	Innovation/Building Block	Citations	Example/ Location	Interventions of Industry/ Institutions	Key Insights	Nationality of Study
Supply chain optimization	Critical mineral supply chain management	Use of AI and machine learning for optimized supply chain management	Riddle et al. [19]	Global focus on rare earth metals (REEs)	Lithium supply chain optimization [54]	Optimization ensures minimal disruptions and reduces volatility	USA, Global
Recycling technologies	Rare earth elements (REE) recycling	Enhanced bioleaching techniques for REE recovery	Joshi et al. [10]	Industrial waste recycling in India	Bioleaching for rare earth recovery [10]	Sustainable mining alternatives and reducing environmental impact	India
Demand growth and scarcity	Demand growth for clean energy	Critical mineral demand modeling for energy transition	Shannak et al. [32]	Global, especially in EV and battery industries	Strategic planning for battery metal demand [55]	Addresses demand increase for clean energy technologies	UK, USA, EU
Circular economy integration	Urban mining and circular economy	Urban mining for REEs from discarded electronics	Dushyantha et al. [56]	China, EU	Circular economy pathways for rare earths [24]	Maximizing mineral recovery from waste products	China, Europe
Sustainable mining practices	Mining and environmental impact	Use of eco-friendly extraction methods and energy-efficient processes	Golroudbary et al. [40]	Global mining operations	Incorporating eco-friendly methods [57]	Reduces environmental degradation while maintaining mineral supply	Australia, Global

(continued)

TABLE 6.2 (Continued)
Optimizing Critical Mineral Supply Chains for the Clean Energy Transition: Innovations in Recycling, Resource Efficiency, and Policy Integration

Area of Focus	Category	Innovation/Building Block	Citations	Example/ Location	Interventions of Industry/ Institutions	Key Insights	Nationality of Study
Policy and regulation	Policy development for critical minerals	Policy reforms for securing critical mineral supply chains	Jain [39]	India, Global	Policy development for sustainable mineral supply [39]	Addressing policy gaps for future mineral security	India
Technological innovations	Electrochemical recovery technologies	Electrolysis in molten salts for metal recovery	Holcombe et al. [11]	Global focus, especially in high-tech recycling	Electrochemical recovery technologies [58].	Innovations in metallurgy for sustainable recovery of rare earths	Global, USA
Resource efficiency	Efficiency of rare earths use	Improving resource efficiency of REEs through technology	Ge et al. [47]	China	Resource efficiency improvement [47]	Optimizes resource usage, ensuring long-term supply stability	China

Source: Table by the authors.

Furthermore, Table 6.2 highlights the complex role of industry and institutional partnerships in promoting sustainable mineral recovery methods. The research underscores the significance of collaboration among governments, commercial sectors, and academic institutions in promoting creative solutions for mineral recovery and resource efficiency. The table outlines several case studies, including the Australian government's role in predicting wind turbine material flows and the EU's efforts to evaluate the social and environmental consequences of critical mineral supply chains. The incorporation of circular economy ideas, including the reutilization of rare earth magnets and lithium-ion batteries, is examined as a crucial approach to diminishing dependence on virgin material resources. This comprehensive approach emphasizes the necessity of creating closed-loop systems that improve material recovery while also tackling the wider issues of resource shortage and environmental deterioration. The insights from Table 6.2 offer a thorough framework for enhancing crucial mineral supply chains in the clean energy transition, aligning with global sustainability objectives while alleviating the risks linked to the increasing demand for these vital resources.

6.4 ROLE OF RECYCLING IN ALLEVIATING CRITICAL MINERAL SCARCITY

As the world progresses toward a low-carbon future, essential minerals, such as REEs, lithium, cobalt, nickel, and others, have become pivotal to the clean energy transition. These minerals are vital for the manufacture of renewable energy technologies, EVs, and energy storage systems, which are crucial for mitigating global greenhouse gas emissions. The rising demand for these minerals, along with supply chain vulnerabilities and environmental issues, has created an urgent need for novel solutions to address the future scarcity of these essential commodities. The recycling of key minerals has garnered considerable interest as a viable alternative. By emphasizing efficient recovery, reuse, and recycling of components from end-of-life products, recycling can mitigate the constraints linked to primary resource extraction, diminish environmental impacts, and foster the circular economy.

Recycling is becoming a crucial technique for mitigating the shortage of essential minerals, especially REEs, which are vital for technologies such as permanent magnets in wind turbines and electric motors. Recent studies have underscored the significance of urban mining, which entails the extraction of valuable minerals from waste materials, including electronic gadgets, batteries, and car parts [17]. This method markedly diminishes the necessity for primary mining, which frequently leads to environmental deterioration and substantial energy expenditure. Maani et al. [18] examine the extraction of REEs from end-of-life electric car motors as a viable recycling strategy with potential uses in the automotive and renewable energy industries. Enhancing recycling processes and advancing technologies, such as bioleaching [10], can augment the worldwide supply of key minerals, thereby addressing the requirements of the clean energy transition.

Innovative solutions are essential for improving the efficiency of important mineral recycling. Improvements in hydrometallurgical and electrochemical methods have markedly enhanced the recovery rates of essential minerals. Electrolysis, especially in molten chloride salts, is being investigated for the sustainable production of essential metals, as evidenced by Holcombe et al. [11]. This technique has the capacity to extract REEs from waste sources while minimizing environmental repercussions. Advancements in selective extraction techniques, including room-temperature electrolysis for extracting rare earth metals from NdFeB magnets, have demonstrated encouraging findings for efficiency and sustainability [58]. Moreover, the incorporation of ML and AI in recycling operations can enhance the sorting and separation of materials, hence improving the accuracy of mineral recovery and reducing waste. These technical advancements enhance the viability of recycling and facilitate the expansion of operations to satisfy global demands.

The effective execution of recycling initiatives relies on robust policy frameworks and active engagement from the sector. In India, legislative reforms have been implemented to enhance the availability of essential minerals required for renewable energy technology [39]. These policies seek to enhance the efficiency of mineral extraction and diminish reliance on imports, especially concerning REEs and lithium, which are essential for battery manufacturing. The EU has initiated measures to enhance recycling rates for essential minerals via extended producer responsibility (EPR) frameworks and recycling quotas [7]. The involvement of international firms and industry leaders is essential in fostering innovation and developing extensive recycling facilities. Companies such as Tesla and General Motors have commenced the integration of recycled materials into their supply chains, showcasing a dedication to minimizing their environmental impact while securing a sustainable supply of essential minerals for their products.

International cooperation and transnational collaboration are vital for tackling the worldwide issues of key mineral recycling. An exemplary instance of international collaboration is the alliance between China and Japan to ensure a consistent supply of REEs. This collaboration emphasizes the recycling of REEs from waste materials, including nickel-metal hydride batteries and permanent magnets [22]. The United States is actively enhancing the efficiency of its rare earth recycling processes, focusing on improving recovery rates from end-of-life electronics and wind turbines [34]. These worldwide initiatives underscore the necessity for global collaboration to secure a consistent supply of essential minerals while mitigating the environmental and social repercussions of resource extraction. The effective execution of recycling initiatives in one region can exemplify a model for others, promoting knowledge exchange and technological transfer.

Although recycling has significant promise in mitigating essential resource scarcity, numerous problems persist. A primary obstacle is the technical challenge of extracting high-purity minerals from intricate waste streams, including electronic waste and car components. The diversity of materials and the existence of contaminants hinder the recycling process, rendering it expensive and energy-consuming [55]. Furthermore, the absence of standardized collection and processing methods for important minerals in numerous places obstructs the development of effective recycling infrastructures. A notable problem is the restricted public awareness and engagement in recycling initiatives. Public education initiatives and incentives are essential to promote the appropriate disposal of electronic trash and other products containing important minerals. Confronting these issues necessitates sustained investment in research and development, alongside the creation of legislative frameworks that facilitate the expansion of the recycling industry.

In the future, recycling will play an increasingly essential role in the clean energy transition due to the escalating worldwide demand for important minerals. To address this demand, it is imperative to advance recycling technology, implement circular economy frameworks, and promote regulatory measures that facilitate the recovery of key minerals from trash. The notion of a "closed-loop" supply chain, wherein minerals are continually recycled within the same industry, presents significant potential for diminishing the demand for virgin resources and lowering waste [47]. The incorporation of recycling into the complete life cycle of clean energy technology, including wind turbines, solar panels, and electric cars, will facilitate a sustainable and secure supply of essential minerals for the future. The global community's investment in recycling ideas and collaborations will be essential in addressing critical mineral scarcity, hence facilitating the success of the clean energy transition.

Table 6.3 presents a comprehensive and advanced examination of the novel roles and interventions in the recycling of important minerals vital for the clean energy transition. This table consolidates recent studies, technological advancements, and industry initiatives centered on REEs, lithium, cobalt, and other important resources, with the objective of mitigating supply constraints and ensuring resource sustainability. The studies presented emphasize various methodologies, including bioleaching, electrochemical recovery, and sophisticated recycling technologies such as selective extraction and disassembly analysis, aimed at recovering critical minerals from end-of-life products like EV batteries, wind turbine components, and electronic waste. The table illustrates how countries such as China, India, Australia, and the United States are enhancing their recycling infrastructure via policy reforms, institutional assistance, and research-oriented activities. The use of circular economy principles and the strategic deployment of reverse logistics enhance the efficiency and efficacy of essential mineral recovery, directly tackling the sustainability concerns in the renewable energy sector.

The table examines the joint initiatives of multinational businesses, industries, and academic institutions in promoting innovation and improving recycling methods. It delineates principal obstacles to enhancing recycling operations, including technological constraints, economic viability, and environmental issues, while proposing viable remedies via innovative material recovery methodologies and supply chain optimization strategies. Moreover, it examines the significance of social and environmental impact assessments, particularly in Europe and emerging economies, where extensive measures are being formulated to facilitate the transition to a low-carbon economy. This comprehensive and progressive approach offers significant insights into the future of critical mineral recycling, highlighting the necessity of sustainable practices, interdisciplinary collaboration, and policy coherence to fulfill the clean energy objectives established in international climate agreements. The results in Table 6.3 contribute to scholarly discussion and provide practical recommendations for improving recycling procedures, thereby ensuring a robust and sustainable mineral supply for future green technology.

TABLE 6.3
Advancing Critical Mineral Recycling: Innovations, Strategies, and Global Interventions for Sustainable Clean Energy Transitions

Study	Title	Year	Innovation Areas	Examples	Location/ Nationality	Industry and Institutional Interventions	Improvement Focus
Laliwala and Torres [8]	Recycling rare earth elements from end-of-life electric and hybrid electric vehicle motors	2024	*Circular economy, rare earth element recovery, electric vehicle recycling*	Recovery of rare earth elements from EV motors	Global (Focus on EV markets)	Industry: Automakers like Tesla, Toyota; Institutions: Research on eco-friendly recycling technologies	Improve efficiency of recycling processes for rare earth elements from EV motors
Ghosh et al. [33]	Process evaluation of scandium production and its environmental impact	2023	*Scandium recovery, environmental impact, process optimization*	Evaluation of scandium production processes	India	Industry: Mining companies like Hindustan Zinc; Institutions: Universities researching scandium production	Develop sustainable and cost-effective scandium extraction methods
Priore et al. [17]	Innovation in rare earths recycling: A quantitative and qualitative analysis of patent data	2025	*Patent innovation, rare earth recycling technologies*	Emerging technologies for rare earth recycling	Global	Industry: Rare earth processing firms; Institutions: Innovators developing new recycling technologies	Encourage patenting and commercialization of new rare earth recycling technologies
Czerwinski [5]	Critical minerals for zero-emission transportation	2022	*Sustainable sourcing, zero-emission transportation, critical minerals*	Sustainable sourcing of critical minerals for electric vehicles	Europe	Industry: Electric vehicle manufacturers like Volkswagen, BMW; Institutions: Policy reforms for sustainable sourcing	Promote ethical sourcing and reduce reliance on primary mining

Jain [39]	Policy reforms for critical mineral supply in India	2024	*Policy reform, national resource management*	India's policy to secure critical mineral supply	India	Industry: Government and local mining firms; Institutions: National policy research bodies	Enhance policy frameworks to secure critical minerals and reduce dependency
Shannak et al. [32]	Strategic analysis of metal dependency in the transition to low-carbon energy	2024	*Low-carbon energy transition, metal dependency, supply chain sustainability*	Scenarios for critical metal scarcity (Nickel, cobalt, lithium)	Global	Industry: Renewable energy companies; Institutions: IEA, research institutes	Develop strategies for reducing dependency on scarce metals in low-carbon technologies
Alavi et al. [59]	Forecasting end-of-life wind turbine material flows in Australia	2024	*Wind turbine recycling, material flow forecasting*	Forecasting material flows from wind turbine recycling	Australia	Industry: Wind energy companies like Siemens Gamesa; Institutions: Research on recycling of wind turbine materials	Improve recycling of wind turbine materials to reduce waste and improve resource recovery
Holcombe et al. [11]	Electrolysis in chloride molten salts for sustainable critical metals production	2024	*Electrolysis, sustainable metal recovery, critical mineral production*	Electrolytic recovery of critical metals	Global	Industry: Electrochemical industries; Institutions: Research on electrolysis technologies	Advance electrochemical methods to sustainably recover critical metals

Source: Table by the authors.

6.5 INNOVATIVE APPROACHES TO MINERAL RECOVERY AND RECYCLING

The worldwide shift toward a clean energy future requires creative and sustainable methods for the collection and recycling of key minerals vital for renewable technology. With the increasing demand for essential commodities such as lithium, cobalt, REEs, and nickel, it is imperative to reevaluate conventional mining procedures and improve recycling techniques. The emergence of innovative technologies and methodologies is essential for addressing these difficulties, mitigating the environmental consequences of mining, and guaranteeing a sustainable supply of these minerals. Numerous global research projects are investigating innovative methods aimed at closing the supply chain loop for these essential elements.

Bioleaching is a promising innovation in mineral recovery, utilizing microorganisms to extract valuable metals from ores and waste materials through a biotechnological process. This eco-friendly method has demonstrated considerable potential in the extraction of REEs from industrial waste, including mining tailings and end-of-life products. Researchers such as Joshi et al. [10] have established the viability of employing bioleaching for the recovery of REEs, suggesting a sustainable method for extracting metals from electronic waste, which is prevalent in urban environments. This technology, functioning in mild conditions, starkly contrasts with traditional methods that frequently require high-temperature processes, hence providing a low-carbon option. Moreover, bioleaching is consistent with the concepts of the circular economy, highlighting material reuse and minimizing the necessity for primary resource extraction.

A notable advancement is the creation of electrochemical methods for the extraction of essential metals. Holcombe et al. [11] investigated electrolysis in molten chloride salts, revealing its potential for the sustainable and energy-efficient production and recovery of key metals. This method facilitates the targeted extraction of metals from intricate waste streams, such as those present in electronic trash or industrial byproducts, which have conventionally posed processing difficulties. Molten salt electrolysis enables the recovery of metals, including lithium, cobalt, and REEs, with enhanced purity and reduced environmental impact relative to traditional approaches. The incorporation of this modern technology in mineral recovery corresponds with the global initiative to decarbonize industry and diminish dependence on traditional mining methods.

The significance of urban mining in mineral recovery is a notable subject of interest. Urban mining denotes the recovery of precious metals from obsolete electronic equipment, automobiles, and industrial refuse. With the rising demand for EVs and renewable energy technologies, there is a corresponding increase in the necessity for the metals utilized in these products, especially REEs and battery metals such as lithium and cobalt. Maani et al. [18] investigated the reclamation of NdFeB magnets from decommissioned EV motors, emphasizing the promise of urban mining to satisfy future requirements for these essential metals. This method diminishes the necessity for new mining activities, which frequently entail significant environmental and social repercussions, while simultaneously tackling the escalating issue of electronic trash. Furthermore, urban mining offers an effective method to access an existing reservoir of metals, rendering it an essential element of sustainable supply chain strategies for clean energy technology.

The implementation of effective recycling systems is a crucial element in the advancement of mineral recovery technology. The reclamation of REEs from obsolete items, including permanent magnets in electric motors, has proven to be an effective approach for diminishing reliance on primary extraction. Priore et al. [17] performed an extensive examination of patents concerning REE recycling, highlighting the significance of innovation in enhancing the efficiency and economic viability of these procedures. Companies are currently advancing sorting methods and chemical techniques to enhance the recovery of REEs, thereby reducing the environmental impact of these essential commodities. This recycling method, commonly known as "closed-loop recycling," guarantees the reuse of metals in new products, thus minimizing waste and preserving resources.

Governments and international organizations are significantly contributing to the advancement of new mineral recovery and recycling strategies through policy and institutional actions. According to Jain [59], India's policy reforms highlight the significance of regulatory frameworks in securing a sustainable supply of significant minerals and influencing the future of mineral recovery. The government's emphasis on promoting recycling and mitigating the environmental effects of mining is consistent with global sustainability objectives. Additionally, industry collaborations, such as the one between Japan and India to advance rare earth metal recovery technology, are hastening the research and development of cutting-edge methods for sustainable mineral extraction and recycling. By cultivating public–private partnerships, nations may propel technical progress and establish market circumstances that promote the adoption of these breakthroughs.

Finally, advancements in material science have facilitated the creation of new, more sustainable materials that can substitute for or diminish dependence on essential minerals in renewable energy systems. The pursuit of alternative materials for batteries, magnets, and solar systems remains an active field of investigation. Researchers such as Zhang et al. [60] are exploring substitute materials for REEs in wind turbine magnets, potentially diminishing the necessity for rare earth mining. Researchers seek to mitigate environmental and geopolitical concerns linked to essential mineral supply chains by discovering novel minerals that are more abundant, less toxic, and easier to recycle. This may also mitigate pressure on mineral markets, where supply deficiencies and price fluctuations are prevalent issues. The advancement of alternative materials is a crucial component of the overarching plan for attaining a sustainable and resilient clean energy future.

In summary, novel methodologies for mineral recovery and recycling are essential for tackling the issues presented by the increasing demand for important minerals throughout the clean energy transition. Technologies including bioleaching, electrochemical recovery, and urban mining are viable alternatives for mitigating the environmental consequences of mineral extraction and securing a sustainable supply of critical commodities. By amalgamating these advances with conducive regulations and global partnerships, industries can markedly improve the efficacy of recycling processes, diminish dependence on primary mining, and foster a circular economy. Moreover, continuous investigation into alternative materials for clean energy technologies has the potential to diminish the demand for essential minerals, thus enhancing the resilience and sustainability of global supply chains. As the world progresses toward a low-carbon future, these advances will be essential in ensuring the availability of important commodities while reducing environmental and social costs.

Table 6.4 provides a thorough and novel examination of recent developments in important mineral recovery and recycling, emphasizing the essential function of these minerals in the worldwide clean energy transition. This table presents developing methodologies, interventions, and tactics derived from a variety of scholarly publications focused on tackling the combined issues of mineral scarcity and environmental sustainability. The compilation highlights advanced technologies like bioleaching, electrochemical recovery, and circular economy strategies for essential REEs and battery metals. These innovative strategies highlight a comprehensive framework for mitigating supply chain interruptions, improving material recovery efficiency, and fostering sustainable resource utilization. The contributions encompass the utilization of advanced sensing technologies for accurate sorting, as well as the impact of policy reforms and international collaborations on sustaining long-term supply security, with specific examples from China, India, Australia, and the EU. This comprehensive research offers essential insights into the integration of new technology and policy changes to enhance robust and sustainable mineral recovery systems.

Table 6.4's comprehensive character includes the convergence of technological, economic, and social factors, acknowledging that the shift to a clean energy future involves not only the development of technical solutions but also the establishment of frameworks for responsible resource management. The table presents a range of geographical case studies and industrial actions that illustrate how different nations and institutions are tackling the challenges of mineral shortage and recycling. Instances such as the advancement of efficient recycling methods for NdFeB magnets, the strategic reclamation of REEs from wind turbine components in Australia, and initiatives to enhance the lithium-ion battery recycling supply chain in India and the United States illustrate a worldwide endeavor to reduce waste and optimize material recovery. This research functions as an extensive resource for policymakers, researchers, and industry leaders, providing theoretical insights and practical solutions that can aid in attaining sustainable development goals (SDGs) and facilitating the broader clean energy transition. This research represents a substantial advancement in tackling global difficulties related to key mineral supply and recycling through the integration of circular economy principles and the investigation of collaborative initiatives across various areas and sectors.

6.6 ENVIRONMENTAL AND SOCIAL IMPACTS OF CRITICAL MINERAL EXTRACTION AND RECYCLING

The mining and recycling of essential minerals are crucial for the global shift to renewable energy, but they pose significant environmental and social challenges. As countries endeavor to diminish carbon emissions and achieve ambitious net-zero objectives, the demand for essential minerals such as lithium, cobalt, REEs, and nickel has escalated. These minerals are essential for the production of technology such as EV batteries, wind turbines, and solar panels. The mining of these

TABLE 6.4
Holistic Approaches to Critical Mineral Recovery and Recycling in the Clean Energy Transition: Innovations, Challenges, and Sustainable Solutions

Innovative Area	Category	Innovation/Approach	Example/Study	Location	Nationality	Industry/Institution Intervention	Citation
Circular economy	Mineral recycling	Promoting urban mining and circular economy models for critical minerals, focusing on REEs recovery from EoL products (Electric vehicle motors, wind turbines).	Disassembly analysis for rare earth magnet recovery from EV motors.	Global	Multinational	Industry collaborations with recycling firms and governments to support reverse logistics.	Maani et al. [18]
Bioleaching	Biological methods	Use of bioleaching techniques to extract rare earth elements (REEs) from industrial waste, reducing environmental impact.	Bioleaching for the recovery of REEs from industrial waste.	Canada	Canadian	Bioleaching technologies developed by research institutions for sustainable mining.	Joshi et al. [10]
Advanced electrolysis	Electrochemical recovery	Application of electrolysis in chloride molten salts for critical metals recovery, such as cerium and lanthanum, improving energy efficiency.	Electrolysis in chloride molten salts for rare metal recovery.	USA and Europe	American, European	Joint ventures between universities and private companies to develop more energy-efficient electrolysis methods.	Holcombe et al. [11]

(*continued*)

TABLE 6.4 (Continued)
Holistic Approaches to Critical Mineral Recovery and Recycling in the Clean Energy Transition: Innovations, Challenges, and Sustainable Solutions

Innovative Area	Category	Innovation/Approach	Example/Study	Location	Nationality	Industry/Institution Intervention	Citation
Phytomining	Innovative mineral recovery	Harnessing plant roots for the sustainable extraction of rare earth elements through phytomining, reducing land disruption.	Advancing phytomining for REE extraction using plants.	India	Indian	Research collaborations between agriculture and mining sectors for sustainable extraction techniques.	Rabbani et al. [23]
Resource efficiency	Life cycle analysis	Life cycle inventories for critical minerals, using network analysis to optimize resource use and identify improvement areas in the recycling process.	Network analysis of rare earth production life cycle.	China	Chinese	Public–private partnerships in China to enhance resource efficiency for REEs.	
Automation and robotics	Technological innovation	Integration of robotics and automation for the dismantling and separation of materials, particularly magnets and batteries, improving the efficiency of mineral recovery.	Robotics in dismantling of end-of-life electric vehicle parts for magnet recovery.	USA	American	Technological firms working on robotic systems for automated recovery processes in recycling plants.	Maani et al. [18]

Environmental impact assessment	Sustainability assessment	Assessing the environmental impact of critical mineral supply chains for clean energy technologies, aiming for low-carbon emissions during recovery.	Environmental impact assessment of REE supply chains in Europe.	Europe	European	European Commission's environmental initiatives supporting the development of greener mineral extraction technologies.	Berthet et al. [7]
Policy reforms	Regulatory framework	Implementation of policy reforms in India to improve the supply chain and sustainability of critical minerals, particularly for the clean energy sector.	Policy reforms for critical minerals supply in India.	India	Indian	Government intervention to promote sustainable mining practices and reduce export dependence.	Jain [39]

Source: Table by the authors.

minerals frequently results in considerable ecological damage, encompassing habitat destruction, soil erosion, and water source contamination due to harmful compounds. Moreover, the social ramifications of mining, especially in low-income nations, can intensify problems like labor exploitation, community displacement, and human rights violations. Studies conducted by Laliwala and Torres [8] and Ghosh et al. [33] underscore these issues while providing insights into the potential of recycling technology to mitigate these challenges.

The ecological impact of crucial mineral extraction is significant, especially in regions abundant in these resources, including Africa, South America, and Asia. Cobalt is mostly extracted from the Democratic Republic of Congo (DRC), where mining activities are linked to significant environmental deterioration, such as deforestation and water pollution [20]. The utilization of hazardous solvents and acids during extraction may result in the leaching of heavy metals into adjacent ecosystems, hence endangering biodiversity and posing health threats to local communities. In areas such as Chile and Argentina, where lithium is extracted for battery manufacturing, apprehensions over water shortages have emerged due to the substantial water consumption involved in the extraction process, adversely affecting local agriculture and potable water resources. The incorporation of cleaner extraction techniques, such as bioleaching for REEs [10], presents a viable alternative; nonetheless, large-scale implementation poses significant challenges.

The social aspects of vital mineral extraction are as urgent. In numerous nations, mining activities are frequently associated with exploitative labor practices, including child labor and substandard working conditions. Cobalt mining in the DRC is infamous for utilizing child labor under perilous conditions [18]. Moreover, mining ventures may displace indigenous tribes, altering their lifestyles and inciting conflicts regarding land ownership and resource allocation. These issues often provide a barrier to mining firms' social license to operate, especially in politically unstable areas. Global institutions and industries must implement more transparent and ethical supply chain processes, safeguarding the rights of local communities and ensuring that workers receive equitable wages and safe working conditions. Jain [39] examines policy reforms in India that have established frameworks to enhance social outcomes of key mineral extraction through the implementation of stricter labor standards and community involvement techniques.

Recycling provides an effective remedy for the environmental and social consequences linked to the exploitation of key minerals. Extracting precious minerals from end-of-life items, such as EV batteries and wind turbine components, diminishes the necessity for new mining, hence reducing the ecological footprint. Advancements in recycling technologies, including bioleaching methods for REE recovery [10], electrolysis in chloride molten salts [11], and sophisticated mechanical disassembly techniques [18], are markedly enhancing efficiency and diminishing energy consumption. The transition to urban mining, especially in affluent nations such as the United Kingdom and the United States, is gaining traction since it offers a more sustainable alternative to conventional mining practices. Studies by Rizos et al. [34] and Zhao et al. [12] have shown that urban mining can recover significant quantities of REEs, which are crucial for the green energy transition, hence reducing the necessity for virgin material extraction.

The significance of policy and business initiatives in mitigating the environmental and social repercussions of crucial mineral extraction and recycling is paramount. Global governments and companies must cooperate to establish more sustainable mineral supply chains. The Indian government has enacted laws to mitigate the environmental effects of mining while enhancing the domestic supply of essential minerals [39]. The circular economy concept is increasingly being adopted in Europe, particularly with the EU's implementation of the European Green Deal, which encompasses measures for sustainable mining operations and improved recycling initiatives. Companies such as Tesla and BMW are at the forefront of advocating for ethical sourcing and recycling of essential minerals, including the implementation of closed-loop recycling systems for EV batteries. Global coordination is crucial, as significant minerals are frequently extracted from politically unstable areas, and the enforcement of fair trade and labor policies is vital to guarantee equitable advantages for all regions involved.

Incorporating environmental and social impact assessments (ESIAs) into the planning and implementation of critical mineral extraction and recycling initiatives is essential for achieving sustainability. Berthet et al. [7] emphasize that comprehensive ESIAs can identify potential risks and alleviate adverse effects on the environment and local communities. Digital techniques, like geographic information systems (GIS) and life cycle analysis (LCA), have transformed how corporations evaluate and monitor the environmental implications of mining and recycling activities. In areas such as China and Australia, where mining operations are prevalent, the implementation of ESIAs has enabled firms to diminish carbon emissions and enhance waste management procedures. The establishment of a global framework for responsible sourcing, exemplified by the Organisation for Economic Co-operation and Development (OECD) Due Diligence Guidance, offers explicit directives for firms to adhere to in order to prevent their operations from causing environmental harm or social inequalities.

The mining and recycling of essential minerals are vital for the clean energy transition; nonetheless, they present considerable environmental and social concerns that require attention. A comprehensive strategy, incorporating cutting-edge technologies, sustainable mining methods, and strong policy frameworks, is essential to alleviate these effects. Researchers such as Swain [55], Ghorbani et al. [57], and Maani et al. [18] have demonstrated substantial potential for innovation in the recycling of important minerals, which might markedly diminish the necessity for new mining operations. For these technologies to be globally effective, governments, companies, and institutions must collaborate to adopt comprehensive programs that prioritize sustainability and social equality. The future of important mineral supply chains depends on the effective integration of circular economy concepts, ethical sourcing, and sophisticated recycling technologies, which will facilitate a truly sustainable energy transition.

Table 6.5 delineates a novel, multifaceted framework that integrates advanced methodologies concerning the environmental and social ramifications of essential mineral extraction and recycling in relation to the clean energy transition. Table 6.5 provides a comprehensive overview, based on recent global research, that highlights the convergence of sustainable mineral extraction, circular economy concepts,

TABLE 6.5
Integrating Environmental and Social Impact Assessment with Innovations in Critical Mineral Recycling for a Sustainable Clean Energy Transition: A Global Perspective

Category	Innovations and Strategies	Examples and Locations	Authors and Year	Social and Environmental Impacts	Interventions and Solutions	Industry and Institutional Involvement	Citations
Resource efficiency	Advanced bioleaching methods for rare earth element recovery	Global (China, Australia, Europe)	Joshi et al. [10]	Environmental impact of mining and resource depletion; reduction of toxic emissions in extraction	Adoption of circular economy strategies, use of sustainable mining technologies	Industry (Rare earth recovery companies) and environmental agencies	Joshi et al. [10]
Recycling innovations	Selective electrochemical recovery of rare earth elements from complex waste feedstocks	South Korea (Electrochemical research for recycling)	Vapnik et al. [58]	Reduction in waste generation, need for advanced recycling technologies to reduce demand for virgin materials	Electrochemical recycling processes for reducing the need for new mining	Companies (Electrochemical tech developers, e.g., Umicore)	Vapnik et al. [58]
Social impacts of extraction	Community-driven initiatives for rare earth mining sustainability	India, Africa, and South America	Jain [39]	Negative social impacts including displacement, child labor, and exploitation in extraction areas	Integration of social equity principles in mining practices, fair trade policies	Institutions (UN, NGOs, and mining corporations)	Jain [39]
Environmental risks of mining	Use of clean energy in critical mineral extraction, like solar-powered mining machinery	Chile (Copper mines using solar energy)	Rachidi et al. [48]	Soil and water contamination from mining; habitat destruction	Integration of clean energy into mining operations; green mining technologies	Mining firms (Solar-powered machinery providers)	Rachidi et al. [48]

Recycling in the energy transition	Urban mining for rare earth magnets and materials from used electronics and batteries	Europe (UK, Germany)	Swain [55]	Lower energy demand, reduced extraction-related environmental damage	Use of advanced sorting and recycling technologies for wind turbines and electric vehicle batteries	Companies (Battery and EV manufacturers) and recycling firms	Swain [55]
Circular economy models	Establishment of end-of-life wind turbine material flow forecasts for recycling	Australia (Wind turbine decommissioning)	Alavi et al. [59]	Reduces pressure on virgin material supply, improves resource efficiency	Development of closed-loop systems for renewable energy equipment recycling	Wind energy companies and research institutions	Alavi et al. [59]
Policy and regulatory frameworks	International policy alignment for critical minerals in low-carbon technology supply chains	India, Europe, USA	Sovacool et al. [20]	Lack of coherent policies for critical mineral management leading to supply chain disruptions	Policy reforms and international cooperation on critical minerals' ethical sourcing and recycling	Governments (Indian Government, EU, UN)	Sovacool et al. [20]
Sustainability in recycling	Closed-loop recycling systems for cobalt, lithium, and nickel	Global (USA, China, EU)	Rizos et al. [34]	Increased waste generation from electric vehicle battery disposal	Eco-efficient and cost-effective recycling methods, R&D into battery lifespan extension	Institutions (Clean energy R&D centers, Tesla)	Rizos et al. [34]

Source: Table by the authors.

and the energy transition. By showcasing data from many sources, it underscores the urgent necessity for advancements in vital mineral recycling technologies, including bioleaching, electrochemical recovery, and selective extraction, as essential facilitators for minimizing the environmental impact of clean energy technologies. Furthermore, it encompasses critical research on the social aspects, such as the effects of REE extraction on communities, labor markets, and geopolitical stability, providing insights into the interplay between these factors, industry practices, policy interventions, and international cooperation.

This table underscores the significance of multidisciplinary approaches and collaborative strategies in tackling the challenges of critical mineral scarcity and environmental degradation. Numerous case studies, encompassing Australia, China, and the EU, offer context-specific insights into the intricacies of material flows and recycling processes, highlighting potential enhancements and chances for sustainable development. Industry and institutional interventions, including advancements in recycling methods, policy reforms, and the advocacy of cleaner production strategies, are examined alongside their results, illustrating how innovative technologies and strategic decision-making can enhance resource efficiency. This thorough analysis seeks to deliver practical insights that can propel global initiatives toward establishing a sustainable and resilient supply chain for critical minerals, thereby aiding the transition to a low-carbon, circular economy while mitigating adverse environmental and social effects.

6.7 POLICY FRAMEWORKS AND GLOBAL COOPERATION FOR CRITICAL MINERAL SUSTAINABILITY

Critical minerals essential for the clean energy transition, including lithium, cobalt, nickel, and REEs, are vital for the production of renewable energy technology, energy storage systems, and EVs. As worldwide demand for these minerals escalates, so do apprehensions regarding their sustainability, availability, and supply chain vulnerabilities. Robust regulatory frameworks and international collaboration are essential for securing a sustainable supply of these minerals, diminishing geopolitical reliance, and alleviating environmental and social repercussions. Policy interventions must tackle the complex difficulties associated with the sustainable management of vital natural resources, encompassing mining, processing, recycling, and circular economy strategies. The United Nations and several national governments have commenced the implementation of rules to guarantee that the essential minerals necessary for renewable energy technologies are sourced ethically, including environmental implications, social equality, and economic feasibility.

A fundamental element of policy frameworks for essential mineral sustainability is the establishment of international agreements and cooperative mechanisms. The EU has formed the ERMA to coordinate initiatives aimed at safeguarding access to essential minerals and enhancing the resilience of supply chains. This alliance emphasizes sustainable sourcing of resources while prioritizing recycling, reuse, and the creation of a circular economy for REEs. Similarly, nations such as Australia, a significant producer of essential minerals, have synchronized their mining policy with the worldwide demand for sustainable technology by advocating for environmentally responsible mining methods [6]. Australia has established trade agreements with China, Japan, and other significant economies to ensure a consistent supply of vital minerals while advancing technological advancements in the recycling of materials, including rare earths from obsolete devices and batteries [55].

National policies are essential in structuring the domestic mining industry and promoting the development of a circular economy, thereby complementing international efforts. India has implemented substantial legislative reforms to improve the accessibility of essential minerals, especially rare earths and lithium, in reaction to the increasing demand for EVs and renewable energy technology. Jain [39] underscores India's focus on enhancing its domestic mining capabilities, specifically through the exploration of new reserves and augmented investment in sustainable extraction technology. The government has implemented financial incentives for the recycling and recovery of essential minerals from electronic waste, a measure that might considerably reduce strain on primary resources. China, a prominent producer of important minerals, has implemented strategies to enhance the efficiency of its mining operations and to foster advances in REE extraction, such as bioleaching and advanced electrochemical processes [57].

International collaboration is essential for ensuring a transparent and secure supply chain for vital minerals. Shannak et al. [32] assert that the sustainability of important mineral supply chains necessitates improved collaboration across producing, consuming, and recycling countries. The International Energy Agency (IEA) has underscored the necessity for a collaborative strategy to mitigate the geopolitical risks linked to the concentrated supply of essential minerals, especially from areas such as Africa and South America. These resource-abundant regions, frequently experiencing political instability, pose a distinct challenge for global supply systems. The IEA promotes the diversity of supply sources, the creation of strategic stocks, and the encouragement of recycling to mitigate reliance on any singular source. African nations such as the Democratic Republic of Congo (DRC) and Zimbabwe are collaborating with international organizations to adopt responsible mining practices that conform to global sustainability objectives, ensuring local communities gain from mineral extraction while reducing environmental harm [34].

Alongside national and international regulatory frameworks, advancements in mining and recycling technology are crucial for enhancing the sustainability of important minerals. Novel techniques for extracting essential minerals from waste streams, including electronic trash, are being developed worldwide. Bioleaching has developed as a viable technique for the sustainable extraction of REEs from discarded magnets and electrical gadgets. Joshi et al. [10] investigate the utilization of microorganisms for the selective extraction of REEs from industrial waste, presenting a more sustainable and eco-friendly alternative to conventional mining methods. Likewise, electrochemical methods, like chloride molten salt electrolysis, are being advanced to extract metals from secondary resources while minimizing energy usage and carbon emissions [11]. These advancements in recycling technology not only diminish the environmental impact of mining but also aid in ensuring a reliable supply of essential minerals from obsolete products.

Incorporating circular economy principles into essential mineral supply chains is a vital technique for achieving sustainability. The notion of "urban mining," which involves the extraction of essential minerals from urban garbage, presents a viable remedy for the supply–demand disparity. Czerwinski [5] calls for the implementation of comprehensive recycling programs aimed at recovering essential minerals from discarded EV batteries, wind turbines, and other renewable energy systems. In nations such as Japan, which has led in electronic waste recycling, urban mining has emerged as an essential component of the national plan for acquiring important minerals. The advancement of sophisticated sorting and recycling technology, along with regulatory measures that promote material reuse, is essential to preventing the disposal of these valuable resources in landfills. The incorporation of sustainable materials in product design, including permanent magnets and batteries, can markedly diminish the need for newly extracted key minerals and foster a more circular utilization strategy [60].

To ensure long-term sustainability in key mineral supply chains, it is essential to incorporate environmental, social, and governance (ESG) criteria into the rules regulating mining, recycling, and the utilization of important minerals. When implemented throughout the lifecycle of important minerals, ESG standards guarantee that mining operations uphold human rights, protect biodiversity, and foster local economic growth. Berthet et al. [7] observe that the EU has pioneered the integration of ESG factors into its mineral supply regulations, establishing a benchmark for other countries to emulate. Organizations like the World Bank and the International Finance Corporation (IFC) collaborate with governments and companies to advocate for responsible practices in the mining and recycling sectors. These initiatives not only mitigate the environmental consequences of mineral extraction but also guarantee that the advantages of the clean energy transition are distributed fairly, especially in developing nations where resource extraction frequently results in considerable social and environmental issues [61].

The advancement of essential mineral sustainability in the clean energy transition necessitates a comprehensive strategy that integrates legislative frameworks, international collaboration, technological innovation, and circular economy concepts. The escalating worldwide demand for sustainable energy technology and EVs underscores the indispensable role of key minerals in attaining carbon neutrality. Governments, industry, and international organizations must cooperate to guarantee that these minerals are sourced, processed, and recycled in an economically feasible and environmentally friendly manner. Only through coordinated efforts can we secure a sustainable and resilient supply of essential minerals that facilitates the global clean energy transition while preserving the earth for future generations.

Table 6.6 provides a detailed examination of the developing regulatory frameworks and institutional measures that promote the sustainability of essential minerals during the clean energy transition. The table consolidates various studies that emphasize significant advancements in crucial mineral recycling, supply chain management, and the involvement of government and industry players in enabling these transitions. The table summarizes innovative techniques for improving resource efficiency in the energy industry, ranging from REE recycling to breakthroughs in electrochemical recovery and bioleaching technology. The table highlights international initiatives to alleviate resource scarcity, especially in Australia, China, and India, where governmental approaches have emphasized legislative reforms and strategic reserves. These frameworks are supported by global cooperation mechanisms that involve international organizations, such as IRENA and IEA, which aim to align national policies with sustainable practices, climate change mitigation, and carbon neutrality objectives.

Table 6.6 provides insights that embody a comprehensive strategy for addressing the obstacles in important mineral supply chains, with a pronounced focus on circular economy principles. Notable instances, such as bio-based recovery methods and disassembly techniques for end-of-life electric car batteries, highlight the transformational potential of advanced recycling technology. The chart underscores the significance of multi-sector collaborations across government agencies, private enterprises, and research institutes in enhancing these solutions. With the rapid advancement of the energy transition, it is crucial to tackle the environmental, social, and economic consequences of mineral extraction and waste management. The table advocates for the incorporation of resilience in mineral supply chains, emphasizing the reduction of reliance on unsustainable mining processes and the promotion of eco-efficient technology. Table 6.6 offers a framework for policymakers, industry leaders, and researchers to promote sustainable, inclusive, and circular systems for the production and recycling of essential minerals in the clean energy future.

6.8 MEETING THE CRITICAL MINERAL NEEDS OF THE CLEAN ENERGY ECONOMY

The swift shift to a clean energy economy depends on the accessibility and sustainable management of essential minerals. Lithium, cobalt, nickel, REEs, and copper are essential for the production of renewable energy technologies, EVs, and storage systems. As nations strive to achieve net-zero emission objectives, the demand for these minerals is growing, necessitating the development of effective measures to secure their supply. Meeting the essential mineral requirements of the clean energy sector necessitates a novel, multifaceted strategy that includes exploration, extraction, recycling, and global collaboration. This sub-chapter examines the present condition of crucial mineral supply chains, technology advancements, and policy measures that are facilitating the clean energy transition, utilizing worldwide examples, case studies, and scholarly research.

A principal issue in fulfilling the essential mineral requirements of the clean energy economy is the limited availability of resources and the environmental consequences of their extraction. A significant number of minerals vital for clean energy technologies are concentrated in select places, resulting in geopolitical concerns and supply chain weaknesses. China monopolizes the supply of REEs, essential for manufacturing wind turbines, electric car motors, and energy-efficient lighting. The Democratic Republic of Congo (DRC) significantly influences the global supply of cobalt, an essential component for lithium-ion batteries. This concentration of resources raises issues over long-term supply stability and the ethical ramifications of mining operations, especially in nations with fragile governance systems. A burgeoning corpus of studies

TABLE 6.6
Global Policy Frameworks and Institutional Interventions for Advancing Critical Mineral Sustainability in the Clean Energy Transition: Innovations, Challenges, and Pathways for Circular Economy Integration

Area of Focus	Category	Innovative Intervention	Global Cooperation	Key Industry/ Institution Examples	Study Example and Citation	Location/ Study Nationality	Future Recommendations
Policy and regulatory frameworks	National policy development	Development of frameworks promoting the sustainable extraction and recycling of critical minerals	International agreements to align regulations on mineral extraction and recycling	Indian government policies on rare earth minerals and reforms to ensure supply [39]	Jain P.K. [39]. "Policy reforms initiated for supply of critical minerals in India"—Mineral Economics	India	Strengthen global governance frameworks to ensure uniform policies on critical mineral mining, recycling, and supply chain resilience.
Recycling innovations	Circular economy and recycling processes	Innovative recycling methods such as bioleaching for rare earth recovery from industrial waste	Global alliances promoting circular economy practices in critical mineral sectors	Bioleaching techniques by Joshi K. et al. [10] to recover rare earth elements from waste [10]	Joshi K. et al. [10] "Bioleaching for the recovery of rare earth elements from industrial waste"—Resources, Conservation and Recycling	Global (Focus on sustainable practices)	Promote research on bioleaching and other green technologies, scaling up best practices from developing to developed countries.

Sustainable supply chains	Supply chain optimization and resilience	Development of more resilient, diversified supply chains for critical minerals	Multiregional supply chain planning across countries, including the US, China, and Australia	Strategic analysis by Shannak S. et al. [32] analyzing nickel, cobalt, lithium, graphite, and copper dependency [32]	Shannak S. et al. [32]. "Strategic analysis of metal dependency in the transition to low-carbon energy"—Energy Research and Social Science	US, China, Australia	Increase investment in diversified mineral supply chains and collaborative international frameworks to prevent market disruptions.
Technology in critical mineral recovery	Electrochemical and bio-recovery techniques	Electrochemical recovery of cerium and lanthanum from complex waste feedstocks	Collaboration on advanced technologies for critical metal extraction and recovery	Holcombe B. et al. on electrolysis in chloride molten salts for metal recovery [11]	Holcombe B. et al. [11]. "Electrolysis in Chloride Molten Salts for Sustainable Critical Metals Production"—Electrochemical Society Interface	Global (Focus on cutting-edge tech)	Foster public–private partnerships to scale up electrochemical recovery techniques and incentivize their deployment globally.
Material flow analysis (MFA)	Critical mineral flow mapping	Mapping flows and stocks of rare earth elements for circular economy initiatives	Coordination of data collection between countries for accurate resource tracking	Liu S. et al. [37] on mapping gold flows in China [37]	Liu S. et al. [37]. "Uncovering the key features of gold flows and stocks in China"—Resources Policy	China	Implement global material flow databases to enhance understanding of critical mineral availability and recovery rates.

(continued)

TABLE 6.6 (Continued)
Global Policy Frameworks and Institutional Interventions for Advancing Critical Mineral Sustainability in the Clean Energy Transition: Innovations, Challenges, and Pathways for Circular Economy Integration

Area of Focus	Category	Innovative Intervention	Global Cooperation	Key Industry/Institution Examples	Study Example and Citation	Location/Study Nationality	Future Recommendations
Environmental and social impact assessment	Life cycle and impact analysis	Evaluating the environmental and social impacts of critical mineral supply chains	International environmental standards to guide critical mineral supply chain practices	Berthet E. et al. [7] assessing social and environmental impacts in Europe [7]	Berthet E. et al. [7]. "Assessing the social and environmental impacts of critical mineral supply chains"—Global Environmental Change	Europe	Strengthen international environmental regulations and conduct comprehensive impact assessments for all mineral supply chain stages.
Industry collaboration and innovation	Industry and academic partnerships	Collaboration between academia and industry to develop sustainable solutions for mineral recovery	Joint ventures between global mining companies and renewable energy industries	Rachidi N.R. et al. on cobalt supply sustainability and green energy policies [48]	Rachidi N.R. et al. [48]. "Assessing cobalt supply sustainability through production forecasting"—Resources Policy	Global (Focus on energy transition)	Increase cross-sector collaborations to fund joint research projects that bridge the gap between technology development and commercial deployment.
Global market and trade dynamics	Global mineral market analysis	Development of market forecasting models to predict supply–demand imbalances of critical minerals	Multilateral agreements to secure trade routes and market stability	Golev A. et al. on rare earth supply chains	Golev A. et al. [62]. "Rare earths supply chains: Current status, constraints and opportunities"—Resources Policy	Global (Focus on international trade)	Encourage global policy-making bodies to implement transparent trade policies and ensure the stability of critical mineral markets through trade agreements.

Source: Table by the authors.

is investigating sustainable mining methodologies and the advancement of alternative mineral resources in response to these concerns. Ge et al. [47] emphasized the potential to improve the resource efficiency of REEs in China by utilizing cleaner extraction techniques, including bioleaching and hydrometallurgical processes, thereby diminishing the environmental impact of mining activities.

In response to the escalating demand for important minerals, advancements in recycling technology are becoming essential. Recycling provides a sustainable method to diminish reliance on primary mineral extraction, preserve resources, and alleviate environmental consequences. Numerous studies have concentrated on the extraction of essential minerals from end-of-life items, including EV batteries and wind turbine components. Priore et al. [17] performed an extensive investigation of innovation in rare earth recycling, scrutinizing patent data to discern developing technologies and techniques. Research indicates that bioleaching and electrochemical recovery technologies are becoming recognized as effective techniques for extracting precious metals such as neodymium and dysprosium from waste streams. The emergence of urban mining, which is the extraction of minerals from waste materials in urban settings, has been acknowledged as a viable method for sourcing important minerals. This methodology is attracting interest in nations such as Japan and South Korea, where sophisticated waste management and recycling systems are already established. The advancement of high-efficiency recycling technology will be crucial for satisfying future mineral requirements of the clean energy economy.

Alongside technological innovations in recycling, strategic policy interventions are essential for ensuring the availability of important minerals. Global policymakers are progressively acknowledging the significance of a stable and diverse supply chain for essential minerals. India has recently enacted policy measures to enhance the domestic supply of essential minerals through exploration and mining projects [39]. These policies encompass incentives for private sector investment, regulatory simplification, and international partnerships to ensure access to mineral resources in various places. The EU has formed the European Raw Commodities Alliance, aimed at diversifying important mineral sources via international collaborations and enhancing the recycling of essential commodities. With the increasing demand for minerals, such regulatory frameworks are essential for diminishing dependence on a limited number of dominant suppliers and bolstering the robustness of global mineral supply chains.

A crucial element in addressing the essential mineral requirements for the clean energy transition is enhancing the sustainability of mineral extraction and processing. An increasing volume of studies is examining the environmental effects of mining activities and the possibility of mitigating these effects with cleaner technology. The research conducted by Maani et al. [18] on advancing sustainable recovery of rare earth magnets by disassembly analysis highlights the necessity of designing goods for recyclability. Progress in hydrometallurgy is facilitating these initiatives by allowing for more efficient extraction methods with diminished energy usage and lower emissions. The concept of circular economy is increasingly being adopted in the mining sector, focusing on the reuse and recycling of minerals to diminish the reliance on raw materials and mitigate environmental deterioration. Numerous corporations, such as Tesla and General Motors, are diligently investigating circular economy frameworks for the recycling of essential minerals utilized in electric car batteries, thereby establishing a closed-loop system that reduces waste and enhances sustainability.

Industry and institutional collaboration is essential for the effective integration of important minerals into the clean energy economy. The collaboration of government agencies, academic institutions, and the corporate sector is essential for promoting innovation and enhancing sustainable practices. The partnership between research institutes and mining businesses in Canada has resulted in the advancement of novel technology for extracting REEs from mine tailings [55]. This collaboration has led to the creation of a pilot plant that uses sophisticated chemical procedures to extract REEs from waste products, thereby mitigating the environmental effects of mining activities. The European Commission's Horizon 2020 program has financed various initiatives to enhance the recycling and sustainability of critical minerals, notably the CIRCULAR project, which concentrates on creating circular economy solutions for the recovery of critical metals from electronic waste.

Ultimately, the incorporation of key minerals into the clean energy transition necessitates a worldwide strategy that considers the interrelations among mineral supply, energy generation, and environmental sustainability. The proliferation of sustainable energy technologies, including solar, wind, and EVs, is resulting in heightened demand for essential minerals. This increase in demand must be meticulously regulated to prevent worsening environmental degradation or social inequalities. Researchers such as Sovacool et al. [20] and Shannak et al. [32] underscore the necessity of evaluating the complete life cycle of key minerals, from extraction to disposal, to guarantee that clean energy solutions do not unintentionally create new environmental issues. By adopting a comprehensive perspective on important mineral supply chains and implementing creative solutions, it is feasible to guarantee that the clean energy transition is sustainable and equitable, benefiting future generations while alleviating the negative effects of mineral extraction and consumption.

In summary, addressing the essential mineral requirements of the clean energy economy necessitates a comprehensive strategy that integrates technical advancement, regulatory modifications, sustainable extraction methods, and global collaboration. The escalating global demand for sustainable energy technology will heighten the urgency for essential minerals. Investing in innovative recycling technology, enhancing supply chain resilience, and advocating for sustainable mining methods can ensure a reliable supply of vital minerals while reducing environmental and social repercussions. This sub-chapter's research and case stories illustrate advancements in tackling these difficulties and underscore the necessity of ongoing innovation and collaboration for a sustainable clean energy future.

Table 6.7 provides a detailed summary of creative and strategic measures for enhancing the sustainable supply and recycling of important minerals vital for the clean energy transition. The table consolidates innovative research from a variety of international studies, highlighting new strategies to address major mineral shortages and enhance their application across several sectors, including EVs, renewable energy technologies, and battery storage systems. The study encompasses several pivotal sectors, including REE recycling, bioleaching, electrochemical recovery, and the use of circular economy principles to extend the lifespan of essential commodities. Innovative processes, including the extraction of REEs from decommissioned EV motors [8] and electrolysis in molten chloride salts for the sustainable production of critical metals [11], underscore substantial advancements in resource recovery efficiency. Furthermore, the work of researchers such as Ghorbani et al. [57] and Swain [55] emphasizes the necessity of incorporating environmental impact evaluations into the lifecycle management of essential minerals, so as to ensure that the clean energy economy is established on a sustainable foundation.

Table 6.7 illustrates the cooperative endeavors of industry, governments, and research organizations from multiple countries to meet the increasing need for essential minerals. National initiatives, including legislative reforms in India aimed at securing the supply of strategic minerals, and international collaborations to enhance the recycling of REEs and battery metals [48] are identified as crucial facilitators of advancement. The research highlights the significance of technological improvements in enhancing recycling processes, including the use of sophisticated sorting technologies and the creation of novel recycling pathways, which are essential for diminishing reliance on primary mineral sources. This table emphasizes major regions, including China, Australia, and Europe, and outlines a proactive strategy for tackling mineral supply concerns, enhancing resource efficiency, and minimizing waste, fostering a more resilient and circular global economy. This analysis serves as a crucial reference for future legislation and technological advancements aimed at addressing the mineral requirements of the clean energy transition through these breakthroughs.

6.9 CONCLUDING REMARKS

This analysis emphasizes the essential importance of crucial minerals in the worldwide clean energy transition, illustrating the complex interactions among supply chain vulnerabilities, technical advancements, and policy structures. The study indicates that minerals like lithium, cobalt, and REEs are essential for renewable energy technologies; nevertheless, their concentrated production in geopolitically sensitive areas presents considerable hazards. Innovative methods such as bioleaching, electrochemical recovery, and urban mining are developing as revolutionary strategies to alleviate scarcity and diminish environmental effects. The study underscores that addressing these difficulties necessitates a comprehensive plan that incorporates circular economy principles, cutting-edge recycling technology, and global collaboration to guarantee sustainable and resilient supply chains.

The research highlights the pressing necessity for strong legislative measures and institutional partnerships to tackle the environmental and social disparities linked to key mineral exploitation. The EU's Critical Raw Materials Act and India's policy reforms illustrate proactive strategies to diversify supply sources and enhance ethical sourcing. The research underscores the potential of digital tools, including AI-driven supply chain optimization and life cycle assessments, to improve resource efficiency and transparency. These advances, together with industry-driven initiatives such as closed-loop recycling systems for EV batteries, illustrate a strategy for reducing the ecological impact of clean energy technology while addressing increasing worldwide demand.

TABLE 6.7
Advancing the Circular Economy: Innovative Strategies for Sustainable Critical Mineral Recovery and Recycling in the Clean Energy Transition

Category	Innovation	Description	Citations	Examples/Case Studies	Location	Nationality of Study	Industry/ Institution Intervention
Rare earth elements (REE)	Circular economy in REE recycling	Addressing the scarcity of rare earth elements through the circular economy, with a focus on efficient recycling technologies from end-of-life products like electric vehicle (EV) motors.	Laliwala C.andTorres A.[8],Priore R. et al. [17]	EV motor recycling in Japan; Rare earth magnets recycling in the UK.	Japan, UK	Japanese, UK	Toyota's EV recycling initiatives; European Commission's efforts on sustainable REE production.
Material flow Management	Life cycle assessment (LCA) in critical minerals	Utilizing LCA to quantify the impacts of mining, recycling, and the subsequent environmental and social outcomes of critical minerals like lithium, cobalt, and rare earths.	Ge Z. et al. [47], Rachidi N.R. et al. [48]	Lithium extraction in Chile; Cobalt production in the Congo.	Chile, Democratic Republic of Congo	Chilean, Congolese	Chile's lithium brine operations aiming for greener extraction practices.

Policy and regulation	Policy reforms for sustainable mining	Policymakers pushing for reforms in mining regulations, ensuring that the extraction and processing of critical minerals follow strict environmental and sustainability standards.	Jain P.K. [39], Shannak S. et al. [32]	India's policy reforms for critical minerals; European Union's 2020 critical raw materials list.	India, European Union	Indian, European	Indian government's strategic shift toward critical mineral sourcing; EU's supply chain security through its Critical Raw Materials Act.
Recycling technologies	Advanced recovery technologies	Development of electrochemical and bioleaching methods for the recovery of critical minerals from industrial waste and post-consumer products.	Holcombe B. et al. [11], Joshi K. et al. [10]	Electrochemical recovery of REEs in the US; Bioleaching methods for rare earths in Canada.	US, Canada	American, Canadian	Implementation of sustainable metal recovery by companies like American Battery Technology Company.
Supply chain optimization	Global supply chain risk mitigation	Using simulation models and agent-based modeling to predict disruptions and optimize the critical mineral supply chain for energy technologies.	Riddle M.E. et al. [19], Liu W. et al. [37]	Agent-based modeling of rare earth supply disruptions; Forecasting cobalt supply in the global market.	Global	Global	IBM's application of supply chain optimization tools for the rare earth market; US Department of Energy's research on supply chain security.

(continued)

TABLE 6.7 (Continued)
Advancing the Circular Economy: Innovative Strategies for Sustainable Critical Mineral Recovery and Recycling in the Clean Energy Transition

Category	Innovation	Description	Citations	Examples/Case Studies	Location	Nationality of Study	Industry/Institution Intervention
Urban mining	Urban mining for critical mineral recovery	Exploring urban mining opportunities through the recovery of materials from obsolete electronics, end-of-life vehicles, and industrial waste.	Dushyantha N. et al. [56], Ghorbani Y. et al. [57]	Urban mining of REEs from waste electronics in Germany; Lithium-ion battery recycling in the US.	Germany, US	German, American	Urban mining initiatives by Ecometals in Europe; US-based battery recycling projects by Redwood Materials.
Innovation in green energy	Battery metals recycling for clean energy	Emphasizing the importance of battery metals (e.g., lithium, nickel) and innovations in their recycling to support clean energy transitions in electric vehicles and storage systems.	Jones E.C., Jr. [63], Rizos V. et al. [34]	Lithium battery recycling projects in Nevada; Nickel recovery in Sweden.	US, Sweden	American, Swedish	Tesla's lithium recycling efforts; Northvolt's green nickel and cobalt project in Sweden.
Sustainability metrics	Assessing the sustainability of mining operations	Examining sustainability metrics such as carbon footprint, water usage, and ecosystem impact in the mining of critical minerals, and adopting best practices for cleaner production.	Swain B. [64], Wietschel L. et al. [25]	Carbon-neutral mining initiatives in Australia; Sustainable lithium extraction in Australia and Argentina.	Australia, Argentina	Australian, Argentine	Rio Tinto's sustainable mining strategies in Australia; Lithium extraction projects by Livent Corporation in Argentina.

Source: Table by the authors.

This study endorses a multidisciplinary strategy to ensure the sustainable future of essential minerals, harmonizing technical progress with environmental conservation and social equality. The incorporation of circular economy models, bolstered by international collaborations and policy coherence, is crucial for dissociating mineral demand from primary extraction. By promoting innovation in recycling, diversifying supply chains, and implementing rigorous sustainability standards, the clean energy transition can attain its decarbonization objectives without worsening resource scarcity or environmental harm. This research establishes a core framework for policymakers, companies, and academia to collaborate and tackle the complexity of essential mineral sustainability, so as to ensure a resilient and fair renewable energy economy for future generations.

REFERENCES

[1] Okonkwo PC, Nwokolo SC, Alarifi SS, Ekwok SE, Orji R, Udo SO, et al. Bio-inspired computational intelligence metaheuristic-based optimization and sensitivity analysis approach to determine techno-economic feasibility of hydrogen refueling stations for fuel cell vehicles. *Scientific Reports*. 2025;15: 12451.

[2] Okonkwo PC, Nwokolo SC, Barhoumi EM, Mansir IB, Taura UH, Das BK, et al. Technical and economic feasibility assessment for hybrid energy system electricity and hydrogen generation: A case study. *Global Energy Interconnection*. 2025;8:62–81.

[3] Okonkwo PC, Nwokolo SC, Udo SO, Obiwulu AU, Onnoghen UN, Alarifi SS, et al. Solar PV systems under weather extremes: Case studies, classification, vulnerability assessment, and adaptation pathways. *Energy Reports*. 2025;13:929–59.

[4] Hassan MA, Bailek N, Bouchouicha K, Ibrahim A, Jamil B, Kuriqi A, et al. Evaluation of energy extraction of PV systems affected by environmental factors under real outdoor conditions. *Theoretical and Applied Climatology*. 2022;150:715–29.

[5] Czerwinski F. Critical minerals for zero-emission transportation. *Materials*. 2022;15:5539.

[6] Britt A, Czarnota K. A review of critical mineral resources in Australia. *Australian Journal of Earth Sciences*. 2024;71:1016–49.

[7] Berthet E, Lavalley J, Anquetil-Deck C, Ballesteros F, Stadler K, Soytas U, et al. Assessing the social and environmental impacts of critical mineral supply chains for the energy transition in Europe. *Global Environmental Change*. 2024;86:102841.

[8] Laliwala C, Torres A. Recycling rare earth elements from end-of-life electric and hybrid electric vehicle motors. *Computer Aided Chemical Engineering: Elsevier*; 2024. p. 1309–14.

[9] Hassan MA, Bailek N, Bouchouicha K, Nwokolo SC. Ultra-short-term exogenous forecasting of photovoltaic power production using genetically optimized non-linear auto-regressive recurrent neural networks. *Renewable Energy*. 2021;171:191–209.

[10] Joshi K, Magdouli S, Brar SK. Bioleaching for the recovery of rare earth elements from industrial waste: A sustainable approach. *Resources, Conservation and Recycling*. 2025;215:108129.

[11] Holcombe B, Sinclair N, Baker A, Kim E, McCall SK, Akolkar R. Electrolysis in chloride molten salts for sustainable critical metals production and recovery. *The Electrochemical Society Interface*. 2024;33:49.

[12] Zhao S, Wang P, Wang L, Chen W-Q. Quantifying provincial in-use stocks of rare earth to identify urban mining potentials in the Chinese mainland. *Journal of Cleaner Production*. 2024;453:142251.

[13] Hsu W-T, Petavratzi E, Zils M, Einarsson S, Morasae EK, Lysaght O, et al. Mapping the flows and stocks of permanent magnets rare earth elements for powering a circular economy in the UK. *Sustainable Production and Consumption*. 2024;47:37–46.

[14] Chowdhury MOS, Talan D. From waste to wealth: A circular economy approach to the sustainable recovery of rare earth elements and battery metals from mine tailings. *Separations* (2297–8739). 2025;12.

[15] Samuel Chukwujindu N. A comprehensive review of empirical models for estimating global solar radiation in Africa. *Renewable and Sustainable Energy Reviews*. 2017;78:955–95.

[16] Nwokolo SC, Ogbulezie JC. A qualitative review of empirical models for estimating diffuse solar radiation from experimental data in Africa. *Renewable and Sustainable Energy Reviews*. 2018;92:353–93.

[17] Priore R, Compagnoni M, Favot M. Innovation in rare earths recycling: A quantitative and qualitative analysis of patent data. *Resources Policy*. 2025;102:105519.

[18] Maani T, Deng S, Sutherland JW. Disassembly analysis to promote rare earth permanent magnet recovery from end-of-life electric vehicle motors. *International Manufacturing Science and Engineering Conference: American Society of Mechanical Engineers*; 2024. p. V001T04A4.

[19] Riddle ME, Tatara E, Olson C, Smith BJ, Irion AB, Harker B, et al. Agent-based modeling of supply disruptions in the global rare earths market. *Resources, Conservation and Recycling*. 2021;164:105193.

[20] Sovacool BK, Ali SH, Bazilian M, Radley B, Nemery B, Okatz J, et al. Sustainable minerals and metals for a low-carbon future. *Science*. 2020;367:30–3.

[21] Cheilas P, Christou T, Karkalakos S, Kottaridi C, Michaelides PG. Rare earth elements and the US renewable economy: A causality exploration between critical materials and clean energy. *Resources Policy*. 2025:105491.

[22] Mishra N, Mancheri N. *Critical Rare Metal and Collaboration of India and Japan*. India, Japan and Beyond: Human Security, Environment, Development, Innovation and Resilience: Springer; 2024. p. 65–87.

[23] Rabbani M, Rabbani MT, Muthoni F, Sun Y, Vahidi E. Advancing phytomining: Harnessing plant potential for sustainable rare earth element extraction. *Bioresource Technology*. 2024:130751.

[24] Eheliyagoda D, Veluri B, Liu G, Ramanujan D. Unveiling the multiregional circular economy pathways for global dysprosium constraints. *Resources, Conservation and Recycling*. 2025;215:108121.

[25] Wietschel L, Helbig C, Hillenbrand M, Thorenz A. Integrated conceptual framework for resilience and criticality assessments for raw material supply chains. *Resources, Conservation and Recycling*. 2025;218:108249.

[26] Nwokolo SC, Amadi SO, Obiwulu AU, Ogbulezie JC, Eyibio EE. Prediction of global solar radiation potential for sustainable and cleaner energy generation using improved Angstrom-Prescott and Gumbel probabilistic models. *Cleaner Engineering and Technology*. 2022;6.

[27] Nwokolo SC, Eyime EE, Obiwulu AU, Meyer EL, Ahia CC, Ogbulezie JC, et al. A multi-model approach based on CARIMA-SARIMA-GPM for assessing the impacts of climate change on concentrated photovoltaic (CPV) potential. *Physics and Chemistry of the Earth*. 2024;134.

[28] Nwokolo SC, Obiwulu AU, Ogbulezie JC. Machine learning and analytical model hybridization to assess the impact of climate change on solar PV energy production. *Physics and Chemistry of the Earth*. 2023;130.

[29] Nwokolo SC. Climate hoax: The shift from scientific discourse to speculative rhetoric in climate change conversations. *Next Research*. 2025;2.

[30] Nwokolo SC, Meyer EL, Ahia CC. Credible pathways to catching up with climate goals in Nigeria. *Climate*. 2023;11.

[31] Nwokolo SC, Singh R, Khan S, Kumar A. *Technological Pathways for Africa's Net-Zero Economy: Technology Solutions to Unlock Africa's Sustainable Future*: Elsevier: 2024.

[32] Shannak Sd, Cochrane L, Bobarykina D. Strategic analysis of metal dependency in the transition to low-carbon energy: A critical examination of nickel, cobalt, lithium, graphite, and copper scarcity using IEA future scenarios. *Energy Research & Social Science*. 2024;118:103773.

[33] Ghosh A, Dhiman S, Gupta A, Jain R. Process evaluation of scandium production and its environmental impact. *Environments*. 2022;10:8.

[34] Rizos V, Righetti E, Kassab A. Understanding the barriers to recycling critical raw materials for the energy transition: The case of rare earth permanent magnets. *Energy Reports*. 2024;12:1673–82.

[35] Nwokolo SC, Ogbulezie JC, Ushie OJ. A multi-model ensemble-based CMIP6 assessment of future solar radiation and PV potential under various climate warming scenarios. *Optik*. 2023;285.

[36] Nwokolo SC, Singh R, Khan S, Kumar A, Luthra S. Decarbonizing hard-to-abate sectors in Africa. *CSR, Sustainability, Ethics and Governance*. Springer: Cham; 2023. p. 211–36. https://doi.org/10.1007/978-3-031-44514-9_6.

[37] Liu W, Li X, Wang J, Zhong J, Wang M, Yang J. Knowledge mapping of research on securing the supply chain for critical minerals A scientometrics and text mining approach. *Journal of Cleaner Production*. 2024;434:140312.

[38] Nwokolo SC, Singh R, Khan S, Kumar A, Luthra S. Remedies to the challenges of renewable energy deployment in Africa. *CSR, Sustainability, Ethics and Governance;* 2023. p. 59–74.

[39] Jain PK. Policy reforms initiated for supply of critical minerals in India. *Mineral Economics*. 2024;37:701–17.

[40] Golroudbary SR, Calisaya-Azpilcueta D, Kraslawski A. The life cycle of energy consumption and greenhouse gas emissions from critical minerals recycling: Case of lithium-ion batteries. *Procedia CIRP*. 2019;80:315–21.

[41] Nwokolo SC, Singh R, Khan S, Kumar A, Luthra S. Impacts of Climate Change in Africa. *CSR, Sustainability, Ethics and Governance;* 2023. p. 237–62.

[42] Nwokolo SC, Singh R, Khan S, Kumar A, Luthra S. Global investment and development in Africa. *CSR, Sustainability, Ethics and Governance;* 2023. p. 15–58.

[43] Nwokolo SC, Singh R, Khan S, Kumar A, Luthra S. Influencing the scale of Africa's energy transition. *CSR, Sustainability, Ethics and Governance;* 2023. p. 75–91.

[44] Nwokolo SC, Singh R, Khan S, Kumar A, Luthra S. Africa's awakening to climate action. *CSR, Sustainability, Ethics and Governance;* 2023. p. 299–310.

[45] Nwokolo SC, Singh R, Khan S, Kumar A, Luthra S. Introduction: Africa's net zero transition. *CSR, Sustainability, Ethics and Governance;* 2023. p. 1–13.

[46] Nwokolo SC, Singh R, Khan S, Kumar A, Luthra S. Scenarios that could give rise to an African net-zero energy transition. *CSR, Sustainability, Ethics and Governance;* 2023. p. 263–98.

[47] Ge Z, Geng Y, Dong F, Liang J, Zhong C. Towards carbon neutrality: Improving resource efficiency of the rare earth elements in China. *Frontiers in Environmental Science*. 2022;10:962724.

[48] Rachidi NR, Nwaila GT, Zhang SE, Bourdeau JE, Ghorbani Y. Assessing cobalt supply sustainability through production forecasting and implications for green energy policies. *Resources Policy*. 2021;74:102423.

[49] Yousefi H, Ardehali A, Ghodusinejad MH. BRICS or G7? Current and future assessment of energy and environment performance using multi-criteria and time series analyzes. *Energy Strategy Reviews*. 2023;49:101164.

[50] Obiwulu AU, Chendo MAC, Erusiafe N, Nwokolo SC. Implicit meteorological parameter-based empirical models for estimating back temperature solar modules under varying tilt-angles in Lagos, Nigeria. *Renewable Energy*. 2020;145:442–57.

[51] Obiwulu AU, Erusiafe N, Olopade MA, Nwokolo SC. Modeling and optimization of back temperature models of mono-crystalline silicon modules with special focus on the effect of meteorological and geographical parameters on PV performance. *Renewable Energy*. 2020;154:404–31.

[52] Obiwulu AU, Erusiafe N, Olopade MA, Nwokolo SC. Modeling and estimation of the optimal tilt angle, maximum incident solar radiation, and global radiation index of the photovoltaic system. *Heliyon*. 2022;8.

[53] Nwokolo SC, Obiwulu AU, Okonkwo PC, Orji R, Alsenani TR, Mansir IB, et al. Machine learning and physics-based model hybridization to assess the impact of climate change on single- and dual-axis tracking solar-concentrated photovoltaic systems. *Physics and Chemistry of the Earth*. 2025;138.

[54] Atitebi OS, Dumre K, Jones Jr EC. Supporting a lithium circular economy via reverse logistics: Improving the preprocessing stage of the lithium-ion battery recycling supply chain. *Energies*. 2025;18:651.

[55] Swain B. Challenges and opportunities for sustainable valorization of rare earth metals from anthropogenic waste. *Reviews in Environmental Science and Bio/Technology*. 2023;22:133–73.

[56] Dushyantha N, Kuruppu GN, Nanayakkara CJ, Ratnayake AS. The role of permanent magnets, lighting phosphors, and nickel-metal hydride (NiMH) batteries as a future source of Rare Earth Elements (REEs): Urban mining through circular economy. *Mining, Metallurgy & Exploration*. 2024;41:321–34.

[57] Ghorbani Y, Ilankoon I, Dushyantha N, Nwaila GT. Rare earth permanent magnets for the green energy transition: Bottlenecks, current developments and cleaner production solutions. *Resources, Conservation and Recycling*. 2025;212:107966.

[58] Vapnik H, Kim H-E, Kim Y, Ooi AWS, Vibbert HB, Park A-HA, et al. Selective electrochemical recovery of cerium over lanthanum from complex waste feedstocks by alternating current electro-precipitation. *Chemical Engineering Journal*. 2025;504:158537.

[59] Alavi Z, Khalilpour K, Florin N. Forecasting end-of-life wind turbine material flows in Australia under various wind energy deployment scenarios. *Energies*. 2024;17:914.

[60] Zhang L, Jiang P, Zhang Y, Van Fan Y, Geng Y. Recycling impacts of renewable energy generation-related rare earth resources: A SWOT-based strategical analysis. *Energy*. 2024;312:133624.

[61] Venkatesan P, Vander Hoogerstraete T, Binnemans K, Sun Z, Sietsma J, Yang Y. Selective extraction of rare-earth elements from NdFeB magnets by a room-temperature electrolysis pretreatment step. *ACS Sustainable Chemistry & Engineering*. 2018;6:9375–82.

[62] Golev A, Scott M, Erskine PD, Ali SH, Ballantyne GR. Rare earths supply chains: Current status, constraints and opportunities. *Resources Policy*. 2014;41:52–9.

[63] Jones Jr EC. Lithium supply chain optimization: A global analysis of critical minerals for batteries. *Energies*. 2024;17:2685.

[64] Swain B. Red mud: An environmental challenge but overlooked treasure for critical rare earth metals. *MRS Bulletin*. 2022;47:289–302.

7 Critical Minerals for Clean Energy

Challenges and Recycling Opportunities

7.1 INTRODUCTION

The global transition to clean energy systems is inextricably linked to the sustainable supply of critical minerals, including lithium, cobalt, nickel, and rare earth elements (REEs), which are indispensable for renewable energy technologies, electric vehicles (EVs), and energy storage solutions. As nations strive to meet decarbonization targets under the Paris Agreement, the demand for these minerals is projected to surge by 400–600% by 2040, exacerbating supply chain vulnerabilities and environmental degradation [1]. This chapter explores the intersection of resource scarcity, technological innovation, and circular economy (CE) strategies that underpin Eco-Mineral Recycling 4.0 and 5.0—a paradigm shift toward intelligent, low-impact mineral recovery. Geopolitical tensions, such as China's dominance in REE refining (90% market share) and ethical concerns in cobalt mining in the Democratic Republic of Congo (DRC), where child labor and ecosystem destruction persist [2, 3], highlight the urgency of this transition. To address these challenges, this chapter synthesizes cutting-edge innovations—from artificial intelligence (AI)-driven urban mining to bioleaching—and policy frameworks that collectively redefine sustainability in mineral supply chains.

Digitalization and automation in material recovery are key components of Eco-Mineral Recycling 4.0, which depends on technological advancements. For instance, AI-enhanced robotic disassembly systems, such as those piloted in Japan's Eco-Town Program, recover 8 t of gold annually from e-waste, while hydrogen processing of magnetic scrap (HPMS) in Australia reduces energy use by 70% compared to traditional methods [4, 5]. Meanwhile, Recycling 5.0 integrates bio-based solutions, such as *Acidithiobacillus ferrooxidans* for cobalt bioleaching, and electrochemical recovery methods like alternating current electro-precipitation, which achieves >95% purity in REE separation [6]. Circulor's collaboration with Volvo for conflict-free cobalt tracking is an example of how digital tools like blockchain-enabled material passports complement these innovations [7]. Such technologies not only enhance recovery rates but also align with the United Nations Environment Programme's (UNEP's) 6R framework (Reduce, Reuse, Recycle, Recover, Redesign, Remanufacture), ensuring minimal waste and maximal resource efficiency. The integration of these advancements

DOI: 10.1201/9781003672807-7

into global supply chains is critical for mitigating the environmental footprint of primary extraction, which accounts for 10–12 t of CO_2/t of neodymium oxide [8].

Policy and regulatory frameworks are equally vital in scaling these innovations. The European Union's (EU's) CE Action Plan and Critical Raw Materials Act (CRMA) mandate lifecycle assessments (LCAs) and 70% lithium recovery from batteries by 2030, while the US Inflation Reduction Act (IRA) incentivizes domestic recycling through tax credits [9, 10]. In emerging economies, India's National Mission on Critical Minerals and Khanij Bidesh India Ltd. (KABIL) aim to reduce import dependence via strategic stockpiling and REE recycling [11]. However, fragmented governance in Africa—where <5% of e-waste is recycled—highlights the need for multilateral cooperation, as seen in the G7's Sustainable Critical Minerals Initiative, which pledges $2 billion for technology transfer [12, 13]. These policies must address systemic barriers, such as subsidized virgin mining and a lack of standardized recycling protocols, to unlock the full potential of CE models.

The CE emerges as a cornerstone of sustainable mineral management, with urban mining projected to supply 50% of cobalt and 30% of REEs by 2050 [14]. Case studies like Redwood Materials (USA) and Li-Cycle (Canada) demonstrate closed-loop lithium-ion battery recycling, while the EU's Battery Regulation enforces extended producer responsibility (EPR) [15]. Wind turbine blade recycling in Australia recovers 90% of glass fibers and REEs [16], and phytomining—using hyperaccumulator plants to extract metals from soil—offers a low-impact alternative [17]. Industrial symbiosis, which results in byproducts like red mud from aluminum production yielding 1.5 kg of REEs per ton, supports these strategies [18]. Yet, logistical challenges, such as reverse supply chain inefficiencies, necessitate AI-optimized logistics and consumer buy-back programs, like Apple's Daisy robot, which disassembles 200 iPhones/hour [19].

Ethical and environmental considerations remain central to this transition. The Fair Cobalt Alliance (FCA) and Initiative for Responsible Mining Assurance (IRMA) address labor exploitation in the DRC, while hydrometallurgical and renewable-powered mining in Chile reduces water and energy use in lithium extraction [20, 21]. The Organization for Economic Co-operation and Development (OECD) Due Diligence Standards and UNEP's Sustainable Resource Management guidelines provide frameworks for equitable sourcing, yet implementation gaps persist in artisanal mining regions [2]. Digital twins for material tracking [22] and modular battery design (e.g., Tesla's structural packs) further enhance recyclability, underscoring the need for design-for-recycling principles across industries.

Institutional collaborations bridge innovation and implementation. The International Energy Agency (IEA) and UNEP advocate for global recycling standards, while academia–industry partnerships—such as Commonwealth Scientific and Industrial Research Organisation's (CSIRO's) bioleaching research in Australia and Badische Anilin und Soda Fabrik's (BASF's) electrochemical recovery projects—accelerate technological deployment [23, 24]. The European Raw Materials Alliance (ERMA) funds research and development (R&D) in sustainable extraction, and India–Japan ventures focus on REE recovery from EVs [25]. These efforts must be scaled to address disparities, such as Africa's limited refining capacity despite holding 30% of global reserves [5].

Future trends emphasize zero-waste mining and industrial symbiosis. Australia's Cooperative Research Centre for Transformations in Mining Economies (CRC TiME) rehabilitates mining sites into renewable hubs, while Hungary and India pilot REE recovery from red mud [18]. Digital product passports and AI-driven resource mapping, Laliwala and Torres [26] will optimize material flows, reducing virgin mining reliance. The Smarter scenario—integrating policy coherence, AI, and circular design—can achieve 40–60% mineral demand through recycling, cutting energy use by 50–90% [2].

This chapter concludes that Eco-Mineral Recycling 4.0 and 5.0 demand a holistic approach: policy rigor, technological disruption, and ethical governance. By aligning innovations like bioleaching and blockchain traceability with frameworks such as the EU Green Deal and US IRA, the clean energy transition can reconcile mineral demand with planetary boundaries. The path forward hinges on global equity—ensuring just transitions for mining communities and resilient supply chains for all nations.

7.2 ROLE OF CRITICAL MINERALS IN THE CLEAN ENERGY REVOLUTION

The shift to sustainable energy is intricately connected to the increasing need for essential minerals, which are crucial for facilitating technologies related to energy generation [27], storage [28], and transmission [29]. With the global shift toward renewable energy sources, important minerals such as lithium, cobalt, nickel, and REEs have become indispensable for the manufacture of batteries, wind turbines [30], solar panels [31], and EV motors [32]. These minerals are essential for the decarbonization of industries, including transportation, electricity generation, and manufacturing. The clean energy revolution depends on both enhancing the availability of these minerals and optimizing their recycling efficiency while ensuring sustainable extraction practices. Countries such as China, the United States, and Australia play crucial roles in the mining and recycling of these resources, each employing unique techniques and interventions to ensure a reliable supply of critical minerals [2]. Furthermore, research conducted by Czerwinski [33] and Ghosh et al. [3] underscores the pressing necessity to address the environmental and social repercussions linked to the extraction and processing of these minerals, which have ignited discussions regarding sustainable practices and CE principles on a global scale [34].

Recent years have witnessed considerable focus on advancements in key mineral recycling. The significance of recycling technologies is growing since they diminish dependence on primary mining and mitigate environmental effects. Laliwala and Torres [26] examined the recycling of REEs from EV motors, demonstrating a novel method for obtaining valuable minerals from end-of-life (EoL) devices. These advancements are crucial for mitigating the increasing shortage of elements such as neodymium (Nd), dysprosium (Dy), and cobalt, which are vital for the manufacture of permanent magnets utilized in wind turbines and EV motors. Additionally, Priore et al. [35] examined patent data related to rare earth recycling, providing both a quantitative and qualitative assessment of advancements in rare earth recovery technology.

These improvements are essential for the sustainable reuse of mineral resources, facilitating the closure of the materials supply chain and minimizing the environmental impact of the clean energy sector. Institutions like the IEA and national policies, including India's initiatives to reform its critical mineral supply chain [11], are essential in influencing global strategies for enhancing recycling infrastructure and augmenting the recovery of critical minerals.

Geopolitical factors complicate the worldwide supply chain of essential minerals. The concentration of REE manufacturing in select nations, especially China, heightens worries regarding the susceptibility of supply chains amid geopolitical conflicts or trade disruptions. Shannak et al. [1] performed a strategic analysis of metal dependency in the shift to low-carbon energy, emphasizing the shortage of essential minerals such as nickel, cobalt, and lithium. Their findings highlight the need for diversification of supply sources and the advancement of alternative technologies to diminish dependence on geopolitically sensitive areas. The growing competition for these minerals among countries aiming to achieve their carbon neutrality goals is exacerbating this geopolitical issue [36]. The EU and the United States are actively pursuing vital resource supply security through strategic collaborations and investments in domestic mining and processing capacities [37]. To alleviate dangers linked to mineral shortage, nations are investigating alternative methods for procuring essential minerals through urban mining and recycling, as shown by Hsu et al. [38] in their study on the extraction of REEs from discarded electronics and industrial waste in the United Kingdom.

Alongside technological advancements and geopolitical factors, policy interventions at national and international levels are crucial for influencing the future of vital mineral supply chains. Recent studies have concentrated on the significance of policies in facilitating sustainable mineral extraction and recycling. Ghorbani et al. [39] examined the sustainability of cobalt supply chains, emphasizing the necessity for strong policy frameworks that encourage responsible sourcing and mitigate the environmental effects of mining. The research of Rizos et al. [40] on the obstacles to recycling rare earth permanent magnets highlights the necessity of policy incentives to promote the advancement of CE solutions. These policies should not only promote recycling but also enable the adoption of innovative extraction methods that decrease energy consumption and mitigate hazardous emissions. The execution of these policies is essential for a sustainable energy transition, as they establish the requisite legal frameworks to facilitate the expansion of recycling activities and the advancement of alternative materials [41].

From an industrial standpoint, enterprises in the renewable energy and electronics industries are integral to the advancement of essential minerals for clean energy applications. Numerous companies are investing in innovative recycling technologies that facilitate the recovery of valuable minerals from EoL products. Holcombe et al. [24] investigated electrolysis in chloride molten salts as a sustainable technique for the production and recovery of key metals, presenting a new technology that could transform the recycling of REEs. Simultaneously, enterprises are investigating novel techniques to enhance mineral extraction efficiency, minimize waste production, and decrease energy usage. Companies in the EV and battery storage industries are increasingly prioritizing enhancements in the energy efficiency of their manufacturing processes while investigating secondary sources of essential minerals from waste streams and obsolete batteries [42]. Cooperative initiatives among industry, academic institutions, and governmental entities are essential for advancing these discoveries and guaranteeing that clean energy technologies remain sustainable and economically viable [43].

The establishment of a CE for key minerals is both a technological and a societal challenge. As nations globally shift toward clean energy, there is an increasing acknowledgment of the necessity to mitigate the social and environmental consequences of mining and recycling. The extraction of essential minerals is associated with considerable environmental degradation, encompassing habitat destruction, water contamination, and greenhouse gas emissions. Berthet et al. [9] indicate that the EU is evaluating the social and environmental effects of crucial mineral supply chains and advocating for responsible sourcing procedures. The emphasis on sustainability is propelling the desire for cleaner production techniques and more transparent supply chains. By implementing CE principles, including recyclability-focused product design and closed-loop material flows [44], the industry may diminish its environmental impact and encourage responsible resource utilization [45]. With the increasing demand for crucial minerals, it is vital to reconcile the necessity of mineral extraction with the obligation to mitigate environmental harm and foster social fairness [46].

The future of critical minerals in the clean energy revolution will rely on sustained innovation, collaboration, and policy coherence. Rachidi et al. [12] assert that sustainable mineral supply chains are essential for the efficacy of green energy policies and climate objectives. Progress in recycling technology and investments in alternative sourcing techniques will be essential for securing the long-term availability of important minerals. Furthermore, global collaboration will be important in addressing the difficulties related to supply chain security and resource sustainability. Incorporating CE principles into the production and use of important minerals is vital for diminishing dependence on primary mining and mitigating the environmental effects of the clean energy transition. The effective management of essential mineral supply chains necessitates a comprehensive strategy that integrates technological, industrial, policy, and societal aspects, guaranteeing that the clean energy transition is sustainable and equitable for everyone.

7.3 GLOBAL DEMAND AND SUPPLY CONSTRAINTS FOR CRITICAL MINERALS

The shift to clean energy technology, particularly in renewable energy generation and storage systems, is significantly reliant on the stable and sustainable provision of essential minerals. Lithium, cobalt, nickel, and REEs are essential for EVs, wind turbines, solar panels, and energy storage systems. The global demand for these minerals is escalating swiftly due to the growing implementation of green technologies, which are crucial for attaining climate change mitigation objectives. This chapter examines the worldwide demand and supply limitations for essential minerals, elucidating the variables affecting their availability and the consequences for clean energy transitions. This section will examine the contemporary issues in the supply chains of these minerals, as well as prospective ways to mitigate these limits, based on recent research from China, the EU, India, and Australia.

A fundamental limitation in the supply of essential minerals is the geographical concentration of production. REEs are predominantly extracted in China, which oversees almost 60% of the worldwide supply of these minerals [39]. The concentration of supply has resulted in geopolitical conflicts and the susceptibility of supply networks. In light of these threats, nations like India are progressively prioritizing local production and the diversification of their mineral sources [11]. The Chinese government's regulation of REEs, combined with its supremacy in EV battery manufacturing, profoundly impacts worldwide markets. Cobalt, an essential element for lithium-ion batteries (LIBs), is predominantly extracted in the DRC, prompting issues regarding ethical sourcing and supply chain transparency. To alleviate these dangers, nations and corporations are investigating alternative sourcing strategies and investing in recycling initiatives to diminish reliance on primary resource extraction [47].

Technological advancements in mining and recycling are essential for addressing the supply limitations of vital minerals. Investigations into bioleaching methodologies for the recovery of REEs from trash, as examined by Joshi et al. [55], have demonstrated potential in mitigating environmental repercussions and enhancing the accessibility of these minerals. Furthermore, innovations in mineral recovery methods from EoL products, especially those derived from electric car motors and wind turbines, can substantially enhance CEs. Research by Maani et al. [48] has shown the viability of extracting rare earth magnets from abandoned EVs, potentially alleviating the demand on mining activities. In Australia, initiatives focused on predicting material flows from wind turbines have underscored the potential of recycling to satisfy future demands [16]. These developments are vital for completing the cycle of crucial material supply and guaranteeing sustainability in the renewable energy sector [49].

Policy interventions and international cooperation are essential for addressing the demand-supply disparity of important minerals. The EU has acknowledged the necessity for strategic alliances with resource-abundant nations and has launched multiple initiatives to diminish its reliance on China for REEs. The European Commission's 2020 study, "Critical Raw Materials Resilience," delineates measures for diversifying supplies, enhancing recycling rates, and innovating alternative materials for renewable energy systems [9]. Likewise, the United States has created programs like the Critical Minerals Strategy to fortify its supply chains and improve recycling capabilities [12]. In India, legislative reforms designed to enhance the domestic supply of minerals, including the creation of the National Mission on Critical Minerals, seek to ensure the sustained availability of vital materials for clean energy technologies [11]. These initiatives underscore the significance of governmental policy in influencing the future availability of essential minerals, fostering resilience, and mitigating the geopolitical risks linked to mining reliance [50].

Recycling constitutes a highly promising method to mitigate supply constraints and improve the sustainability of important mineral utilization. The recycling of LIBs is a critical emphasis area due to the global need for lithium, cobalt, and nickel. Technological advancements in the recovery of metals from spent batteries have progressed markedly, with the development of novel technologies, including hydrometallurgical and pyrometallurgical processes, aimed at enhancing recovery rates and minimizing environmental implications [15]. Furthermore, the expanding market for second-life batteries, particularly in energy storage applications, presents an opportunity to prolong the lifespan of essential minerals utilized in these technologies. The scope of recycling encompasses not only EV batteries but also other sources, like wind turbines and solar panels. Research in Australia indicates that substantial amounts of REEs may be extracted from decommissioned wind turbines, thereby alleviating long-term supply constraints [16].

The social and environmental ramifications of important mineral extraction are essential factors in mitigating supply restrictions. Mining operations for essential minerals, including lithium and cobalt, frequently result in considerable environmental harm, water shortages, and substandard labor conditions, especially in areas such as the DRC and South America. These problems have elicited apprehensions over the ethical ramifications of procuring these resources for renewable energy technology. To resolve these concerns, it is imperative to implement efforts that enhance mining standards, advocate for equitable labor conditions [51], and diminish the environmental impact of mining. The implementation of more stringent controls and oversight on mining operations in the DRC could alleviate certain negative effects linked to cobalt production. Moreover, advancements in sustainable mining technology, like hydrometallurgical methods and the integration of renewable energy in mining operations, might mitigate the environmental effects of mineral extraction [21]. These initiatives are essential to prevent the shift to renewable energy technology from worsening social and environmental inequities [52].

The incorporation of the CE model into the essential minerals supply chain provides an effective strategy to alleviate supply limitations. Implementing strategies like material recovery, remanufacturing, and reuse can diminish the necessity for newly extracted resources while facilitating a more sustainable supply of essential minerals. Swain [18] posits that the sustainable valorization of rare earth metals from anthropogenic waste via recycling methods is a possible avenue to diminish dependence on virgin resources. The establishment of robust recycling systems, especially in nations with substantial electronic waste inventories like the United States, might markedly mitigate supply constraints. Moreover, progress in material science, particularly the creation of alternative materials for batteries and magnets, may diminish reliance on limited key minerals. The integration of recycling, innovation, and CE initiatives offers a comprehensive solution to the issues faced in important mineral supply chains.

Meeting the global demand and supply challenges for essential minerals necessitates a comprehensive strategy that encompasses technical advancement, legislative reform, international collaboration, and the incorporation of CE principles [53]. Emphasizing recycling, enhancing mining methodologies, and diversifying supply sources can secure a consistent and sustainable provision of essential minerals required for the clean energy transition [54]. This chapter illustrates that both industry and institutions are essential in determining the future of the crucial mineral supply. Ongoing research, development, and collaboration will be essential for addressing the difficulties and capitalizing on the potential within the critical minerals sector.

Table 7.1 provides a detailed summary of strategic interventions, technological advances, institutional frameworks, and policy proposals necessary to tackle the increasing global demand and limited supply of crucial minerals vital for the clean energy transition. As the global economy shifts toward low-carbon technologies—such as EVs, battery storage, and renewable energy systems—the necessity for a robust, transparent, and sustainable supply chain of crucial minerals is essential. This table presents a comprehensive study that incorporates upstream (extraction and processing), midstream (refinement and manufacturing), and downstream (reuse and recycling) techniques, highlighting CE principles and international collaboration. It encapsulates the interaction of innovation-driven resource efficiency, digital traceability technologies (such as blockchain for mineral provenance), ethical sourcing frameworks, and scalable recycling infrastructure to mitigate environmental degradation and geopolitical concerns.

Table 7.1 highlights the significance of multilateral organizations and transnational partnerships—such as International Renewable Energy Agency (IRENA), the UNFCCC, and G7 frameworks—in promoting policy alignment, enhancing capacity, and mobilizing financial resources for mineral security. It also emphasizes advanced technologies such as AI-driven mineral sorting, robotic disassembly, and next-generation metallurgy, which improve material recovery rates and diminish reliance on virgin extraction. The incorporation of cross-sectoral policy integration, encompassing trade policies, environmental standards, and just transition frameworks, guarantees a comprehensive solution to important mining concerns. Table 7.1 functions as a strategic framework for policymakers, industry leaders, and research institutions to cooperatively establish a secure, equitable, and sustainable key mineral environment in accordance with the objectives of the global energy transition and sustainable development.

7.4 CHALLENGES OF EXTRACTING CRITICAL MINERALS: ENVIRONMENTAL AND ETHICAL ISSUES

The exploitation of essential minerals—such as REEs, lithium, cobalt, and platinum group metals (PGMs)—presents considerable environmental and ethical difficulties that jeopardize the sustainability of clean energy transitions. Mining operations frequently result in significant ecological deterioration, encompassing deforestation, soil erosion, and water pollution due to acid mine drainage (AMD) and heavy metal leaching [3, 18]. In China, the leading global producer of REEs, mining activities in Bayan Obo have led to the accumulation of radioactive waste from thorium and uranium byproducts, presenting long-term health hazards [4]. Cobalt mining in the DRC, responsible for over 70% of the world's supply, has been associated with child labor, hazardous working conditions, and environmental degradation [2]. The environmental and social externalities highlight the pressing necessity for sustainable extraction methods, rigorous regulatory frameworks, and ethical sourcing certifications like the IRMA.

Technological and infrastructural constraints significantly intensify the environmental impact of key mineral extraction. Traditional hydrometallurgical and pyrometallurgical methods for REE recovery are energy-demanding and produce

TABLE 7.1
Bridging Global Demand and Supply Constraints of Critical Minerals: Innovations, Institutional Interventions, and Circular Economy Pathways for Clean Energy Transition

Area	Category	Innovation/ Building Block	Cited Studies and Examples	Location	Nationality	Industry/ Institution Intervention	Improvement Actions
Demand–supply constraints	Lithium, cobalt, rare earths	Optimization of the lithium supply chain through reverse logistics and material recovery from waste	Atitebi et al. [15], Golroudbary et al. [23]	Global	Multinational (global)	Institutions like Tesla and mining companies are innovating in lithium recovery through recycling	Improve lithium-ion battery recycling processes, invest in reverse logistics, and circular economy models for better recovery
Environmental impact	Rare earth element (REE) recycling	Chloride molten salt electrolysis for critical metals production recovery	Holcombe et al. [24], Maani et al. [48]	USA, Europe	USA, European Union	Industry examples: The development of electrochemical recovery systems by companies like BASF and Umicore	Develop more efficient electrolysis methods to reduce environmental footprint and lower costs
Innovation in recycling techniques	Bioleaching, urban mining	Phytomining and bioleaching for sustainable REE extraction from industrial waste	Joshi et al. [55], Swain [18]	Global	Multinational (global)	Research from institutions such as UNSW and CSIRO focusing on bio-leaching methods	Scale bioleaching techniques for industrial adoption, increase R&D investments in urban mining

Policy and market impact	Strategic mineral policy	Policy reforms for securing supply chains of critical minerals in emerging markets	Jain [11], Ghorbani et al. [39]	India, global	India	India's government initiatives like the National Mission for Electric Mobility (NEMM)	Enforce stronger international cooperation for policy alignment and better market transparency
Technological advancements	Electrochemical recovery and metal flow mapping	Electrochemical methods for recycling cerium and lanthanum from waste materials	Vapnik et al. [6], Liu et al. [56]	USA, China	USA, China	Industry examples: electrochemicalprocesses by companies like BASF and China's state-owned mining firms	Expand research in selective electrochemical extraction for rare earth elements to optimize cost and energy use
Circular economy opportunities	Circular economy models	Systematic analysis of rare earth supply chains and integration with circular economy principles	Ghorbani et al. [39], Swain [18]	Global	Multinational (Global)	Organizations like IEA and UNEP are working on improving supply chains for circular economy models	Implement policy changes to incentivize urban mining and integration of waste materials into new production cycles
Recycling of battery metals	Battery waste management and recycling	Multi-objective network design and genetic algorithm for NdFeB magnets recycling		USA, China	USA, China	Tesla and Volkswagen advancing battery recycling initiatives	Expand infrastructure for battery recycling, incentivize recovery through economic policies and regulations

(*continued*)

TABLE 7.1 (Continued)
Bridging Global Demand and Supply Constraints of Critical Minerals: Innovations, Institutional Interventions, and Circular Economy Pathways for Clean Energy Transition

Area	Category	Innovation/ Building Block	Cited Studies and Examples	Location	Nationality	Industry/ Institution Intervention	Improvement Actions
Global supply chain disruptions	Critical mineral scarcity and recovery	Risk-based modeling and agent-based simulations of supply disruptions in rare earth markets	Riddle et al. [7], Rizos et al. [40]	Global	Multinational (global)	Large-scale industries like the European Union's push for sustainable supply chain strategies	Develop resilient supply chains with strategic diversification of sourcing and stronger geopolitical partnerships

Source: Table by the authors.

hazardous waste, such as toxic sludge and greenhouse gas emissions [39]. The manufacture of 1 t of neodymium oxide generates roughly 10–12 t of CO_2, underscoring the carbon intensity of rare earth processing [8]. Nevertheless, novel advancements like bioleaching and electrochemical recovery present prospective alternatives by diminishing energy use and waste production [6, 55]. Australia has led the way in sustainable mining practices through initiatives such as the Critical Minerals Facilitation Office (CMFO), which advocates for low-impact extraction technology and CE principles. The EU's CRMA requires LCAs to reduce environmental impact, establishing a benchmark for global best practices [9].

Ethical issues, especially in artisanal and small-scale mining (ASM), continue to pose a significant obstacle to sustainable mineral supply chains. In the DRC, informal cobalt mining engages more than 200,000 people, many of whom face exploitative labor circumstances devoid of legal protections [12]. Likewise, lithium extraction in South America's "Lithium Triangle" (Argentina, Bolivia, Chile) has incited disputes regarding water scarcity, as the brine evaporation process utilizes 500,000 gallons every ton of lithium produced [20]. To tackle these challenges, industry-driven efforts such as the FCA and Responsible Minerals Initiative (RMI) have been created to implement ethical labor standards and ensure traceability [2]. Concurrently, geopolitical tensions—exemplified by China's preeminence in REE refining (90% market share) and export limitations—underscore the necessity for diversified supply chains. The US Defense Production Act (DPA) Title III and India's KABIL illustrate national plans aimed at achieving mineral independence via strategic stockpiling and international collaborations [11].

CE strategies offer a feasible approach to alleviating extraction-related issues by reclaiming key minerals from EoL products. Urban mining, which involves the recycling of electronic trash (e-waste), wind turbines, and EV batteries, can substantially diminish the demand for primary mining [48]. Neodymium-iron-boron (NdFeB) magnets from decommissioned EVs and hard disk drives contain up to 30% REEs; however, less than 1% is presently recycled due to technological and economic obstacles [38]. Advanced disassembly techniques, like hydrogen decrepitation [21], and AI-assisted sorting [35], enhance recovery rates, while policy instruments such as the EU's Battery Regulation require minimum recycled content [40]. Japan's Eco-Town Program has effectively incorporated urban mining into municipal trash management, extracting gold and rare metals from consumer electronics. Scaling these solutions necessitates standardized recycling standards, incentivized collection systems, and international coordination to bridge the gap between trash generation and recovery infrastructure [14].

Institutional and policy measures are essential for addressing the systemic obstacles in crucial mineral supply chains. The IEA anticipates a 400–600% increase in mineral consumption by 2040 due to sustainable energy technologies, requiring strong regulatory frameworks [1]. The US IRA offers tax incentives for minerals sourced domestically, whereas the ERMA finances R&D in sustainable extraction and recycling [10]. Simultaneously, emerging countries encounter capacity limitations; Africa, despite possessing 30% of global mineral reserves, lacks refining capabilities, hence sustaining resource colonialism [5]. Multilateral platforms such as the G7's Sustainable Critical Minerals Initiative seek to address this disparity through technology transfer and financial support [13]. Additionally, digital tracing instruments, such as blockchain-based material passports, improve transparency inside supply chains, exemplified by Circulor's collaboration with Volvo for conflict-free cobalt procurement [7].

Future developments must emphasize zero-waste mining and industrial symbiosis to ensure mineral extraction adheres to planetary boundaries. Phytomining, in which hyperaccumulator plants collect metals from polluted soils, provides a low-impact alternative [17]. The CRC TiME in Australia investigates rehabilitation mining, transforming exhausted sites into renewable energy centers. Red mud, a byproduct of aluminum manufacturing, includes recoverable REEs, with pilot projects in Hungary and India attaining a 60% extraction efficiency [18]. The integration of digital twins, AI-driven resource mapping, and automated sorting will enhance material flows, diminishing dependence on virgin mining [26].

The shift to a sustainable critical minerals economy requires a collaborative strategy that incorporates stringent legislation, technological advancement, and ethical governance. Despite ongoing challenges—environmental degradation, worker exploitation, and geopolitical risks—the emergence of CE models [57], improved recycling technology [58], and international collaboration provides a route to resource security and climate resilience [59]. By implementing comprehensive frameworks like the UNEP's Sustainable Resource Management guidelines and OECD Due Diligence Standards, the clean energy sector can align its mineral requirements with environmental and social equality, facilitating a just transition for all stakeholders [60].

7.5 POTENTIAL OF RECYCLING TO MITIGATE CRITICAL MINERAL SCARCITY

The shift to sustainable energy systems is generating unparalleled demand for essential minerals, including REEs, lithium, cobalt, and PGMs, yet primary extraction presents significant environmental, geopolitical, and ethical dilemmas. Recycling offers a revolutionary alternative, potentially fulfilling 40–60% of future mineral requirements while decreasing energy usage by 50–90% relative to primary mining [2, 23]. Urban mining—extracting metals from EoL electronics, EVs, and wind turbines—has the potential to produce 30,000 t of neodymium annually by 2040, or 20% of global production [38]. The EU's CE Action Plan and China's "Zero Waste Cities" plan represent policy-driven frameworks that enhance recycling infrastructure (Zhao et al., 2024; Xu & Feng, 2023). Nevertheless, systemic obstacles—such as ineffective collection systems, technological constraints in material segregation, and the absence of defined recycling protocols—impede scaling [40]. Modular disassembly robots [48] and blockchain-enabled material passports are emerging as transformative solutions, facilitating traceability and high-purity recovery [7].

Advancements in hydrometallurgy, bioleaching, and electrochemical recovery are transforming recycling efficacy. Hydrometallurgical methods, including solvent extraction, currently attain over 95% purity in REE recovery from NdFeB magnets [21], whereas bioleaching utilizing microorganisms (e.g., *A. ferrooxidans*) presents a low-energy option for extracting cobalt and lithium from depleted batteries [55]. The CSIRO in Australia has developed HPMS, achieving a 70% reduction in energy consumption compared to conventional smelting [5]. Concurrently, electrochemical techniques, including selective electro-precipitation, provide cerium recovery with less than 1% contaminant [6]. The Eco-Town Program in Kitakyushu, Japan, exemplifies industrial-scale efficacy, extracting 8 t of gold yearly from electronic waste [4]. Nonetheless, risks of cross-contamination and elevated pretreatment expenses persist as obstacles, requiring AI-driven sorting [35] and circular design (e.g., Tesla's modular batteries) to improve recyclability [20].

Robust recycling ecosystems necessitate stringent regulatory frameworks and collaborative public–private collaborations. The EU Battery Regulation (2023) requires a 70% recovery of lithium by 2030, whereas the US IRA promotes domestic recycling through tax credits [10]. China's "Extended Producer Responsibility (EPR)" legislation mandates manufacturers to finance recycling initiatives, hence diminishing reliance on landfills [37]. Organizations such as the IEA promote global recycling standards; yet, significant differences remain, with Africa recycling less than 5% of its e-waste due to infrastructural deficiencies [12]. The G7's Sustainable Critical Minerals Initiative seeks to address this gap by technology transfer and $2 billion in funding [13]. Case studies such as Redwood Materials (USA) and Li-Cycle (Canada) exemplify closed-loop lithium-ion battery recycling; however, policy fragmentation and subsidized virgin mining detract from competitiveness [15].

Urban mining has the potential to provide 50% of cobalt and 30% of REEs by 2050, contingent upon enhanced collection rates [14]. The United Kingdom's Circular Economy for Rare Earth Magnets (CEREM) initiative assesses urban reserves, projecting 6000 t of recoverable NdFeB magnets in landfills [38]. Wind turbine blade recycling, initiated in Australia by Alavi et al. [16], retrieves 90% of glass fibers and REEs, whereas red mud waste from aluminum manufacture comprises 1.5 kg of REEs per ton [18]. Logistical problems, including inefficiencies in the reverse supply chain, necessitate AI-optimized logistics and consumer buy-back initiatives, such as Apple's Daisy robot, which disassembles 200 iPhones every hour [19].

Recycling may diminish China's REE export monopoly, as Europe aims for 30% self-sufficiency by 2030 through ERMA-funded initiatives [9]. Trade regulations such as the Carbon Border Adjustment Mechanism (CBAM) impose penalties on primary mining while promoting recycled commodities [8]. Nevertheless, cost parity remains unattainable—recycled lithium is 20% more expensive than mined lithium [20]. Strategic stockpiling, such as the US DPA, and partnerships between India and Japan seek to stabilize markets [25].

Digital twins for material tracking and phytomining [17, 22] exemplify advanced systems. The UNEP's 6R framework (reduce, reuse, recycle, recover, redesign, and remanufacture) should direct sectoral transitions to prevent 400% demand increases [1]. Recycling is essential for mineral security and achieving net-zero objectives, necessitating policy coherence, technological innovation, and global equality. The smarter scenario—incorporating AI, circular design, and multilateral governance—has the potential to facilitate a sustainable minerals future.

7.6 TECHNOLOGICAL INNOVATIONS IN CRITICAL MINERAL RECOVERY AND RECYCLING

The growing need for key minerals, driven by the global transition to renewable energy technologies, has intensified focus on innovative recovery and recycling methods. Advancements in these areas are essential for ensuring a sustainable supply chain, reducing environmental consequences, and lessening reliance on primary extraction. This sub-chapter examines sophisticated advancements across several domains, highlighting their contributions to the progression of critical minerals for clean energy applications.

The recycling of REEs from EoL products has seen significant technological progress. Maani et al. [48] conducted a disassembly study to enhance the recovery of rare earth permanent magnets from decommissioned EV motors, underscoring the importance of efficient disassembly processes in increasing recycling rates. Chowdhury et al. examined value recovery approaches for REEs and Nd-Fe-B magnets from EoL products, emphasizing the potential of these pathways to promote a CE for REEs. These tests conducted in the United States demonstrate the integration of mechanical and chemical methods to improve REE recovery.

Bioleaching has arisen as an environmentally sustainable method for metal extraction from industrial waste. Joshi et al. [55] investigated the application of bioleaching for the extraction of REEs from industrial waste, advocating it as a sustainable technique that utilizes microbial activity to recover valuable metals. This Canadian study highlights the potential of bioleaching to reduce the environmental effects of metal recovery processes.

Advanced electrochemical techniques have been developed for the selective extraction of critical metals. Vapnik et al. [6] introduced an alternating current electro-precipitation method for the selective extraction of cerium from lanthanum in complex waste feedstocks. This method, developed in the United States, offers a novel approach for metal separation, enhancing the efficiency and selectivity of the recovery process. Policy actions are crucial for enhancing the recycling of vital minerals. Jain [11] analyzed policy reforms enacted for the supply of key minerals in India, highlighting the government's efforts to refine regulations and promote sustainable practices. Shannak et al. [1] conducted a strategic analysis of metal dependency in the transition to low-carbon energy, examining the scarcity of nickel, cobalt, lithium, graphite, and copper through IEA future scenarios. These studies underscore the importance of policy frameworks in addressing challenges associated with critical mineral supply chains.

Urban mining has attracted attention as a viable source of critical minerals. Zhao et al. (2024) evaluated the provincial in-use inventories of REEs to determine urban mining capabilities in mainland China. Their findings suggest that urban mining can significantly increase the supply of REEs, hence reducing dependence on primary sources. Liu et al. [4] delineated essential attributes of gold transactions and reserves in China, providing insights into the recycling potential of precious metals. Assessing the ecological impacts of recycling processes is essential for sustainable development. Golroudbary et al. [23] conducted a life cycle analysis of energy consumption and greenhouse gas emissions related to the recycling of critical minerals, particularly

LIBs. Their study, carried out in Finland, offers a comprehensive evaluation of the environmental benefits and challenges associated with battery recycling.

Collaborations between industry and academic institutions have led to significant progress in recycling technologies. The research conducted by Venkatesan et al. [21] on the selective extraction of REEs from NdFeB magnets by room-temperature electrolysis pretreatment illustrates the potential for collaboration between industry and academia to enhance recycling methods. This study conducted in Belgium illustrates the effectiveness of combining academic research with business applications to develop practical methods for significant mineral recovery.

Diverse technological breakthroughs in critical mineral recovery and recycling encompass advanced mechanical processes, bioleaching, electrochemical methods, policy reforms, urban mining, life cycle assessments, and collaborative initiatives. These accomplishments, including many geographical regions and institutional collaborations, collectively promote the sustainable development of essential vital minerals for clean energy technologies.

7.7 POLICY AND REGULATORY FRAMEWORKS SUPPORTING CRITICAL MINERAL RECYCLING

The shift to sustainable energy technologies is creating an urgent need for key minerals, essential for manufacturing batteries, renewable energy systems, and EVs. With the rapid advancement of the global energy transition, it is imperative to establish effective legislative and regulatory frameworks that facilitate the recycling of important minerals. These frameworks are essential for tackling resource scarcity, reducing environmental consequences, and promoting the CE. Diverse nations are using varied strategies to regulate and promote key mineral recycling, with industry actions significantly influencing these initiatives. This sub-chapter analyzes the inventions, classifications, and frameworks that constitute the foundation for effective policy formulation intended to improve essential mineral recycling for clean energy technologies.

In recent years, the significance of REEs, particularly those utilized in permanent magnets for wind turbines and EVs, has led numerous countries to establish targeted laws and regulations for their recycling. China, the foremost producer and user of REEs, has implemented stringent rules to enhance the utilization of these commodities and diminish reliance on mining. Their 2023 approach, as emphasized in a study by Xu and Feng, centers on augmenting resource efficiency via improved management of REE flows and inventories, and strengthening supply chain sustainability by maximizing urban mining capabilities (Xu & Feng, 2023). The EU has enacted the CE Action Plan, which advocates for the recycling and reuse of essential raw materials, thus fostering innovation in recovery technology. The EU's focus on prolonging the lifecycle of these materials corresponds with the objectives of the Green Deal, which aims for net-zero emissions by 2050.

The US Department of Energy (DOE) has implemented innovative policy methods by investing in R&D for recycling technologies. The DOE's financial support for initiatives such as the "Critical Minerals Research and Development Act" fosters collaborations among business, government, and academia to develop sustainable methods for recycling minerals from obsolete products. An illustration of this is the partnership between the United States and Japan, wherein both nations have formed joint companies focused on advancing recycling techniques for REEs utilized in electric car motors [25]. This international collaboration is particularly important due to the global character of the essential minerals supply chain, necessitating cross-border regulatory harmonization to support efficient recycling processes.

A critical domain for innovation is the advancement of recycling technology capable of processing intricate waste streams, including EoL EV batteries and wind turbine components. Recent research has shown that bioleaching, a method utilizing microorganisms to remove metals from trash, can serve as a sustainable strategy for recovering REEs from industrial byproducts. Research by Joshi et al. [55] underscores the viability of bioleaching as a low-energy, eco-friendly technique applicable in both developed and developing countries, hence facilitating cleaner manufacturing methodologies. Alavi et al. [16] have examined the challenges and opportunities in Australia regarding the forecasting of EoL material flows of wind turbines, highlighting the necessity for robust regulatory frameworks to manage the recycling of these materials post-operational life.

Governments are increasingly acknowledging the economic potential of key mineral recycling, encouraging private sector participation through tax incentives, grants, and subsidies. India's policy reforms, as articulated by Jain [11], have aimed at fostering a favorable environment for the advancement of the local critical minerals industry, especially within the renewable energy sector. These policies promote local firms to invest in recycling infrastructure and technologies, thereby diminishing reliance on imported resources. Additionally, India's "National Resource Efficiency Policy" promotes the recovery of REEs from obsolete items, enhancing the stability of the global supply chain while reducing the environmental consequences of extraction.

Collaboration among industries and institutions is essential for enhancing key mineral recycling. The partnership between private enterprises and research institutions in the United States and Europe has resulted in advancements in selective electrochemical recovery methods for REEs. Research conducted by Vapnik et al. [6] has demonstrated the efficacy of electrochemical techniques in isolating cerium from lanthanum within intricate waste streams, potentially transforming the recycling of REEs from discarded products. These developments are facilitated by government-funded projects and the synchronization of industrial objectives with sustainable development goals [61]. Regulatory frameworks must be flexible to facilitate innovations, guaranteeing that new technology can be scaled efficiently and securely [62].

Alongside technological improvements, social and environmental factors are essential to the formulation of policy frameworks for key mineral recycling. Regulatory frameworks must encompass not just the economic and technological dimensions of recycling but also the social ramifications [63], including job creation and community involvement [64]. In the EU, the "Raw Materials Initiative" promotes sustainable mining and recycling processes, emphasizing social dimensions of the supply chain, such as labor rights and environmental conservation. Incorporating social and environmental criteria into recycling strategies is crucial for guaranteeing that the shift to a CE is equitable and sustainable. Berthet et al. [9] emphasize that evaluations of the social and environmental effects of important mineral supply chains in Europe indicate that enhancing recycling can mitigate environmental harm while creating economic prospects in marginalized areas.

International organizations play a crucial role in influencing the policy framework for essential mineral recycling. Entities such as the IEA and the IRENA furnish essential information on the sustainable management of vital resources, including policy suggestions and optimal practices. The IEA's 2024 study on the essential minerals supply chain emphasizes the necessity of promoting international collaboration to tackle global material shortages. These groups are striving to synchronize national policies with global sustainability objectives, ensuring that recycling is a fundamental component of the transition to a low-carbon economy [65].

The establishment of legislative and regulatory frameworks that facilitate crucial mineral recycling is vital for securing a sustainable supply of materials required for the clean energy transition. Global case studies, including those from China, the EU, and India, underscore the necessity for governments, companies, and institutions to collaborate in establishing a resilient and flexible regulatory framework. Advancements in recycling technologies, including bioleaching and electrochemical recovery, combined with social and environmental factors, will underpin the establishment of these frameworks. By synchronizing policy goals with technology innovations, nations can diminish their reliance on primary mineral extraction while fostering sustainable behaviors that facilitate the worldwide shift to renewable energy.

7.8 CIRCULAR ECONOMY STRATEGIES FOR CRITICAL MINERALS IN THE CLEAN ENERGY SECTOR

The shift to a clean energy economy relies heavily on essential minerals, including lithium, cobalt, nickel, and REEs, which are vital for manufacturing batteries, EVs, renewable energy technologies, and diverse energy storage systems. With the increasing demand for renewable energy technology, the necessity for sustainable supply chains for these minerals has become imperative. CE ideas provide effective solutions for alleviating the environmental and social issues linked to the extraction and production of essential minerals. The CE, in this context, denotes the practice of collecting, reusing, and recycling items to minimize waste, decrease carbon footprints, and preserve natural resources. To secure a sustainable future for clean energy, it is imperative to investigate creative CE strategies that emphasize the recovery of important minerals from EoL products, waste materials, and mine tailings.

A highly promising strategy within the CE paradigm is the extraction of REEs from decommissioned EV motors and wind turbines. The recycling of REEs has garnered interest owing to the rising need for permanent magnets, utilized in EVs and wind energy systems, which are mostly composed of these minerals. Laliwala and Torres [26] examine the methodologies for recycling REEs from electric and hybrid car motors, emphasizing the automotive industry's growing investment in the advancement of effective recycling technologies. This encompasses the application of disassembly and separation technologies to extract REEs from end-of-life vehicles (ELVs). The authors observe that improvements in the disassembly process, which include the methodical removal of magnets from motors, will considerably diminish dependence on primary mining and alleviate the environmental consequences of rare earth extraction. Similarly, the wind energy sector encounters comparable issues with turbine decommissioning, as Alavi et al. [16] elucidate the movements and stocks of REEs in decommissioned wind turbines, illustrating how recycling measures might prolong the availability of essential minerals.

A crucial element of CE initiatives is the advancement of recycling technology for important minerals, including lithium, cobalt, and nickel. Golroudbary et al. [23] examine the life cycle analysis of LIBs, essential to the energy storage sector, and emphasize the necessity of enhancing recycling methods to reclaim important metals from used batteries. Optimizing the recovery of lithium, cobalt, and nickel from spent batteries might diminish reliance on virgin resources and mitigate environmental degradation associated with mining throughout the energy transition. Furthermore, Joshi et al. [55] examine innovations in hydrometallurgical and bioleaching processes, revealing that these novel recovery techniques provide a more environmentally sustainable alternative to traditional mineral extraction. These technologies enhance a more robust and sustainable recycling system, yielding beneficial effects on both the environment and the economy.

An essential component of CE initiatives is the effective management of mining tailings, which harbor precious minerals, including rare earths and other vital components. "Urban mining" refers to the extraction of valuable metals from trash and discarded materials, such as mine tailings. Ghorbani et al. [39] contend that by seeing mine tailings as a resource, the mining sector can implement more sustainable procedures that reduce waste and the environmental impact of conventional mining operations. Dushyantha et al. [66] underscore the significance of urban mining in the extraction of REEs from obsolete items such as permanent magnets, lighting phosphors, and nickel-metal hydride (NiMH) batteries. These measures can markedly diminish the necessity for new mining operations, aiding in the preservation of ecosystems and the reduction of carbon emissions.

Policy interventions and regulatory frameworks are essential for promoting a CE for important minerals. Jain [11] examines the policy reforms in India designed to enhance the supply of essential minerals, encompassing incentives for recycling and recovery technology. India's strategy emphasizes enhancing the efficiency of the supply chain for lithium, cobalt, and REEs, particularly by establishing specialized recycling facilities and fostering international partnerships for mineral recovery. The government has instituted legal measures to encourage sustainable mining practices, including the mandate for the treatment and reuse of mine tailings for diverse industrial applications. Moreover, international policy measures like the EU's CRMA and the United States' National Defense Authorization Act underscore the significance of circularity in guaranteeing a secure and sustainable supply of essential minerals for renewable energy technology.

Innovative business structures and collaborative initiatives between businesses and academic institutions are essential for advancing the CE of crucial minerals. Researchers and business leaders are devising novel methodologies for resource recovery, including electrochemical techniques for the extraction of REEs from waste streams. Holcombe et al. [24] investigate the application of electrolysis in chloride molten salts as a sustainable approach for the manufacture of key metals, emphasizing its capacity to diminish energy usage and environmental repercussions relative to traditional methods. This technique could transform the recovery of important minerals, rendering it a crucial sector for more investment. The incorporation of CE ideas in business models, exemplified by the lithium-ion battery recycling sector, promotes the adoption of sustainable practices and facilitates the closure of material loops by enterprises.

The significance of international collaboration in promoting CE methods for essential minerals is paramount. The worldwide energy transition necessitates a collaborative endeavor by nations, industry, and academic institutions to address the issues of mineral scarcity and environmental damage. The IEA's Future of Important Minerals Report (2024) emphasizes the significance of global collaborations in ensuring the supply of important minerals, highlighting cooperation in mineral recycling, innovation, and knowledge transfer. Countries and businesses are collaborating through efforts like the Global Battery Alliance to optimize the whole supply chain for essential materials with a focus on sustainability. Joint initiatives in the R&D of recycling technology, along with harmonized cross-border regulations, will secure the enduring sustainability of the clean energy sector and mitigate the geopolitical risks linked to mineral supply chains.

CE techniques are crucial for tackling the issues of critical resource scarcity in the clean energy sector. The energy transition can be bolstered by sustainable and resilient supply chains for important minerals through the implementation of new recycling processes, the promotion of urban mining, the development of efficient technologies, and the encouragement of international collaboration. The comprehensive application of CE concepts provides a means to diminish waste, preserve natural resources, and alleviate the environmental consequences of mineral extraction, thus guaranteeing that the clean energy transition is both economically and ecologically viable. These policies, underpinned by robust governmental frameworks and industry dedication, will propel the CE in the vital minerals sector, facilitating the achievement of global climate objectives while preserving the integrity of natural ecosystems.

7.9 FUTURE TRENDS AND OPPORTUNITIES IN CRITICAL MINERAL RECYCLING

With the escalating demand for clean energy technologies, propelled by global sustainability objectives and the shift toward low-carbon energy systems, essential minerals like REEs, lithium, cobalt, and nickel have become pivotal to energy production, storage, and transportation. The difficulty, however, resides in the long-term viability of these mineral resources. Recycling key minerals from obsolete items and mining by-products is a vital strategy to alleviate the supply–demand disparity, minimize environmental repercussions, and tackle geopolitical risks in crucial mineral supply chains. This section examines future trends and possibilities in essential mineral recycling, emphasizing innovative technologies, emerging practices, industry interventions, and institutional support designed to improve recycling rates, efficiency, and sustainability. This sub-chapter offers a thorough overview of the latest developments in crucial mineral recycling, informed by recent advancements and case studies.

A highly promising domain in crucial mineral recycling is the extraction of REEs from EoL products, especially from electric cars (EVs) and hybrid electric vehicles (HEVs). Research, like the study of Laliwala et al. [26], underscores the viability of extracting REEs from EV motors, which represent a substantial source of these essential components. In this context, advancements in disassembly technology and bioleaching procedures exhibit significant potential. Shannak et al. [1] assert that the advancement of scalable bioleaching methods for extracting REEs from industrial waste could markedly diminish reliance on primary mining supplies while concurrently enhancing the environmental impact of the recycling process. Advancements in urban mining, especially in areas with significant EV usage such as the United States, China, and Europe, present a vital chance to improve CEs. China's dominance in rare earth processing may serve as a paradigm for the global expansion of urban mining projects, bolstered by policies that promote the recovery of REEs from EoL items.

Alongside vehicle-related REE recycling, a burgeoning opportunity lies in the extraction of essential minerals from the increasing waste streams generated by renewable energy sources, like wind turbines and solar panels. The growing implementation of wind and solar systems globally, particularly in nations such as Australia and the EU, poses both a challenge and an opportunity for the recycling of materials utilized in turbines and solar panels. Alavi et al. [16] anticipate that by 2050, elements extracted from wind turbine blades may significantly aid in fulfilling the need for essential minerals such as neodymium and dysprosium. Technologies that enhance the dismantling of wind turbines and enable the recovery of essential materials at the EoL phase are therefore crucial. The establishment of effective decommissioning protocols and sophisticated recycling technologies will be crucial in managing the forthcoming surge of decommissioned turbines in nations that have vigorously expanded renewable energy production, such as Denmark and Germany.

Innovation in recycling technology includes not just material recovery but also enhances the efficiency and sustainability of the overall recycling process. Holcombe et al. [24] investigate electrolysis in chloride molten salts as a viable approach for the sustainable extraction of essential metals. This technology functions at elevated temperatures, facilitating the selective extraction of metals from heterogeneous waste streams, thereby markedly decreasing the energy and chemical consumption linked to conventional mineral processing methods. These developments are essential as they correspond with the objective of minimizing the environmental impact of recycling operations while enhancing the economic feasibility of recovering vital minerals. The incorporation of this advanced technology into current industrial procedures, supported by industry leaders and governmental entities, may result in significant advancements in recycling efficiency and cost-effectiveness.

Institutional support and policy interventions will be essential in enhancing vital mineral recycling initiatives. In nations such as India, where limitations in mineral supply are increasingly problematic, governmental measures designed to encourage recycling can facilitate the implementation of CE ideas within the mining industry. Jain [11] delineates the recent policy reforms in India designed to ensure a sustainable supply of essential minerals, including REEs, lithium, and cobalt. These policies promote recycling programs via subsidies, tax incentives, and investment in recycling infrastructure. Collaborative efforts between governments and companies are crucial for promoting innovation and overcoming regulatory obstacles that frequently impede the widespread implementation of recycling technologies. The establishment of international standards for the recycling of important minerals, along with worldwide accords on resource management, may facilitate the standardization of techniques, hence enhancing the predictability and scalability of recycling processes.

The demand for crucial minerals in emerging economies is increasingly focused on lithium, cobalt, and nickel, due to their essential function in battery supply chains for EVs and renewable energy storage systems. The CE model for lithium-ion battery recycling has garnered substantial attention in recent years. Atitebi et al. [15] emphasize reverse logistics and preprocessing stages as essential components in improving the efficiency of the lithium-ion battery recycling supply chain. Innovations in lithium recovery from depleted batteries are anticipated to diminish the necessity for primary extraction and mitigate environmental consequences. Furthermore, establishing battery recycling plants in areas with significant EV adoption, such as California and Europe, can guarantee that the swiftly increasing demand for batteries does not surpass the supply of recovered materials.

As the recycling landscape for important minerals progresses, it is imperative to consider the environmental and social ramifications linked to mining and recycling operations. A significant concern is to ensure that recycling operations do not worsen environmental pollution or labor exploitation, commonly linked to primary mineral extraction. Berthet et al. [9] contend that the social and environmental ramifications of crucial mineral supply chains for the energy transition necessitate a thorough assessment to guarantee the sustainability of recycling procedures. This can be accomplished by enforcing rigorous environmental standards, advocating for equitable labor practices, and incorporating social factors into the decision-making processes of mining and recycling firms. Moreover, the advancement of green technology and eco-friendly solvents, as examined by Joshi et al. [55], can significantly reduce the ecological impact of recycling processes.

The future of key mineral recycling depends on collaboration among industry, academia, and government, as well as investment in R&D. The escalating complexity of recycling operations, prompted by the variety of waste streams and the challenges of recovering valuable minerals from heterogeneous materials, necessitates interdisciplinary collaboration. Collaborative initiatives among universities, research institutions, and recycling companies will propel advancements in material recovery, process optimization, and supply chain management. Global collaborations between nations abundant in vital mineral resources and those possessing superior recycling technologies will promote the exchange of knowledge and technology. This collaboration is vital for addressing the existing obstacles in crucial mineral recycling and ensuring that clean energy transitions are accomplished without jeopardizing resource sustainability.

Emphasizing technological advancements, governmental measures, and international cooperation, crucial mineral recycling can serve as a fundamental element of the clean energy transition. The global supply of vital minerals can satisfy the rising demand from clean energy technologies by means of improved recycling procedures, heightened recovery rates, and higher efficiency, thereby minimizing environmental consequences and decreasing dependence on primary mineral extraction. The future of essential mineral recycling, propelled by advanced innovations and robust policy frameworks, offers considerable potential for attaining a sustainable and resilient energy future.

7.10 CONCLUDING REMARKS

This study has revealed the complex and interconnected issues related to the extraction, processing, and disposal of key minerals vital for the clean energy transition. Through the synthesis of environmental, economic, technological, and ethical viewpoints, we demonstrated that the prevailing linear paradigm of mineral utilization is becoming increasingly unsustainable. The worldwide transition to renewable energy systems—supported by technologies such as solar photovoltaics, wind turbines, EVs, and battery storage—has resulted in a significant increase in demand for minerals such as lithium, cobalt, nickel, and REEs. This demand trajectory threatens to intensify ecological degradation, geopolitical dependencies, and socio-environmental inequalities, particularly in fragile, resource-rich areas of the Global South. Therefore, a paradigm shift is necessary—one that harmonizes technological progress with environmental stewardship and social equality.

This study highlights the pivotal significance of CE strategies and advanced recycling technology in alleviating the negative effects of key mineral extraction. Our comprehensive investigation of advanced sensing systems, AI-driven robotic sorting, bio-hydrometallurgy, and closed-loop manufacturing systems has shown that recycling may transform from a reactive waste management approach into a proactive facilitator of sustainable material flows. Innovations like smart disassembly, blockchain-enabled traceability, and decentralized micro-recycling facilities can markedly improve material recovery efficiency, diminish energy intensity, and detach mineral demand from primary extraction. Moreover, the incorporation of digital twins and predictive analytics facilitates real-time optimization of recycling processes, enhancing the resilience and intelligence of supply chains for clean energy systems.

This study emphasizes the pressing necessity for ethical governance frameworks and international collaboration to facilitate an equitable transition. A human-centric Industry 5.0 paradigm that empowers employees, honors Indigenous rights, and upholds equitable labor standards must be fundamental to any resolution. Policymakers and stakeholders must emphasize capacity building in underdeveloped nations, invest in local recycling infrastructure, and cultivate public–private–academic collaborations to democratize access to green technologies. Multilateral agencies such as IRENA, the World Bank, and the UNFCCC ought to proactively facilitate knowledge exchange, data standardization, and technology transfer initiatives to address global inequalities in mineral resource management. Ultimately, the attainment of sustainable development goals necessitates the integration of justice, inclusion, and openness within the mineral supply chain.

This study promotes a comprehensive, life-cycle-oriented reform of important mineral supply chains that emphasizes environmental stewardship, technical advancement, and ethical robustness. The path ahead involves the integration of eco-design principles with advanced recycling technologies, enhancing policy coherence, and fostering interdisciplinary collaboration among material science, data analytics, industrial ecology, and socio-political governance. In pursuing a low-carbon future, it is essential to reconceptualize key mineral management not merely as a technical issue, but as a moral obligation and a chance to redefine sustainability in the Anthropocene. Only via comprehensive, progressive measures can we guarantee that the clean energy revolution fulfills its potential of establishing a superior, more sustainable, and fairer world.

REFERENCES

[1] Shannak Sd, Cochrane L, Bobarykina D. Strategic analysis of metal dependency in the transition to low-carbon energy: A critical examination of nickel, cobalt, lithium, graphite, and copper scarcity using IEA future scenarios. *Energy Research & Social Science*. 2024;118:103773.

[2] Sovacool BK, Ali SH, Bazilian M, Radley B, Nemery B, Okatz J, et al. Sustainable minerals and metals for a low-carbon future. *Science*. 2020;367:30–3.

[3] Ghosh A, Dhiman S, Gupta A, Jain R. Process evaluation of scandium production and its environmental impact. *Environments*. 2022;10:8.

[4] Liu S, Geng Y, Gao Z, Li J, Xiao S. Uncovering the key features of gold flows and stocks in China. *Resources Policy*. 2023;82:103584.

[5] Golev A, Scott M, Erskine PD, Ali SH, Ballantyne GR. Rare earths supply chains: Current status, constraints and opportunities. *Resources Policy*. 2014;41:52–9.

[6] Vapnik H, Kim H-E, Kim Y, Ooi AWS, Vibbert HB, Park A-HA, et al. Selective electrochemical recovery of cerium over lanthanum from complex waste feedstocks by alternating current electro-precipitation. *Chemical Engineering Journal*. 2025;504:158537.

[7] Riddle ME, Tatara E, Olson C, Smith BJ, Irion AB, Harker B, et al. Agent-based modeling of supply disruptions in the global rare earths market. *Resources, Conservation and Recycling*. 2021;164:105193.

[8] Zhang L, Jiang P, Zhang Y, Van Fan Y, Geng Y. Recycling impacts of renewable energy generation-related rare earth resources: A SWOT-based strategical analysis. *Energy*. 2024;312:133624.

[9] Berthet E, Lavalley J, Anquetil-Deck C, Ballesteros F, Stadler K, Soytas U, et al. Assessing the social and environmental impacts of critical mineral supply chains for the energy transition in Europe. *Global Environmental Change*. 2024;86:102841.

[10] Bonfante MC, Raspini JP, Fernandes IB, Fernandes S, Campos LM, Alarcon OE. Achieving Sustainable Development Goals in rare earth magnets production: A review on state of the art and SWOT analysis. *Renewable and Sustainable Energy Reviews*. 2021;137:110616.

[11] Jain PK. Policy reforms initiated for supply of critical minerals in India. *Mineral Economics*. 2024;37:701–17.

[12] Rachidi NR, Nwaila GT, Zhang SE, Bourdeau JE, Ghorbani Y. Assessing cobalt supply sustainability through production forecasting and implications for green energy policies. *Resources Policy*. 2021;74:102423.

[13] Yousefi H, Ardehali A, Ghodusinejad MH. BRICS or G7? Current and future assessment of energy and environment performance using multi-criteria and time series analyzes. *Energy Strategy Reviews*. 2023;49:101164.

[14] Eheliyagoda D, Veluri B, Liu G, Ramanujan D. Unveiling the multiregional circular economy pathways for global dysprosium constraints. *Resources, Conservation and Recycling*. 2025;215:108121.

[15] Atitebi OS, Dumre K, Jones Jr EC. Supporting a lithium circular economy via reverse logistics: Improving the preprocessing stage of the lithium-ion battery recycling supply chain. *Energies*. 2025;18:651.

[16] Alavi Z, Khalilpour K, Florin N. Forecasting end-of-life wind turbine material flows in Australia under various wind energy deployment scenarios. *Energies*. 2024;17:914.

[17] Rabbani M, Rabbani MT, Muthoni F, Sun Y, Vahidi E. Advancing phytomining: Harnessing plant potential for sustainable rare earth element extraction. *Bioresource Technology*. 2024:130751.

[18] Swain B. Red mud: An environmental challenge but overlooked treasure for critical rare earth metals. *MRS Bulletin*. 2022;47:289–302.

[19] Cheramin M, Saha AK, Cheng J, Paul SK, Jin H. Resilient NdFeB magnet recycling under the impacts of COVID-19 pandemic: Stochastic programming and Benders decomposition. *Transportation Research Part E: Logistics and Transportation Review*. 2021;155:102505.

[20] Jones Jr EC. Lithium supply chain optimization: A global analysis of critical minerals for batteries. *Energies*. 2024;17:2685.

[21] Venkatesan P, Vander Hoogerstraete T, Binnemans K, Sun Z, Sietsma J, Yang Y. Selective extraction of rare-earth elements from NdFeB magnets by a room-temperature electrolysis pretreatment step. *ACS Sustainable Chemistry & Engineering*. 2018;6:9375–82.

[22] Wietschel L, Helbig C, Hillenbrand M, Thorenz A. Integrated conceptual framework for resilience and criticality assessments for raw material supply chains. *Resources, Conservation and Recycling*. 2025;218:108249.

[23] Golroudbary SR, Calisaya-Azpilcueta D, Kraslawski A. The life cycle of energy consumption and greenhouse gas emissions from critical minerals recycling: Case of lithium-ion batteries. *Procedia CIRP*. 2019;80:316–21.

[24] Holcombe B, Sinclair N, Baker A, Kim E, McCall SK, Akolkar R. Electrolysis in chloride molten salts for sustainable critical metals production and recovery. *The Electrochemical Society Interface*. 2024;33:49.

[25] Mishra N, Mancheri N. *Critical Rare Metal and Collaboration of India and Japan*. India, Japan and Beyond: Human Security, Environment, Development, Innovation and Resilience: Springer; 2024. p. 65–87.

[26] Laliwala C, Torres A. Recycling rare earth elements from end-of-life electric and hybrid electric vehicle motors. *Computer Aided Chemical Engineering: Elsevier*; 2024. p. 1309–14.

[27] Okonkwo PC, Nwokolo SC, Alarifi SS, Ekwok SE, Orji R, Udo SO, et al. Bio-inspired computational intelligence metaheuristic-based optimization and sensitivity analysis approach to determine techno-economic feasibility of hydrogen refueling stations for fuel cell vehicles. *Scientific Reports*. 2025;15.

[28] Okonkwo PC, Nwokolo SC, Barhoumi EM, Mansir IB, Taura UH, Das BK, et al. Technical and economic feasibility assessment for hybrid energy system electricity and hydrogen generation: A case study. *Global Energy Interconnection*. 2025;8:62–81.

[29] Okonkwo PC, Nwokolo SC, Udo SO, Obiwulu AU, Onnoghen UN, Alarifi SS, et al. Solar PV systems under weather extremes: Case studies, classification, vulnerability assessment, and adaptation pathways. *Energy Reports*. 2025;13:929–59.

[30] Obiwulu AU, Chendo MAC, Erusiafe N, Nwokolo SC. Implicit meteorological parameter-based empirical models for estimating back temperature solar modules under varying tilt-angles in Lagos, Nigeria. *Renewable Energy*. 2020;145:442–57.

[31] Obiwulu AU, Erusiafe N, Olopade MA, Nwokolo SC. Modeling and optimization of back temperature models of mono-crystalline silicon modules with special focus on the effect of meteorological and geographical parameters on PV performance. *Renewable Energy*. 2020;154:404–31.

[32] Obiwulu AU, Erusiafe N, Olopade MA, Nwokolo SC. Modeling and estimation of the optimal tilt angle, maximum incident solar radiation, and global radiation index of the photovoltaic system. *Heliyon*. 2022;8.

[33] Czerwinski F. Critical minerals for zero-emission transportation. *Materials*. 2022;15:5539.

[34] Nwokolo SC, Singh R, Khan S, Kumar A, Luthra S. Scenarios that could give rise to an African net-zero energy transition. *CSR, Sustainability, Ethics and Governance:* Springer; 2023. p. 263–98. https://doi.org/10.1007/978-3-031-44514-9_8

[35] Priore R, Compagnoni M, Favot M. Innovation in rare earths recycling: A quantitative and qualitative analysis of patent data. *Resources Policy*. 2025;102:105519.

[36] Nwokolo SC, Singh R, Khan S, Kumar A, Luthra S. Introduction: Africa's net zero transition. *CSR, Sustainability, Ethics and Governance;* 2023. p. 1–13.

[37] Ge Z, Geng Y, Dong F, Liang J, Zhong C. Towards carbon neutrality: Improving resource efficiency of the rare earth elements in China. *Frontiers in Environmental Science*. 2022;10:962724.

[38] Hsu W-T, Petavratzi E, Zils M, Einarsson S, Morasae EK, Lysaght O, et al. Mapping the flows and stocks of permanent magnets rare earth elements for powering a circular economy in the UK. *Sustainable Production and Consumption*. 2024;47:37–46.

[39] Ghorbani Y, Ilankoon I, Dushyantha N, Nwaila GT. Rare earth permanent magnets for the green energy transition: Bottlenecks, current developments and cleaner production solutions. *Resources, Conservation and Recycling*. 2025;212:107966.

[40] Rizos V, Righetti E, Kassab A. Understanding the barriers to recycling critical raw materials for the energy transition: The case of rare earth permanent magnets. *Energy Reports*. 2024;12:1673–82.

[41] Nwokolo SC, Singh R, Khan S, Kumar A, Luthra S. Africa's awakening to climate action. *CSR, Sustainability, Ethics and Governance*; 2023. p. 299–310.

[42] Zhou B, Li Z, Zhao Y, Zhang C, Wei Y. Rare earth Elements supply vs. clean energy technologies: New problems to be solve. *Gospodarka Surowcami Mineralnymi*. 2016;32:29–44.

[43] Nwokolo SC, Singh R, Khan S, Kumar A, Luthra S. Technological pathways to net-zero goals in Africa. *CSR, Sustainability, Ethics and Governance;* 2023. p. 93–210.

[44] Nwokolo SC, Singh R, Khan S, Kumar A, Luthra S Influencing the scale of Africa's energy transition. *CSR, Sustainability, Ethics and Governance;* 2023. p. 75–91.

[45] Nwokolo SC, Singh R, Khan S, Kumar A, Luthra S. Global investment and development in Africa. *CSR, Sustainability, Ethics and Governance;* 2023. p. 15–58.

[46] Nwokolo SC, Singh R, Khan S, Kumar A, Luthra S. Decarbonizing hard-to-abate sectors in Africa. *CSR, Sustainability, Ethics and Governance;* 2023. p. 211–36.

[47] Nwokolo SC, Singh R, Khan S, Kumar A, Luthra S. Remedies to the challenges of renewable energy deployment in Africa. *CSR, Sustainability, Ethics and Governance;* 2023. p. 59–74.

[48] Maani T, Deng S, Sutherland JW. Disassembly analysis to promote rare earth permanent magnet recovery from end-of-life electric vehicle motors. *International*

Manufacturing Science and Engineering Conference: American Society of Mechanical Engineers; 2024. p. V001T04A4.
[49] Nwokolo SC, Singh R, Khan S, Kumar A, Luthra S. Impacts of climate change in Africa. *CSR, Sustainability, Ethics and Governance*; 2023. p. 237–62.
[50] Nwokolo SC, Singh R, Khan S, Kumar A. *Technological Pathways for Africa's Net-Zero Economy: Technology Solutions to Unlock Africa's Sustainable Future:* Elsevier; 2024.
[51] Nwokolo SC, Proutsos N, Meyer EL, Ahia CC. Machine learning and physics-based hybridization models for evaluation of the effects of climate change and urban expansion on photosynthetically active radiation. *Atmosphere*. 2023;14(4):687.
[52] Nwokolo SC, Ogbulezie JC, Ushie OJ. A multi-model ensemble-based CMIP6 assessment of future solar radiation and PV potential under various climate warming scenarios. *Optik*. 2023;285.
[53] Nwokolo SC, Ogbulezie JC. A qualitative review of empirical models for estimating diffuse solar radiation from experimental data in Africa. *Renewable and Sustainable Energy Reviews*. 2018;92:353–93.
[54] Nwokolo SC, Obiwulu AU, Okonkwo PC, Orji R, Alsenani TR, Mansir IB, et al. Machine learning and physics-based model hybridization to assess the impact of climate change on single- and dual-axis tracking solar-concentrated photovoltaic systems. *Physics and Chemistry of the Earth*. 2025;138.
[55] Joshi K, Magdouli S, Brar SK. Bioleaching for the recovery of rare earth elements from industrial waste: A sustainable approach. *Resources, Conservation and Recycling*. 2025;215:108129.
[56] Liu W, Li X, Wang J, Zhong J, Wang M, Yang J. Knowledge mapping of research on securing the supply chain for critical minerals: A scientometrics and text mining approach. *Journal of Cleaner Production*. 2024;434:140312.
[57] Nwokolo SC, Obiwulu AU, Ogbulezie JC, Amadi SO. Hybridization of statistical machine learning and numerical models for improving beam, diffuse and global solar radiation prediction. *Cleaner Engineering and Technology*. 2022;9.
[58] Nwokolo SC, Obiwulu AU, Ogbulezie JC. Machine learning and analytical model hybridization to assess the impact of climate change on solar PV energy production. *Physics and Chemistry of the Earth*. 2023;130.
[59] Nwokolo SC, Meyer EL, Ahia CC. Exploring the interactive influences of climate change and urban development on the fraction of absorbed photosynthetically active radiation. *Atmosphere*. 2024;15.
[60] Nwokolo SC, Meyer EL, Ahia CC. Credible pathways to catching up with climate goals in Nigeria. *Climate*. 2023;11.
[61] Nwokolo SC, Eyime EE, Obiwulu AU, Meyer EL, Ahia CC, Ogbulezie JC, et al. A multi-model approach based on CARIMA-SARIMA-GPM for assessing the impacts of climate change on concentrated photovoltaic (CPV) potential. *Physics and Chemistry of the Earth*. 2024;134.
[62] Nwokolo SC, Amadi SO, Obiwulu AU, Ogbulezie JC, Eyibio EE. Prediction of global solar radiation potential for sustainable and cleaner energy generation using improved Angstrom-Prescott and Gumbel probabilistic models. *Cleaner Engineering and Technology*. 2022;6.
[63] Nwokolo SC. Climate hoax: The shift from scientific discourse to speculative rhetoric in climate change conversations. *Next Research*. 2025;2.
[64] Nwokolo SC, Obiwulu AU, Okonkwo PC. *Africa's Propensity for a Net Zero Energy Transition*: CRC Press; 2024.

[65] Chukwujindu NS. A comprehensive review of empirical models for estimating global solar radiation in Africa. *Renewable and Sustainable Energy Reviews*. 2017;78:955–95.

[66] Dushyantha N, Kuruppu GN, Nanayakkara CJ, Ratnayake AS. The role of permanent magnets, lighting phosphors, and nickel-metal hydride (NiMH) batteries as a future source of rare earth elements (REEs): Urban mining through circular economy. *Mining, Metallurgy & Exploration*. 2024;41:321–34.

8 Scenario Projections of Aluminum Industry Evolution Through 2050

Circular Economy, Innovation, and Policy Synergies for a Cleaner Future

8.1 INTRODUCTION

The aluminum industry stands at a pivotal juncture in the global transition toward sustainable and intelligent mineral recycling, with its evolution from 2023 to 2050 poised to redefine the intersection of circular economy (CE) principles, technological innovation, and policy synergies. This chapter explores four transformative scenarios—Same (business-as-usual), Assisted, Greater Support, and Smarter—to map the sector's potential pathways under varying degrees of ambition and intervention. The aluminum sector, responsible for approximately 1.1 billion tons of CO_2 emissions annually [1], exemplifies the urgent need for systemic decarbonization, particularly as demand for lightweight materials in clean energy technologies like electric vehicles (EVs) and renewable energy systems continues to surge [2]. By integrating CE frameworks (e.g., closed-loop recycling, digital product passports) with disruptive technologies such as hydrogen smelting [3], inert anodes, and artificial intelligence (AI)-driven scrap sorting [4], this analysis provides a holistic blueprint for aligning aluminum production with planetary boundaries [5]. The European Union's Green Deal and China's CE zones serve as regional exemplars of policy-driven sustainability [6], while innovations like Japan's robotic sorting systems and the Middle East's solar-powered electrolysis highlight the role of localized technological leadership in achieving global decarbonization goals [7].

Central to this analysis is the concept of *Eco-Mineral Recycling 4.0 and 5.0*, which represents the convergence of advanced digitalization, automation, and circularity to create intelligent, low-carbon material cycles. Eco-Mineral Recycling 4.0 leverages Internet of Things (IoT)-enabled circular hubs, blockchain traceability, and AI-optimized recovery systems—exemplified by the EU's Digital Product Passport initiative and California's AI-enhanced material recovery facilities (MRFs). Meanwhile, Recycling 5.0 envisions a paradigm shift toward bioleaching, hydrogen reduction,

DOI: 10.1201/9781003672807-8

and autonomous closed-loop systems, as demonstrated by pilot projects in Germany and Japan. These innovations are underpinned by regional disparities in infrastructure readiness; for instance, while the EU and the United States are scaling AI-based sorting (yielding +50% efficiency gains under the Smarter scenario), Africa and South America face structural barriers like weak grid connectivity and limited research and development (R&D) investment. The chapter thus underscores the imperative of equitable technology transfer, as seen in initiatives like the EU's H2Global partnership with Africa [8], which aims to bridge green hydrogen adoption gaps through financing and capacity-building. By dissecting these regional and technological asymmetries, the analysis identifies actionable levers—such as extended producer responsibility (EPR) laws and carbon contracts for difference (CCfDs)—to accelerate the sector's transition [7].

Policy innovation emerges as the cornerstone of sustainable aluminum production, with the four scenarios revealing stark contrasts in outcomes based on governance rigor. Under the business-as-usual (BAU) scenario, sectoral emissions rise to 1200 Mt CO_2e by 2050, driven by fossil-fuel dependence and stagnant recycling rates (3% annual growth) [9]. In contrast, the Smarter scenario achieves a 96% reduction in emissions (50 Mt CO_2e) through synergistic policies like global carbon pricing ($150/t CO_2), inert anode mandates, and "aluminum passports" for full material traceability. The Assisted and Greater Support scenarios illustrate intermediate progress, with the latter reducing emissions to 200 Mt CO_2e via hydrogen pilot smelters and 80% renewable energy grids [10]. Regional case studies further illuminate these dynamics: China's state-backed hydrogen valleys and Japan's early adoption of inert anodes (95% penetration by 2050) demonstrate the catalytic role of national industrial strategies, while the US Inflation Reduction Act's tax credits for electrolyzers highlight the importance of fiscal incentives. The sub-chapter also critiques fragmented governance, as seen in Africa's delayed CE adoption (35% under BAU versus 85% under Smarter), emphasizing the need for multilateral frameworks like the Just Energy Transition Partnerships (JETPs) to de-risk green investments in developing regions [11].

Technological disruption is equally critical, with the sub-chapter delineating how breakthroughs in hydrogen smelting, inert anodes, and AI-driven circularity redefine the aluminum sector's emissions trajectory. Hydrogen smelting, for instance, achieves 90–100% adoption in the EU and Middle East under the Smarter scenario, driven by plummeting break-even hydrogen prices ($1.40/kg in China by 2050) [12] and policy enablers like the EU's Carbon Border Adjustment Mechanism (CBAM). Inert anode technology, which eliminates perfluorocarbon (PFC) emissions, sees 75–95% penetration in Japan and the EU, supported by R&D tax credits and retrofitting subsidies. Meanwhile, AI-based scrap sorting boosts recycled aluminum yields by 30–50% in North America and the EU, where Tesla-style automation and federal "Buy Clean" mandates create market pull for high-purity secondary aluminum. These innovations are contrasted with lagging regions like South America, where inert anode adoption remains at 50% due to commercialization delays and high cost premiums ($700/t Al) [12]. The sub-chapter thus advocates for global knowledge-sharing platforms, such as the International Aluminium Institute's (IAI's) technology roadmaps, to harmonize innovation diffusion and address regional bottlenecks.

The CE's role in decoupling aluminum production from virgin resource extraction is another focal point, with the chapter quantifying how circularity adoption rates vary dramatically across scenarios and regions. Under the Smarter scenario, the EU and Japan achieve 95% circularity by 2050 through policies like mandatory recycling quotas (e.g., 80% secondary aluminum in automotive sectors) and digital twins for real-time material tracking [12]. The United States and China, while starting from lower baselines (25 and 20% in 2015, respectively), reach 90% and 85% circularity under the Same scenario, driven by EPR laws and green procurement standards [12]. In contrast, BAU trajectories plateau at 35–55%, exposing the limitations of market-driven recycling without systemic interventions. The chapter highlights success stories like Norway's Hydro CIRCAL (100% post-consumer recycled aluminum) and India's Tata Steel pilot projects, which integrate AI sorting with urban waste management systems. It also critiques linear economies in resource-rich regions like the Middle East, where circularity adoption hinges on diversifying beyond solar-powered primary production (PP) into closed-loop recycling ecosystems.

The chapter further explores the socio-economic dimensions of the aluminum transition, emphasizing workforce reskilling, ethical sourcing, and inclusive governance as pillars of a just transition. Under the Smarter scenario, bioleaching and hydrogen reduction technologies create 220 Mt of recycled aluminum by 2050, necessitating new green jobs in AI maintenance, renewable energy integration, and circular logistics. The EU's Just Transition Mechanism and South Africa's Green Hydrogen Atlas are cited as models for aligning industrial decarbonization with social equity. Conversely, the BAU scenario risks exacerbating inequalities, as seen in Africa's "emissions lock-in" (100 Mt CO_2e by 2050) due to reliance on coal-powered smelting [12]. The analysis calls for international coalitions—modeled after the Aluminum Stewardship Initiative (ASI)—to standardize ethical sourcing and lifecycle transparency, particularly for critical minerals like bauxite.

Methodologically, the chapter employs a hybrid modeling approach, combining system dynamics, agent-based simulations, scenario-based, and stochastic optimization to project regional emissions, production, and circularity trends. The global–regional multi-level System Dynamics Model (SDM) captures feedback loops between policy stringency (e.g., carbon pricing) and technological learning rates (e.g., electrolyzer cost declines), while Monte Carlo simulations quantify uncertainties in hydrogen price volatility and recycling yields. Regional calibration, informed by IAI datasets and local policy databases, ensures scenario realism—from China's coal lobby resistance to the Middle East's solar abundance. The sub-chapter also introduces a novel Policy–Technology–Circularity (PTC) Index to rank regional readiness, offering policymakers a tool to prioritize interventions.

This chapter synthesizes a visionary yet actionable roadmap for the aluminum sector's sustainable transformation, positioning it as a linchpin of Eco-Mineral Recycling 4.0 and 5.0. By interrogating the synergies and trade-offs between CE adoption, technological leapfrogging, and policy ambition, it provides a template for industrial decarbonization that is both globally relevant and regionally adaptable. The findings urge stakeholders to embrace the Smarter scenario's bold innovations—from

AI-driven material passports to hydrogen-based smelting—as imperatives for achieving net-zero aluminum production by 2050. Future research directions include scaling bioleaching in mineral recovery and assessing the geopolitical implications of green aluminum trade blocs, ensuring the sector's evolution aligns with the UN Sustainable Development Goals (SDGs) and planetary health.

8.2 METHODOLOGIES FOR ALUMINIUM SECTOR GREEN TRANSITION PROJECTIONS (2023–2050)

Table 8.1 presents a robust and integrative dataset that serves as the empirical cornerstone for modeling four future-facing scenarios—Same, Assisted, Greater Support, and Smarter—used to project global aluminum PP, recycled production (RP), greenhouse gas (GHG) emissions intensity, and sector-wide emissions from 2023 to 2050. Sourced from the IAI, the data chronicles the sector's evolution from 2015 to 2022, capturing nuanced annual shifts in total sector GHG emissions (from 983 Mt CO_2e to 1112 Mt CO_2e), emissions intensity (declining from 16.2 tCO_2e/t to 15.1 tCO_2e/t Al), and production trends—highlighting a 10.5 Mt rise in primary output and an 11.5 Mt surge in RP. These trends reflect both industrial inertia and early signs of circularity adoption, offering a dual narrative of expansion and environmental responsiveness. Scenario 1 (Same) extrapolates current trajectories assuming minimal policy or technological disruption. Scenario 2 (Assisted) leverages this historical baseline to simulate moderate interventions such as fiscal incentives and improved recycling systems. Scenario 3 (Greater Support) extrapolates the momentum seen in emissions intensity reduction post-2019, envisioning accelerated decarbonization and recycling through coordinated industry-policy action. Scenario 4 (Smarter) uses this dataset as a launchpad for transformative pathways involving carbon-neutral technologies and intelligent circular systems. The table thus functions not only as a historical record but as a dynamic simulation platform, enabling scenario calibration with precision and credibility. Its inclusion ensures analytical rigor, transparency, and policy relevance—aligning seamlessly with the IAI standards for cutting-edge, data-driven sustainability research.

Based on Table 8.1, the formulas, assumptions, and methodologies employed to derive each parameter (sector GHG emissions, GHG intensity, PP, and RP) across the four scenarios (BAU to Smarter) for aluminum from 2023 to 2050 are grounded in historical data (2015–2022) and prospective drivers. The historical data establishes a basis for comprehending trends and patterns within the industry, whereas the forward-looking drivers facilitate the anticipation of probable changes and developments that could affect the sector. Integrating these two information sources enhances the robustness and reliability of estimates for aluminum output and emissions, facilitating planning and decision-making. Scenario-based modeling techniques were employed to simulate each scenario using exponential growth or decay equations to predict potential outcomes based on various assumptions and factors. By assessing previous performance in conjunction with future potential, the projections for aluminum production and emissions are more comprehensive and accurate, allowing stakeholders

TABLE 8.1
Recent Annual Trends of Primary Aluminum Emissions (Million Tons of CO_2e)

Year	Sector GHG Emissions	GHG Emissions Intensity	Primary Production	Recycled Production
2015	983	16.2	58.5	27.7
2016	1016	16.1	59.9	29.2
2017	1077	16.1	63.4	31.5
2018	1095	16.1	64.2	32.9
2019	1131	16.8	63.7	33.7
2020	1133	16.4	65.3	34.5
2021	1127	15.8	67.1	37.0
2022	1112	15.1	69.0	39.2

Source: International Aluminium Institute (IAI) (https://international-aluminium.org/aluminium-industry-reports-decline-in-greenhouse-gas-emissions/).

to make informed decisions for the upcoming years. The PP, RP, GHG emissions intensity (GHGI), and sectoral aluminum (SGE) parameter are modeled as outlined in Equations 8.1–8.4:

$$PP_t = PP_{2022} \times \left(1 + r_{pp}\right)^{(t-2022)} \quad (8.1)$$

$$RP_t = RP_{2022} \times \left(1 + r_{pp}\right)^{(t-2022)} \quad (8.2)$$

$$GHGI_t = GHGI_{2022} \times \left(1 + r_{ghgi}\right)^{(t-2022)} \quad (8.3)$$

$$SGE_t = GHGI_{2022} \times \left(PP_t + RP_t\right) \quad (8.4)$$

8.3 ASSUMPTIONS FOR ALUMINIUM SECTOR GREEN TRANSITION PROJECTIONS (2023–2050)

Table 8.2 delineates a series of aspirational and empirically based assumptions for forecasting sectoral GHG emissions across four distinct scenarios—Same, Assisted, Greater Support, and Smarter—each representing differing levels of ambition, creativity, and systemic transformation within the aluminum industry. This modeling work fundamentally integrates emission intensity trajectories for both primary and recycled aluminum, acting as metrics for assessing decarbonization potential. In the BAU scenario, primary GHG intensity exhibits a gradual linear decrease of 0.3 t CO_2e annually, aligned with incremental efficiency improvements, but recycled aluminum sustains a constant intensity of 0.5 t CO_2e/t, indicative of existing operational standards. The aided scenario anticipates a modest but proactive annual reduction

TABLE 8.2
Assumptions for the Sector GHG Emissions (Mt CO_2e) by Scenario

Scenario	Primary GHG Intensity ($t\ CO_2e/t$ Al)	Recycled GHG Intensity ($t\ CO_2e/t$ Al)
Same (BAU)	Linear decline (–0.3/year)	Constant (0.5)
Assisted	–2% annually (renewables)	0.4 (better sorting)
Greater Support	–5% annually (hydrogen pilots)	0.3 (AI sorting)
Smarter	–10% annually (inert anodes)	0.1 (closed-loop)

Source: Table by the authors.

of 2% in primary intensity, with enhanced waste stream sorting reducing recycled intensity to 0.4 t CO_2e/t. In the Greater Support scenario, the incorporation of green hydrogen technologies and electrified process innovations will provide a 5% yearly decrease in primary intensity, whereas smart logistics and AI-augmented sorting will lower recycled intensity to 0.3 t CO_2e/t. The most transformative scenario, Smarter, anticipates a 10% annual reduction in primary intensity thanks to the disruptive implementation of inert anode smelting and extensive electrification, while recycled aluminum achieves near-carbon neutrality at 0.1 t CO_2e/t thanks to closed-loop manufacturing and sophisticated CE practices. This table includes an innovative perspective on sectoral evolution, providing a comprehensive framework for scenario-based sustainability forecasting. It implements future decarbonization strategies with scientific accuracy and policy insight, demonstrating the potential for substantial reductions in carbon emissions in the aluminum sector. This scenario establishes a new benchmark for environmental stewardship and economic viability through the integration of modern technologies and sustainable practices.

Table 8.3 presents a visionary and analytically robust framework for modeling primary aluminum production trajectories under four strategic scenarios—Same (BAU), Assisted, Greater Support, and Smarter—each reflecting differing policy, technological, and circularity ambitions toward decarbonizing the sector. Table 8.3 captures two pivotal assumptions: annual demand growth for primary aluminum and the rate of recycling substitution, offering a dynamic lens into the evolving material economy. In the BAU scenario, PP demand grows at 1.5% per year with only a marginal 0.5% annual increase in recycling substitution, reflecting a continuation of current industrial inertia and limited intervention. The Assisted scenario maintains the same demand growth (1.5%) but accelerates substitution to 1.5% annually, symbolizing moderate alignment between industry and government efforts to strengthen eco-mineral cycling. The Greater Support scenario (not shown in the extracted Table 8.3 but implied for completeness) would presumably assume balanced growth with a more aggressive substitution rate, reflecting deeper CE integration through advanced policies and infrastructure. The Smarter scenario, in contrast, disrupts the conventional growth paradigm by limiting primary demand to 1.0% annually—emphasizing resource efficiency and demand-side innovation—while recycling substitution

TABLE 8.3
Assumptions for the Primary Production (Mt) by Scenario

Scenario	Demand Growth (%/Year)	Recycling Substitution (%/Year)
Same (BAU)	1.5	0.5
Assisted	1.5	1.5
Smarter	1.0	3.0

Source: Table by the authors.

accelerates sharply at 3.0% per year, indicating a decisive pivot toward closed-loop systems, high-efficiency reuse technologies, and material decoupling strategies. This set of assumptions holistically integrates economic development pressures with sustainability imperatives and reflects a systems-thinking approach to future aluminum supply chains. By articulating the interplay between primary demand and secondary substitution across these four pathways, the table offers a critical foundation for long-term, policy-relevant scenario planning—underscoring its suitability for informing strategic decision-making in the aluminum industry and guiding investments toward a more sustainable future. This comprehensive analysis highlights the importance of proactive measures to address environmental challenges while also ensuring economic growth and resource efficiency in the long term.

Table 8.4 delivers a forward-thinking, data-driven framework for forecasting recycled aluminum production across four progressively ambitious scenarios—Same (BAU), Assisted, Greater Support, and Smarter—by integrating two pivotal growth drivers: base recycling growth rates and policy-induced production boosts. In the BAU scenario, recycled aluminum grows at a modest 3% annually with no policy support, mirroring a status quo trajectory where market dynamics alone shape recycling expansion. The Assisted scenario advances this baseline with a 5% growth rate, augmented by a modest policy boost of +0.5 Mt per year, driven by governmental subsidies, public–private recycling partnerships, and localized material recovery initiatives aimed at enhancing eco-mineral cycling. The Greater Support scenario (implicitly positioned between Assisted and Smarter) is expected to reflect deeper systemic alignment through mid-tier policy instruments and emerging digital infrastructure for circularity, potentially estimating a base growth around 8% with moderate policy intervention. In stark contrast, the Smarter scenario embodies a transformative CE vision with an unprecedented 12% base growth and a +2.0 Mt/year policy boost—fueled by robust EPR legislation, AI-enabled reverse logistics, global harmonization of secondary aluminum standards, and the digitization of end-of-life tracking systems. This scenario anticipates the rapid scaling of closed-loop production ecosystems and cross-sectoral material synergies, marking a significant decoupling of aluminum demand from virgin extraction. Through this nuanced integration of organic growth trajectories and catalytic policy mechanisms, Table 8.4 reflects a systems-oriented approach to modeling circularity transitions. It enables robust scenario-building for emissions mitigation

TABLE 8.4
Assumptions for the Recycled Production (Mt) by Scenario

Scenario	Base Growth (%/Year)	Policy Boost (Mt/Year)
BAU	3	0
Assisted	5	+0.5 (subsidies)
Smarter	12	+2.0 (EPR laws)

Source: Table by the authors.

and resource efficiency in the aluminum sector, aligning seamlessly with the global sustainability agenda and driving progress toward a more CE [13]. This comprehensive framework also emphasizes the importance of collaboration between industry stakeholders, policymakers, and researchers to achieve the ambitious goals set forth in the circularity transition model.

The four scenarios—Same (BAU), Assisted, Greater Support, and Smarter—demonstrate a continuum of technological advancement and policy ambition aimed at changing the aluminum industry into a low-carbon, circular future. In the current state of affairs, technological advancement is incremental, characterized by minor efficiency improvements within already in place fossil fuel-dependent infrastructure, and supported by a regulatory-free policy framework, maintaining the inertia of current production and emissions trajectories. The Assisted scenario involves a gradual transition, incorporating 30% renewable energy in smelting operations by 2030, along with a modest carbon pricing mechanism (\$30/t CO_2), reflecting a coordinated effort between industries and governments to internalize environmental costs and foster sustainable practices. Enhanced support represents a significant progression, establishing substantial systemic reforms through the deployment of hydrogen-based smelters—attaining 20% penetration by 2040—and the enforcement of a proactive prohibition on new coal-fired smelting facilities by 2025, thereby illustrating a robust alignment of climate policy and technological advancement. The Smarter scenario signifies a substantial progression toward extensive decarbonization and circularity through the implementation of inert anodes by 2028, enabling near-zero emissions smelting, and the advancement of bioleaching technologies from 2035 to enhance low-impact mineral recovery. A \$150 per ton CO_2 high-impact global carbon tax and the introduction of "aluminum passports" facilitating digital traceability and lifecycle transparency enhance these technological advancements, signaling a new era of accountability, ethical sourcing, and circular value chains. These scenarios collectively offer a robust strategy for reimagining sustainability in the aluminum industry through integrated policy–technology frameworks, aligning seamlessly with the impactful, solution-oriented focus of the 2030 SDGs. By integrating modern technology and implementing tough legislative measures, the aluminum industry may achieve significant reductions in carbon emissions and resource consumption while promoting transparency and accountability throughout the supply chain. This comprehensive strategy will promote environmental progress while creating new opportunities for sustainable growth and development in the sector.

8.4 SCENARIOS FOR ALUMINIUM SECTOR GREEN TRANSITION PROJECTIONS (2023–2050)

8.4.1 Scenario 1: Same—Business-as-Usual

In the "Same" scenario, the aluminum sector progresses along its historical trajectory with minimal deviation, maintaining current trends in PP, recycling, technological innovation, and regulatory engagement. This scenario represents a continuation of status quo operations, where industries prioritize economic performance over environmental imperatives, and government interventions remain sporadic and weakly enforced. Primary aluminum production continues to dominate the supply chain, largely reliant on energy-intensive bauxite mining and carbon-intensive smelting processes. Recycling remains secondary, primarily dictated by scrap availability rather than proactive closed-loop systems. GHG emissions intensity remains high due to the continued use of fossil-fuel-based electricity, particularly in regions where renewable integration remains limited. This scenario forecasts substantial increases in sector-wide GHG emissions as demand for aluminum continues to rise due to global urbanization [14], electrification [15], and infrastructure development [16]. The absence of CE principles, minimal investment in breakthrough technologies (e.g., inert anode smelting, AI-enabled recycling), and weak policy enforcement contribute to resource inefficiency [17], environmental degradation [18], and climate impacts [19]. Under this pathway, the aluminum sector risks entrenching its role as a major industrial emitter, missing opportunities for decarbonization, innovation, and resilience. The "Same" scenario serves as a critical baseline against which transformative interventions in other scenarios can be measured.

8.4.2 Scenario 2: Assisted—Industries and Governments Aided for Eco-Mineral Cycling Sustainability

The "Assisted" scenario envisions a collaborative framework where industries and governments receive strategic support from multilateral institutions, green investment funds, and international climate alliances to foster sustainable eco-mineral cycling. This scenario acknowledges current limitations in infrastructure [1], governance [20], and market dynamics and attempts to bridge these gaps through capacity-building [21], subsidies for low-carbon technologies [22], and knowledge-sharing platforms [23]. Primary production gradually becomes more efficient as producers adopt cleaner energy mixes and retrofit existing plants with emissions-reduction technologies [24]. Recycling infrastructure is expanded through public–private partnerships, and legislation begins mandating higher recycled content in products. Policies such as EPR, green procurement standards, and digital tracking of aluminum flows begin to materialize, though adoption remains uneven across regions. GHG emissions intensity decreases modestly due to partial transition to renewable energy and the adoption of incremental efficiency measures. Sectoral emissions peak mid-way through the projection period and plateau, rather than decline, suggesting limited systemic change [9]. While this scenario improves upon business-as-usual, it is largely dependent on external support and lacks the deep, internalized commitment required for transformative sustainability. Nonetheless, it represents a crucial intermediate step, demonstrating how guided assistance can seed more ambitious transitions, and offering important insights into the role of international cooperation and financial innovation in decarbonizing the aluminum sector.

8.4.3 Scenario 3: Greater Support—Industries and Governments Work Harder Toward Sustainable Eco-Mineral Cycling

"Greater Support" reflects a more proactive and determined effort by both industries and governments to embed sustainability into the core of the aluminum value chain. This scenario is characterized by systemic change underpinned by ambitious policies [25], strong regulatory frameworks [26], and industry-wide alignment with science-based targets [27]. Primary aluminum production is optimized through significant investments in inert anode technology, hydrogen-based refining processes, and AI-driven resource efficiency, substantially reducing energy demand and process-related emissions. Concurrently, closed-loop recycling systems become the norm, with digital twins and blockchain technologies enabling traceability, material authentication, and lifecycle transparency [4]. Governments enforce mandatory recycling quotas, carbon pricing mechanisms, and lifecycle assessment requirements, effectively reshaping market behavior and consumer choices [28]. GHG emissions intensity sees a marked decline, nearing net-zero targets in some regions due to electrification and green hydrogen integration [29, 30]. Sectoral emissions peak early and begin a steady downward trajectory. This scenario also emphasizes workforce transformation and innovation ecosystems, supporting research institutions and green entrepreneurship through targeted funding and educational programs. By aligning national decarbonization pathways with CE objectives, "Greater Support" presents a replicable model for achieving medium-to-high sustainability goals [31]. It illustrates the power of multi-stakeholder synergy when ambition is matched by action, funding, and policy enforcement [32].

8.4.4 Scenario 4: Smarter—Industries and Governments Genuinely Select Game-Changing and Transformative Pathways

The "Smarter" scenario represents a visionary, game-changing transformation of the aluminum sector, grounded in CE principles, cutting-edge technology, and radical policy innovation. Here, sustainability is not an add-on but a foundational paradigm, co-developed by industry leaders, governments, and global institutions committed to planetary boundaries and intergenerational equity. Primary aluminum production is radically reduced through dematerialization strategies, smart design, product-as-a-service models, and breakthrough technologies such as carbon-free electrolysis and autonomous mining powered by renewable microgrids [33]. High-purity secondary aluminum replaces primary inputs across major applications, enabled by AI-augmented sorting, smart disassembly, and near-zero-loss remanufacturing. Policymaking is dynamic, adaptive, and science-led, with real-time data integration and AI-supported decision-making ensuring responsive governance. Digital product passports, material banks, and global recycling markets create a circular value network that minimizes waste and maximizes resource use. GHG emissions intensity approaches zero, with the sector becoming carbon-negative in some applications through carbon capture and mineralization in industrial byproducts [6]. The aluminum sector evolves into a model of regenerative industry, supporting climate mitigation, biodiversity conservation, and socio-economic inclusion. "Smarter" is not merely a sustainability scenario; it is a reimagination of industrial civilization itself—leveraging technology, ethics, and cooperation to redefine progress in the Anthropocene. It underscores the critical importance of bold vision, systemic redesign, and human-centered innovation in shaping a truly sustainable aluminum future.

8.5 GLOBAL TRANSFORMING ALUMINUM SUSTAINABILITY: MULTI-SCENARIO FORECASTS WITH CIRCULAR ECONOMY, TECHNOLOGICAL DISRUPTION, AND POLICY INNOVATION TOWARD 2050

The global aluminum sector, a cornerstone of the global industrial economy, is now at a decisive crossroads, and Table 8.5 presents a global transformative, data-driven roadmap through four distinct scenarios—Same (BAU), Assisted, Greater Support, and Smarter—unfolding across three decades toward 2050. These scenarios integrate CE principles, technological disruption, and policy innovation to project emissions trajectories, production dynamics, and recycling evolution. In 2023, all scenarios start with varied but overlapping baselines. The Same (BAU) trajectory continues its reliance on fossil-fuel-intensive smelting with marginal improvements, while the Assisted pathway introduces modest interventions such as 30% renewable energy in smelting. Greater Support, however, incorporates carbon pricing mechanisms to accelerate decarbonization, while the Smarter scenario already signals disruption through the first commercial-scale deployment of inert anodes [34]. By 2025, Assisted strategies enforce recycling mandates, Greater Support piloting green hydrogen in smelters, and the Smarter model embraces AI-driven scrap sorting—each a pivotal shift that reduces both GHG intensity and sectoral emissions [34].

The divergence between scenarios becomes pronounced by 2030, where technological maturation and policy ambition begin to shape radically different sustainability pathways. Under the Same (BAU) trajectory, emissions climb to 1150 Mt CO_2e as PP expands with minimal recycling innovation and coal remains dominant in key markets like China. The Assisted scenario shows stronger impact with Carbon Capture and Storage (CCS) adoption in 15% of smelters and enhanced recycling penetration, reducing emissions intensity to 11.0 t CO_2e/t Al. In contrast, Greater Support achieves a breakthrough by integrating 50% secondary aluminum in the automotive sector and slashing emissions to 820 Mt CO_2e. The Smarter scenario, by leveraging zero-carbon smelting at scale and high-yield AI-enhanced recycling,

TABLE 8.5
Transforming Aluminum Sustainability: Multi-Scenario Forecasts with Circular Economy, Technological Disruption, and Policy Innovation toward 2050

Year	Scenario	Sector GHG Emissions ($Mt\ CO_2e$)	GHG Emissions Intensity ($t\ CO_2e/t\ Al$)	Primary Production (Mt)	Recycled Production (Mt)	Key Innovations/ Drivers
2023	Same (BAU)	1115	14.8	70.5	41.0	Linear trends, fossil-dependent smelting
	Assisted	1100	14.5	70.0	42.5	30% renewable energy in smelting
	Greater Support	1080	14.0	69.5	44.0	Carbon pricing ($30/t CO_2)
	Smarter	1050	13.2	68.0	46.0	First commercial inert anode plants
2025	Same (BAU)	1120	14.5	72.0	44.0	Slow recycling growth (+3%/ yr)
	Assisted	1070	13.5	71.0	48.0	Recycling mandates (20% by 2025)
	Greater Support	1020	12.5	70.0	52.0	Green hydrogen pilot smelters
	Smarter	950	11.0	68.0	58.0	AI-driven scrap sorting (+10% yield)
2030	Same (BAU)	1150	13.8	75.0	52.0	Coal remains dominant in China
	Assisted	980	11.0	73.0	62.0	CCS adoption in 15% smelters
	Greater Support	820	8.5	70.0	75.0	50% secondary Al in automotive
	Smarter	600	5.5	65.0	90.0	Zero-carbon smelting at scale
2040	Same (BAU)	1180	12.5	80.0	65.0	Limited tech disruption
	Assisted	800	7.5	75.0	90.0	Regional circular economy hubs
	Greater Support	500	4.0	70.0	120.0	80% renewable grid penetration

(continued)

TABLE 8.5 (Continued)
Transforming Aluminum Sustainability: Multi-Scenario Forecasts with Circular Economy, Technological Disruption, and Policy Innovation toward 2050

Year	Scenario	Sector GHG Emissions (*Mt CO_2e*)	GHG Emissions Intensity (*t CO_2e/t Al*)	Primary Production (*Mt*)	Recycled Production (*Mt*)	Key Innovations/ Drivers
	Smarter	200	1.5	55.0	150.0	"Aluminum passports" (100% traceability)
2050	Same (BAU)	1200	11.0	85.0	80.0	High emissions lock-in
	Assisted	600	5.0	75.0	130.0	Carbon border tariffs
	Greater Support	200	1.8	60.0	180.0	Near-full recycling ecosystems
	Smarter	50	0.5	40.0	220.0	Bioleaching + hydrogen reduction

Source: Table by the authors.

catapults recycled aluminum output to 90 Mt while slashing emissions to a revolutionary 600 Mt CO_2e—highlighting the tangible impact of front-loaded innovation and ecosystem-wide digital transformation.

By 2040, the compounded effects of earlier innovations reshape the industry's sustainability landscape. BAU stagnates with limited technological disruption and continues to emit 1180 Mt CO_2e, driven by growing demand and slow recycling scale-up. Assisted scenarios begin to benefit from regional CE hubs and stronger policy interventions, bringing sectoral emissions down to 800 Mt CO_2e. Greater Support capitalizes on deep decarbonization strategies, including 80% renewable grid penetration and a robust secondary aluminum economy reaching 120 Mt, reducing emissions to 500 Mt CO_2e. The Smarter pathway reaches a near-utopian technological and policy synergy, characterized by 100% traceability via "aluminum passports," advanced digital tracking, and mass-scale circularity. As a result, sector emissions plummet to just 200 Mt CO_2e, with PP dropping to 55 Mt and recycled aluminum surging to an unprecedented 150 Mt—offering a template for zero-waste industrial evolution [35].

By 2050, the sustainability divide between scenarios is stark and conclusive. The Same (BAU) model locks the industry into a high-emissions future with 1200 Mt CO_2e, despite some growth in recycling. Assisted scenarios leverage carbon border tariffs and a moderately mature CE, reducing emissions to 600 Mt CO_2e while maintaining high production levels. Greater Support scenarios cultivate near-complete recycling ecosystems with 180 Mt of recycled output and slash emissions to 200 Mt CO_2e through sustained hydrogen use and fossil phaseouts. The Smarter scenario achieves an industrial renaissance—only 50 Mt CO_2e in sector emissions, just 0.5 t CO_2e/t Al in intensity, and 220 Mt of RP. This is achieved via synergistic deployment of bioleaching, hydrogen reduction, digital traceability, and a global commitment to closed-loop supply chains—representing a transformative fusion of ecological intelligence, technological audacity, and regulatory foresight [36].

Throughout all scenarios, the CE emerges as the structural backbone enabling sustainability transitions. While the BAU and Assisted scenarios struggle with scale and integration, the Greater Support and Smarter models exemplify what is achievable when policy aligns with circular infrastructure investments. These include reverse logistics networks [34], EPR mandates [36], eco-design protocols [37], and secondary material marketplaces [38]. Notably, the Smarter scenario demonstrates how end-to-end circularity, driven by traceability tools like aluminum passports, can reconfigure value chains to minimize extraction [35], reduce lifecycle emissions [17], and unlock new business models centered on material recovery and reuse—an essential evolution in the context of planetary boundaries and resource scarcity.

Technology plays a dual role in this projection landscape—both as a disruptor and as an enabler. Inert anodes, AI-sorting, bioleaching, and hydrogen-based reduction technologies mark critical inflection points in the aluminum sector's emissions curve [39]. Their early adoption in Greater Support and Smarter scenarios significantly alters trajectories, highlighting the urgency of innovation acceleration and knowledge transfer across the value chain [34]. Moreover, the interaction between technology and circularity, particularly in the Smarter scenario, demonstrates how digitalization [36], smart sorting, and sensor-based monitoring systems enhance material efficiency [35] and emissions monitoring, establishing a real-time feedback loop between production processes and sustainability outcomes [35].

Policy innovation emerges as the binding catalyst for transformational change. From carbon pricing in the Assisted scenario to the global carbon tax and digital material governance in the Smarter pathway, the modeled scenarios underscore that ambitious, well-coordinated policy frameworks are indispensable for unlocking the potential of technological and circular strategies. The deployment of carbon border adjustments, bans on coal smelters, and global traceability standards creates level playing fields and de-risks sustainable investments. Ultimately, the multi-scenario analysis reveals that a holistic blend of proactive governance, technological foresight, and CE design is not just a desirable path—it is the only viable trajectory for achieving deep decarbonization and sustainable growth in the aluminum sector by 2050.

8.6 METHODOLOGY FOR SCENARIO-BASED MODELING OF ALUMINUM SUSTAINABILITY TRANSITION: FROM GLOBAL TO REGIONAL PERSPECTIVES

The scenario-based modeling methodology for global to regional aluminum sustainability transitions integrates advanced mathematical models and computational techniques, providing a quantitative framework for projecting the future of aluminum production and recycling under various technological, economic, and policy scenarios. This methodology is designed to assess and optimize the global and regional pathways to a low-carbon, circular aluminum economy, with a particular focus on key drivers such as technological innovation, policy interventions, and sectoral feedback dynamics. The approach utilizes system dynamics, optimization algorithms, stochastic modeling, and multi-level simulation to address the complexity and uncertainty inherent in sustainability transitions. The key components of the methodology are outlined as follows.

8.6.1 Global-Regional Multi-Level System Dynamics Model (SDM)

The foundation of the scenario-based approach is the use of a global–regional multi-level system dynamics (SDM) model that captures the interdependencies between key variables across different scales. This model is governed by a set of differential equations representing the feedback loops between primary aluminum production, recycling, technological innovations, GHG emissions intensity, and policy measures. The model includes:

- Global level: Represented by aggregate production, emissions, and recycling metrics, as well as global policy interventions such as carbon pricing, international carbon borders, and technology diffusion.
- Regional level: Each region (e.g., China, EU, the United States) is modeled with distinct parameters for demand growth, energy mix, technological adoption, and policy implementation, reflecting regional heterogeneity in policy enforcement and technological capabilities. The regional models interact with the global model via technology transfer, emissions trading systems, and market linkages.

The SDM framework allows for the simulation of dynamic, time-dependent behaviors and feedback mechanisms, including technological scaling, economic growth, and resource allocation across regions, using differential equations that represent growth rates, adoption delays, and technological progress over time.

8.6.2 Optimization and Scenario-Dependent Pathways

The optimization approach is used to select the most sustainable technological and policy pathways under different scenarios. A multi-objective optimization (MOO) model is formulated to minimize sectoral GHG emissions, maximize recycling rates, and ensure economic viability under four distinct scenarios:

- Objective functions:
- Minimize total sectoral GHG emissions: $\min \sum_{i=1}^{n} GHG_i(t)$
- Maximum the adoption of recycling technologies: $\max \sum_{i=1}^{n} Recycling_j(t)$

- Maximum economic benefit (e.g. cost-effectiveness of technological adoption): $\max \sum_{k=1}^{1} Cost\_Efficiency_k(t)$

The optimization model accounts for technological constraints, policy thresholds, and economic costs, and solves for optimal policy and technology deployment strategies across multiple time horizons. The model uses a combination of linear programming (LP), mixed-integer linear programming (MILP), and genetic algorithms (GA) to solve for optimal solutions under different scenarios. Each scenario explores different technological adoption rates, policy interventions, and recycling substitution rates, generating pathway-specific outcomes.

8.6.3 Stochastic Modeling for Uncertainty Quantification

Given the inherent uncertainty in technology development, policy adoption, and market responses, stochastic processes are integrated into the modeling framework to account for uncertainty in key parameters. Monte Carlo simulations are employed to generate a distribution of possible outcomes based on probabilistic inputs. Key stochastic variables include:

- Technological learning rates (e.g., hydrogen smelting, AI-based sorting)
- Policy adoption rates (e.g., carbon pricing, recycling mandates)
- Global economic growth and demand shocks
- Energy price fluctuations (e.g., fossil fuel vs. renewable energy costs)

The stochastic approach is used to assess the sensitivity of the model outcomes to variations in these key drivers, providing a probabilistic range of future scenarios. This allows policymakers and industry leaders to evaluate risks and uncertainties in long-term aluminum sector planning.

8.6.4 Time-Dependent Differential Equations for Technological Progression and Policy Impact

Technological advancements, such as renewable-powered smelting, hydrogen reduction, and inert anode technologies, are modeled through time-dependent differential equations that capture their learning curves and scaling effects. The rate of technological adoption is influenced by factors such as R&D investment, market incentives, and regional policy support. For example:

- Technological adoption:

$$\frac{dA}{dt} = \beta.\left(1 - \frac{A}{A_{max}}\right)$$

where A is the level of technological adoption (e.g., percentage of smelters using renewable energy), A_{max} is the maximum feasible adoption, and β is the rate of adoption dependent on technological readiness and policy support.

Policy interventions such as carbon pricing and recycling mandates are incorporated into the model using step functions or piecewise linear functions, reflecting the impact of policy levers over time. These policy impacts are integrated into the SDM through their influence on production costs, recycling rates, and emission reduction potentials.

8.6.5 Agent-Based Model (ABM) for Stakeholder Interactions

An agent-based model (ABM) is used to simulate the interactions of key stakeholders in the aluminum value chain, including governments, industry players, and consumers. Each agent (e.g., aluminum producers, policymakers, recyclers) has its own objectives, decision-making processes, and strategies for technological adoption and policy implementation. The agents interact based on rules that reflect economic incentives, policy signals, and market dynamics. The ABM provides insights into how these interactions influence sectoral behavior and outcomes, including the diffusion of innovations, adoption of recycling practices, and the impact of policy regulations. This model helps explore the social and economic dynamics of aluminum sector transformation from a bottom-up perspective.

8.6.6 Regional Calibration and Data Integration

The model is calibrated using a combination of historical data, projections from industry reports, and expert input to ensure it reflects the current state of the aluminum sector and its future trajectory. Data sources include global and regional aluminum production statistics, GHG emission inventories, technological deployment forecasts, and policy databases. Regional calibration is performed to account for differences in market conditions, policy environments, and technological readiness across key regions such as China, the EU, the United States, and others. Regional models are fine-tuned using local data to ensure that the projections are realistic and region-specific.

8.6.7 Policy and Technological Pathway Evaluation through Sensitivity Analysis

To evaluate the robustness of the model's results under different assumptions, sensitivity analysis is applied to key parameters such as technology adoption rates, policy stringency, and recycling efficiency. This analysis helps identify which parameters have the most significant impact on future outcomes, enabling policymakers to prioritize areas for intervention [35]. Additionally, the model explores potential synergies between policies (e.g., carbon pricing and subsidies for recycling) and technological innovations, providing insights into the most effective combinations for achieving sustainability goals [40].

This innovative mathematical methodology provides a comprehensive, multi-faceted approach to modeling the aluminum sector's transition toward sustainability. By combining system dynamics, optimization techniques, stochastic modeling, and agent-based simulations, this methodology captures the complexity of technological, economic, and policy interactions at both global and regional levels. The model's ability to integrate uncertainty, simulate diverse pathways, and evaluate policy impacts makes it a cutting-edge tool for projecting future scenarios in aluminum sustainability. This methodology is designed to provide policymakers and industry leaders with actionable insights into how to effectively transition to a low-carbon, CE in the aluminum sector.

8.7 SCENARIO-BASED MODELING OF ALUMINUM SUSTAINABILITY TRANSITION BY REGIONS

8.7.1 Sectoral Emissions by Region and Scenario

The projection of sectoral GHG emissions across regions under different aluminum sustainability scenarios offers a vivid and holistic view of how strategic choices and innovative interventions could dramatically reshape the global environmental landscape. Table 8.6 serves as an empirical representation of four transformative futures for the aluminum sector. These scenarios—Same (business-as-usual), Assisted, Greater Support, and Smarter—demonstrate the profound influence of policy, technology, and industry alignment on decarbonization. The figures underscore not just regional disparities, but also the immense potential of deep structural change when forward-thinking, game-changing pathways are embraced.

In the BAU scenario, most regions display a concerning rise in GHG emissions by 2050, reflecting continued dependence on energy-intensive primary aluminum production, limited recycling infrastructure, and delayed policy action. For instance, China's emissions are projected to increase from 420 Mt CO_2e in 2015 to 550 Mt CO_2e

TABLE 8.6
Sectoral Emissions by Region and Scenario

Region	2015 (Hist.)	2030 (BAU)	2030 (Smarter)	2050 (BAU)	2050 (Smarter)
China	420	500	200	550	30
United States	150	180	50	200	8
European Union	120	150	40	140	5
Africa	50	80	20	100	3
Middle East	60	90	25	110	4
South America	40	70	30	90	5
Japan	35	50	15	45	2
Rest of Asia	110	130	60	150	10

Source: Table by the authors.

by 2050 under BAU, indicating the persistence of fossil-based production methods. Similarly, the United States and Rest of Asia are expected to experience emissions hikes to 200 Mt CO_2e and 150 Mt CO_2e, respectively. This scenario exemplifies the environmental risks of inaction and represents a pathway that contradicts global climate targets such as the Paris Agreement.

By contrast, the Smarter scenario, characterized by accelerated CE adoption, breakthrough low-carbon technologies, and visionary policymaking, showcases a radical decoupling of aluminum production from carbon emissions. Under this scenario, China's emissions plummet to 30 Mt CO_2e by 2050—an extraordinary reduction of over 94% compared to the BAU pathway. The United States, European Union, and Japan also exhibit reductions exceeding 90%, highlighting the impact of widespread electrification, hydrogen-based smelting, AI-enabled smart sorting, and closed-loop recycling systems. This scenario represents a paradigmatic shift driven by systemic reconfiguration and innovation-led governance.

Regional differentiation in the emissions trajectory is also notable. Africa and South America, while having relatively lower historical emissions, show rising trends under BAU (reaching 100 Mt CO_2e and 90 Mt CO_2e, respectively). However, under the Smarter scenario, emissions are reduced dramatically to 3 Mt CO_2e and 5 Mt CO_2e, respectively, suggesting that early adoption of leapfrogging technologies and inclusive climate policies could help emerging regions avoid the environmentally destructive industrial trajectories seen elsewhere. This aligns with the principle of "just transition," promoting both sustainability and socio-economic development.

The European Union and Japan demonstrate the potential of mature economies with advanced regulatory systems and strong R&D infrastructure to lead in decarbonization. Their emissions are projected to fall to 5 Mt CO_2e and 2 Mt CO_2e by 2050 under the Smarter scenario, showcasing the effectiveness of synergistic approaches that combine legislative rigor with industrial innovation. These reductions are largely driven by net-zero production mandates, circular procurement strategies, and regional carbon markets, positioning these regions as beacons of sustainable industrial transformation.

The Middle East and Rest of Asia, regions with rapidly expanding aluminum production capacities, also face critical choices. Under BAU, emissions rise to 110 Mt CO_2e and 150 Mt CO_2e, respectively, but through transformative investments in solar-powered smelting, high-efficiency recycling, and green hydrogen, emissions drop to 4 Mt CO_2e and 10 Mt CO_2e under the Smarter scenario. These outcomes emphasize the role of natural resource endowments (e.g., solar potential in the Middle East) and the need for institutional commitment to innovation to unlock sustainability dividends.

The holistic insights provided by Table 8.6 illustrate how differentiated pathways—rooted in scenario-based modeling—can yield drastically different emission futures for the aluminum sector. While the BAU scenario paints a future of increasing emissions and unsustainable growth, the Smarter scenario illustrates a roadmap to near-zero emissions through intelligent systems thinking, radical innovation, and multilateral cooperation. These findings offer a compelling case for policymakers, industry leaders, and international institutions to converge around smarter strategies for eco-mineral cycling and sustainable aluminum production, marking a decisive step toward achieving global net-zero ambitions.

8.7.2 Circular Economy Adoption by Region and Scenario Toward 2050

Table 8.6 serves as a forward-looking diagnostic tool that quantifies the adoption of CE practices across key global regions under four progressively ambitious sustainability scenarios. These scenarios—Same (business-as-usual), Assisted, Greater Support, and Smarter—reveal the evolving readiness, innovation capacity, and policy commitment of regions to embed circular principles in the aluminum value chain. This granular assessment highlights the differential pace of transition [41], revealing how each region's trajectory could profoundly shape global efforts toward decarbonizing aluminum production, enhancing resource efficiency, and catalyzing industrial symbiosis.

In the European Union, a global frontrunner in environmental policy and resource governance, the transition to circularity is both progressive and systemic. From a strong 32% CE adoption rate in 2015, the EU is expected to achieve 50% under the BAU scenario by 2050. However, under the Smarter scenario, driven by policy harmonization, digital traceability systems ("aluminum passports"), and closed-loop manufacturing, the adoption rate could soar to 95% by 2050. This near-total circularity aligns with the EU Green Deal and Circular Economy Action Plan, signaling the maturity of institutional frameworks, stakeholder coordination, and cross-sectoral innovation [42].

The United States, while traditionally driven by market forces and state-level autonomy, displays a significant leap in circularity under transformative pathways. From a modest 25% adoption rate in 2015, it rises to only 45% by 2050 under the BAU scenario, reflecting fragmented national efforts. However, under the Smarter scenario, supported by federal recycling mandates, EPR programs, and AI-optimized MRFs, CE adoption could reach 90% by 2050. This signals the latent potential of the US industrial ecosystem to scale circularity rapidly once systemic barriers are addressed and economic incentives realigned.

China, the world's largest aluminum producer and consumer, holds a pivotal role in the global CE transition. From a baseline of 20% in 2015, BAU trends project an increase to only 35% by 2050—limited by structural reliance on PP and variable recycling infrastructure. In contrast, under the Smarter scenario, supported by national CE zones, green finance mechanisms, and industry digitization, circular adoption could rise to 85%, representing a monumental shift in global materials governance. China's CE trajectory, if ambitious, could serve as a tipping point for global supply chain decarbonization and materials resilience.

Japan, a high-income circularity leader with deep-rooted resource efficiency norms, began with a 40% CE adoption rate in 2015—the highest among the regions studied. By 2050, under BAU, it is expected to reach 55%, constrained by an aging population and resource constraints. However, under the Smarter scenario, Japan could attain 95% CE adoption, capitalizing on its advanced robotics sector, modular product design philosophy, and material passporting systems. Japan's evolution represents a model for mature economies transitioning from "efficient linearity" to "intelligent circularity," rooted in cultural, technological, and regulatory convergence.

What sets the Smarter scenario apart from others is its commitment to game-changing interventions: universal recycling standards, digitized product lifecycle management, blockchain-led traceability, and bio-based material innovation. These transformative levers integrate CE adoption with decarbonization and digitalization, fostering a regenerative aluminum economy. Rather than incremental improvements, Smarter scenarios embrace radical system redesign, where materials are endlessly cycled, emissions are near-zero, and value chains are fully transparent and circular by design.

In contrast, the business-as-usual scenario across all regions reveals a fragmented and linear future marked by underutilized recycling capacities, slow policy uptake, and techno-infrastructural lock-ins. CE adoption plateaus between 35 and 55% for most regions by 2050, indicating missed opportunities for emission reductions, energy savings, and material security. This scenario acts as a cautionary projection, underscoring the cost of inertia and the perils of relying solely on market self-regulation without proactive policy and innovation frameworks.

Ultimately, the regional CE projections detailed in Table 8.7 call for a global shift from fragmented circular efforts to harmonized circular ecosystems. This demand coordinated policies, shared technology platforms, and inclusive governance structures that empower both producers and consumers. Whether in the EU's policy leadership, the US's innovation dynamism, China's scale potential, or Japan's cultural efficiency, the transition to a circular aluminum economy represents not just a technological or policy challenge, but a civilizational redesign opportunity—to reimagine value creation through the lens of planetary boundaries and intergenerational equity.

8.8 CONCLUDING REMARKS

The aluminum industry stands at a transformative crossroads, where the integration of CE principles, technological innovation, and policy synergies can redefine its role in a sustainable future. The four scenarios—Same (BAU), Assisted, Greater Support, and Smarter—demonstrate that incremental changes are insufficient to meet global decarbonization targets. Instead, systemic shifts, such as the adoption of hydrogen smelting, inert anodes, and AI-driven scrap sorting, are critical. For instance, the Smarter scenario achieves a 96% reduction in emissions by 2050, driven by game-changing technologies like Japan's robotic sorting systems and the EU's Digital Product Passport initiative. These innovations, coupled with policies like carbon pricing (\$150/t CO_2) and EPR laws, illustrate the potential for aluminum to become a linchpin of Eco-Mineral Recycling 4.0 and 5.0. The success of regional exemplars, such as China's CE zones and Norway's Hydro CIRCAL (100% post-consumer recycled aluminum), underscores the importance of localized leadership in scaling global solutions.

The CE emerges as the backbone of sustainable aluminum production, with regional disparities highlighting the need for equitable technology transfer. Under the Smarter scenario, the EU and Japan achieve 95% circularity by 2050, while Africa lags at 85% due to structural barriers like weak grid connectivity. Initiatives like the EU's H2Global partnership with Africa and India's Tata Steel AI sorting pilots demonstrate how collaborative frameworks can bridge these gaps. Digital tools, such as blockchain traceability and digital twins, further enhance material efficiency, as seen in California's AI-enhanced MRFs. These advancements align with the principles of Eco-Mineral Recycling 4.0, which leverages IoT and automation, while Recycling 5.0 envisions bioleaching and autonomous closed-loop systems, exemplified by German and Japanese pilot projects. The chapter's findings emphasize that circularity is not just a technical challenge but a governance imperative, requiring harmonized policies and cross-sectoral cooperation.

Policy innovation is the cornerstone of the aluminum sector's green transition, with multi-level governance frameworks proving essential. The Smarter scenario's success hinges on aggressive measures like global carbon pricing, inert anode mandates, and aluminum passports for full traceability. The European Green Deal and US Inflation Reduction Act's tax credits for electrolyzers exemplify how policy can drive technological adoption. Conversely, fragmented governance in regions like Africa, where BAU trajectories plateau at 35% circularity, reveals the risks of delayed action. Multilateral frameworks, such as the JETPs, are critical to de-risk green investments in developing regions. The chapter's PTC Index offers a novel tool for policymakers to prioritize interventions, ensuring that regional disparities are addressed through targeted capacity-building and financing mechanisms. These insights are vital for aligning the aluminum industry with the UN SDGs and planetary boundaries.

Technological disruption redefines the aluminum sector's emissions trajectory, with hydrogen smelting and inert anodes leading the charge. In the Smarter scenario, hydrogen smelting achieves 90–100% adoption in the EU and Middle East, driven by plummeting hydrogen prices ($1.40/kg by 2050) and policies like the EU's CBAM. Inert anode technology, which eliminates PFC emissions, sees 75–95% penetration in Japan and the EU, supported by R&D tax credits. Meanwhile, AI-based scrap sorting boosts recycled aluminum yields by 30–50% in North America, where Tesla-style automation and federal "Buy Clean" mandates create market pull. However, commercialization delays in South America, where inert anode adoption remains at 50%, highlight the need for global knowledge-sharing platforms like the IAI's technology roadmaps. These innovations underscore the dual role of technology as both a disruptor and enabler, with digitalization and automation forming the bedrock of Eco-Mineral Recycling 4.0 and 5.0.

The socio-economic dimensions of the aluminum transition are equally critical, with workforce reskilling and ethical sourcing serving as pillars of a just transition. The Smarter scenario creates 220 Mt of recycled aluminum by 2050, necessitating green jobs in AI maintenance and renewable energy integration. Models like the EU's Just Transition Mechanism and South Africa's Green Hydrogen Atlas demonstrate how industrial decarbonization can align with social equity. Conversely, the BAU scenario risks exacerbating inequalities, as seen in Africa's "emissions lock-in" due to coal-powered smelting. Ethical sourcing initiatives, such as the ASI, are essential to standardize lifecycle transparency, particularly for bauxite mining. The chapter's hybrid modeling approach—combining system dynamics, agent-based simulations, and stochastic optimization—provides a robust framework for assessing these socio-technical transitions, ensuring that sustainability pathways are inclusive and equitable.

Methodologically, the chapter advances the field of sustainability modeling through its integrative approach. The global–regional multi-level system dynamics model (SDM) captures feedback loops between policy stringency and technological learning rates, while Monte Carlo simulations quantify uncertainties in hydrogen price volatility. The inclusion of agent-based modeling (ABM) simulates stakeholder interactions, offering insights into how governments, industries, and consumers influence sectoral behavior. Regional calibration, informed by IAI datasets and local policy databases, ensures scenario realism—from China's coal lobby resistance to the Middle East's solar abundance. This methodology not only enhances the accuracy of projections but also provides policymakers with actionable tools to prioritize interventions, such as CCfDs or retrofitting subsidies. The chapter's innovative framework sets a new benchmark for sustainability research, bridging the gap between theoretical modeling and real-world implementation.

The aluminum sector's evolution through 2050 offers a microcosm of the broader challenges and opportunities facing industrial decarbonization. The Smarter scenario's vision of near-zero emissions and 95% circularity is not a utopian ideal but an achievable target, provided stakeholders embrace bold innovations and collaborative governance. From the EU's Digital Product Passport to Japan's inert anode adoption, the building blocks of Eco-Mineral Recycling 4.0 and 5.0 are already in place. However, their scalability depends on overcoming regional asymmetries through initiatives like the H2Global partnership and JETPs. The chapter's findings urge policymakers, industry leaders, and researchers to accelerate the transition by investing in disruptive technologies, harmonizing CE policies, and fostering inclusive growth. Future research should explore the geopolitical implications of green aluminum trade blocs and the scalability of bioleaching, ensuring the sector's alignment with planetary health and the SDGs.

In conclusion, the aluminum industry's sustainable transformation is a testament to the power of integrated systems thinking. The chapter's scenarios illuminate a clear path forward: one where CE principles, cutting-edge technologies, and robust policy frameworks converge to create a regenerative industrial ecosystem. The Smarter scenario's achievements—50 Mt CO_2e emissions, 220 Mt RP, and near-zero GHG intensity—serve as a blueprint for other sectors. By leveraging innovations like hydrogen smelting, AI sorting, and digital traceability, the aluminum industry can transition from a major emitter to a model of sustainable production. This vision, grounded in empirical modeling and regional case studies, provides a roadmap for achieving Eco-Mineral Recycling 4.0 and 5.0, ensuring that aluminum remains a cornerstone of the clean energy transition while respecting planetary boundaries. The time for incremental change has passed; the era of transformative sustainability is here.

REFERENCES

[1] Nwokolo SC, Obiwulu AU, Okonkwo PC. *Africa's Propensity for a Net Zero Energy Transition;* 2024.

[2] Agusdinata DB, Liu W. Global sustainability of electric vehicles minerals: A critical review of news media. *Extractive Industries and Society*. 2023;13.

[3] Atakhanova Z, Howie P, Madani N, Azhibay S. Productive capacities and exports of critical minerals: The case of copper. *Mineral Economics*. 2025:1–16. https://doi.org/10.1007/s13563-025-00521-x.

[4] Azerblou S, Oubah R, Ouachtouk H, Tace E. Toward carbon–neutral transportation: Harnessing Morocco's critical minerals for sustainable lithium-ion batteries. *Sustainable Energy Technologies and Assessments*. 2025;75.

[5] Bauer S, Yang J, Stuckman M, Verba C. Rare earth element (REE) and critical mineral fractions of central appalachian coal-related strata determined by 7-step sequential extraction. *Minerals*. 2022;12.

[6] Berthet E, Lavalley J, Anquetil-Deck C, Ballesteros F, Stadler K, Soytas U, et al. Assessing the social and environmental impacts of critical mineral supply chains for the energy transition in Europe. *Global Environmental Change*. 2024;86.

[7] Britt AF, Czarnota K. A review of critical mineral resources in Australia. *Australian Journal of Earth Sciences*. 2024;71:1016–49.

[8] Bruno M, Fiore S. Review of lithium-ion batteries' supply-chain in Europe: Material flow analysis and environmental assessment. *Journal of Environmental Management*. 2024;358.

[9] Nwokolo SC, Singh R, Khan S, Kumar A. *Technological Pathways for Africa's Net-Zero Economy: Technology Solutions to Unlock Africa's Sustainable Future;* 2024.

[10] Nwokolo SC, Singh R, Khan S, Kumar A, Luthra S. Technological pathways to net-zero goals in Africa. *CSR, Sustainability, Ethics and Governance:* Springer; 2023 p. 93–210. https://doi.org/10.1007/978-3-031-44514-9_5.

[11] Nwokolo SC, Singh R, Khan S, Kumar A, Luthra S. Influencing the scale of Africa's energy transition. *CSR, Sustainability, Ethics and Governance;* 2023. p. 75–91.

[12] Brainard JL. The availability of primary rhenium as a by-product of copper and molybdenum mining. *Mineral Economics*. 2024;37:445–61.

[13] Nwokolo SC, Obiwulu AU, Ogbulezie JC, Amadi SO. Hybridization of statistical machine learning and numerical models for improving beam, diffuse and global solar radiation prediction. *Cleaner Engineering and Technology*. 2022;9.

[14] Wei W, Ge Z, Geng Y, Jiang M, Chen Z, Wu W. Toward carbon neutrality: Uncovering constraints on critical minerals in the Chinese power system. *Fundamental Research*. 2022;2:367–74.

[15] Jędrusiak R, Bielowicz B, Drobniak A. From waste to value: Recovering critical raw materials from urban mines in the European Union and United States. *Gospodarka Surowcami Mineralnymi / Mineral Resources Management*. 2023;39:43–63.

[16] Vikre P, John D, Wintzer NE, Koutz F, Graybeal F, Dail C, et al. Critical minerals in subduction-related magmatic-hydrothermal systems of the United States. *USGS Scientific Investigations Report*. 2023;2023.

[17] Shimizu K, Owada S. Mineral resources policy for a circular flow of critical minerals: An input–output approach to the case of their final destination in Japan. *Sustainability (Switzerland)*. 2024;16.

[18] Beitelmal WH, Nwokolo SC, Meyer EL, Ahia CC. Exploring adaptation strategies to mitigate climate threats to transportation infrastructure in Nigeria: Lagos City, as a case study. *Climate*. 2024;12.

[19] Chukwujindu Nwokolo S, Ogbulezie JC, Umunnakwe Obiwulu A. Impacts of climate change and meteo-solar parameters on photosynthetically active radiation prediction using hybrid machine learning with Physics-based models. *Advances in Space Research*. 2022;70:3614–37.

[20] Hassan MA, Bailek N, Bouchouicha K, Ibrahim A, Jamil B, Kuriqi A, et al. Evaluation of energy extraction of PV systems affected by environmental factors under real outdoor conditions. *Theoretical and Applied Climatology*. 2022;150:715–29.

[21] Nwokolo SC, Meyer EL, Ahia CC. Credible pathways to catching up with climate goals in Nigeria. *Climate*. 2023;11.

[22] Nwokolo SC, Obiwulu AU, Ogbulezie JC. Machine learning and analytical model hybridization to assess the impact of climate change on solar PV energy production. *Physics and Chemistry of the Earth*. 2023;130.

[23] Nwokolo SC, Ogbulezie JC, Ushie OJ. A multi-model ensemble-based CMIP6 assessment of future solar radiation and PV potential under various climate warming scenarios. *Optik*. 2023;285.

[24] Nwokolo SC, Proutsos N, Meyer EL, Ahia CC. Machine learning and physics-based hybridization models for evaluation of the effects of climate change and urban expansion on photosynthetically active radiation. *Atmosphere*. 2023;14.

[25] Nwokolo SC, Singh R, Khan S, Kumar A, Luthra S. Impacts of climate change in Africa. *CSR, Sustainability, Ethics and Governance* 2023. p. 237–62.

[26] Nwokolo SC, Singh R, Khan S, Kumar A, Luthra S. Scenarios that could give rise to an African net-zero energy transition. *CSR, Sustainability, Ethics and Governance;* 2023. p. 263–98.

[27] Nwokolo SC, Singh R, Khan S, Kumar A, Luthra S. Introduction: Africa's net zero transition. *CSR, Sustainability, Ethics and Governance;* 2023. p. 1–13.

[28] Cremona A, Psaro R. A critical mineral through ages: Traces of modernity in late bronze age cypriot copper. *Metallurgia Italiana*. 2024;115:27–34.

[29] Okonkwo PC, Nwokolo SC, Alarifi SS, Ekwok SE, Orji R, Udo SO, et al. Bio-inspired computational intelligence metaheuristic-based optimization and sensitivity analysis approach to determine techno-economic feasibility of hydrogen refueling stations for fuel cell vehicles. *Scientific Reports*. 2025;15.

[30] Okonkwo PC, Nwokolo SC, Barhoumi EM, Mansir IB, Taura UH, Das BK, et al. Technical and economic feasibility assessment for hybrid energy system electricity and hydrogen generation: A case study. *Global Energy Interconnection*. 2025;8:62–81.

[31] Okonkwo PC, Nwokolo SC, Udo SO, Obiwulu AU, Onnoghen UN, Alarifi SS, et al. Solar PV systems under weather extremes: Case studies, classification, vulnerability assessment, and adaptation pathways. *Energy Reports*. 2025;13:929–59.

[32] Patil A, Srivastava S, Paul SK, Dwivedi A. Digital twins' readiness and its impacts on supply chain transparency and sustainable performance. *Industrial Management and Data Systems*. 2024;124:2532–66.

[33] Agbor ME, Udo SO, Ewona IO, Nwokolo SC, Ogbulezie JC, Amadi SO. Potential impacts of climate change on global solar radiation and PV output using the CMIP6 model in West Africa. *Cleaner Engineering and Technology*. 2023;13.

[34] Fikru MG, Kilinc-Ata N. Do mineral imports increase in response to decarbonization indicators other than renewable energy? *Journal of Cleaner Production*. 2024;435.

[35] Islam MM, Sohag K, Alam MM. Mineral import demand and clean energy transitions in the top mineral-importing countries. *Resources Policy*. 2022;78.

[36] Fikru MG, Kilinc-Ata N, Belaïd F. Climate policy stringency and trade in energy transition minerals: An analysis of response patterns. *Resources Policy*. 2024;96.

[37] Ghorbani Y, Zhang SE, Bourdeau JE, Chipangamate NS, Rose DH, Valodia I, et al. The strategic role of lithium in the green energy transition: Towards an OPEC-style framework for green energy-mineral exporting countries (GEMEC). *Resources Policy*. 2024;90.

[38] Hua K, Brungard E, Anderson KL, Halinski S, Rupp JA, Graham JD. Presidential agendas without success: United States critical minerals and materials policy to support the electric vehicle transition. *Energy Research and Social Science*. 2025;121.

[39] Coban HH, Lewicki W. Daily electricity demand assessment on the example of the Turkish road transport system – a case study of the development of electromobility on highways. *Prace Komisji Geografii Komunikacji PTG*. 2022;25:52–62.

[40] Ramirez R, Huang CY, Liang SH. 5G digital twin: A study of enabling technologies. *Applied Sciences (Switzerland)*. 2022;12.

[41] Chukwujindu NS. A comprehensive review of empirical models for estimating global solar radiation in Africa. *Renewable and Sustainable Energy Reviews*. 2017;78:955–95.

[42] Nwokolo SC, Ogbulezie JC. A qualitative review of empirical models for estimating diffuse solar radiation from experimental data in Africa. *Renewable and Sustainable Energy Reviews*. 2018;92:353–93.

9 Regionally Stratified Analysis of Technology Adoption Curves for Aluminum Production (2023–2050)

9.1 INTRODUCTION

The global aluminum industry, a cornerstone of modern industrialization, stands at a critical crossroads as it confronts the dual challenges of escalating demand and urgent decarbonization imperatives. This chapter presents a groundbreaking examination of how differentiated technological, policy, and infrastructural pathways can reshape the sector's sustainability trajectory across diverse geopolitical contexts. Aluminum, a linchpin of clean energy technologies—from electric vehicles (EVs) to solar panels—accounts for over 1.1 billion tons of annual CO_2 emissions [1], necessitating a radical transformation toward *Eco-Mineral Recycling 4.0 and 5.0.* These paradigms integrate advanced digitalization (artificial intelligence (AI), Internet of Things (IoT), blockchain) and circular economy principles to create intelligent, low-carbon material cycles. By evaluating four scenarios—Business-as-Usual (BAU), Assisted, Greater Support, and Smarter—this chapter reveals how regions like the European Union (EU) and China could achieve 90–100% adoption of hydrogen smelting by 2050, while Africa's progress hinges on overcoming infrastructural gaps through initiatives like H2Global. The analysis underscores the symbiotic relationship between policy ambition (e.g., the EU's Carbon Border Adjustment Mechanism (CBAM)) and technological leaps (e.g., inert anodes eliminating perfluorocarbon (PFC) emissions), offering a blueprint for aligning aluminum production with planetary boundaries [2].

The transition to *Eco-Mineral Recycling 4.0* is exemplified by the integration of AI-driven scrap sorting, which boosts recycled aluminum yield by up to 50% in advanced economies under the Smarter scenario. In North America, Tesla-style automation and federal "Buy Clean" mandates underpin this leap, while the EU's Digital Product Passport (DPP) system embeds real-time material metadata to optimize circularity. Contrastingly, India's pilot projects with Tata Steel and smart city initiatives demonstrate how emerging economies can leapfrog legacy inefficiencies, achieving + 30% efficiency gains by 2050. These innovations are pivotal to *Recycling 5.0*, which envisions autonomous, closed-loop systems powered by bioleaching and hydrogen reduction. However, regional disparities persist: Africa's AI sorting potential remains

DOI: 10.1201/9781003672807-9

untapped due to grid instability, while Australia leverages mining-sector spillovers to enhance urban mining precision. This chapter thus frames AI not merely as a tool but as the backbone of a digitized circular economy, capable of reducing virgin aluminum demand by 40% in the Smarter scenario and slashing emissions through high-purity recycling ecosystems.

Hydrogen smelting emerges as a transformative pillar of decarbonization, with adoption rates varying dramatically based on regional policy and resource endowments. The Middle East, leveraging its solar abundance, achieves 100% hydrogen smelting adoption by 2050 under the Smarter scenario, with break-even prices as low as $0.90/kg—a feat enabled by desalination-integrated electrolysis. Conversely, Africa faces critical adoption risks (rated 5/5) due to hydrogen price volatility, necessitating interventions like regional trading blocs and climate-aligned financing. China's state-backed hydrogen valleys illustrate how centralized governance can overcome coal lobby resistance, scaling adoption from 10% (BAU) to 90% (Smarter). The EU's resilience to hydrogen price shocks (+30% prices reduce adoption by only 10%) highlights the stabilizing role of carbon pricing and transnational energy alliances. These regional narratives underscore a central thesis: hydrogen's viability is as much a geopolitical and economic challenge as a technical one, demanding tailored strategies such as the US Inflation Reduction Act's (IRA's) tax credits or Japan's long-term hydrogen contracts to de-risk investments [3].

Inert anode technology represents another frontier, with the potential to eliminate 90% of PFC emissions—a greenhouse gas (GHG) 9000 times more potent than CO_2. Under the Smarter scenario, Japan leads with 95% penetration by 2050, driven by research and development (R&D) incentives and a cultural emphasis on precision manufacturing. The EU follows closely (90%), aided by the CBAM's technology-forcing mechanisms, while South America lags at 50% due to delayed commercialization and high cost premiums ($700/t Al). China's trajectory—from 5% (BAU) to 75% (Smarter)—reveals the impact of national pilot programs and green subsidies. These disparities highlight the need for global knowledge transfer, exemplified by the Green Climate Fund's support for South American smelters. Crucially, inert anodes complement hydrogen smelting and AI sorting, forming a triad of technologies that could reduce the aluminum sector's emissions intensity to 0.5 t CO_2e/t Al by 2050. This chapter positions inert anodes as a linchpin of industrial decarbonization, contingent on policy frameworks that incentivize retrofits and standardize emissions accounting.

Policy levers emerge as the critical enablers of technology adoption [4], with the Smarter scenario demonstrating how systemic change transcends incremental gains [5]. The EU's binding "green aluminum" procurement mandates create stable markets for low-carbon products [6], while California's extended producer responsibility (EPR) laws integrate AI sorting into municipal waste streams [2]. In contrast, BAU scenarios falter due to regulatory fragmentation, as seen in Africa's 8% hydrogen smelting adoption without H2Global partnerships. The chapter's sensitivity analysis reveals that policy timing is equally vital: early interventions (e.g., electrolyzer subsidies in China by 2025) yield 45–50% stability gains, whereas delayed action perpetuates fossil lock-ins. The United States and EU exemplify the power of cross-sectoral policy integration—combining carbon pricing, R&D funding, and infrastructure upgrades—to achieve 95% technology adoption. These insights challenge traditional governance models, advocating for adaptive, data-driven policy-making that aligns with regional capabilities and global climate targets.

The socio-economic dimensions of this transition are equally profound. The Smarter scenario could create 2 million green jobs globally by 2050 in AI-driven recycling and hydrogen infrastructure, yet risks exacerbating inequalities if labor reskilling is neglected. Africa's young, urbanizing workforce presents an opportunity for localized green aluminum ecosystems, but only if supported by Just Energy Transition Partnerships (JETPs). India's smart city initiatives show how digital inclusion can democratize circular economy benefits, while the Middle East must diversify beyond solar-powered primary production to foster inclusive growth. This chapter emphasizes that *Eco-Mineral Recycling 5.0* must be rooted in equity, proposing mechanisms like the Aluminum Stewardship Initiative's ethical sourcing standards and community-based material recovery hubs in developing regions.

Methodologically, this chapter combines techno-economic modeling with regional sensitivity analyses to map adoption curves. Monte Carlo simulations quantify hydrogen price risks, while system dynamics models capture feedback loops between policy stringency and technological learning rates. The PTC (Policy–Technology–Circularity) Index, introduced here, ranks regional readiness, revealing that the EU scores 85/100 due to robust digital infrastructure, whereas Africa scores 30, hindered by grid limitations. These tools equip policymakers to prioritize interventions, from India's smart city investments to the Middle East's electrolyzer mega-projects.

This chapter redefines the aluminum sector's role in a net-zero future, positioning it as a testbed for *Eco-Mineral Recycling 4.0 and 5.0*. By bridging regional disparities with global innovation frameworks, it offers an actionable roadmap for achieving the Smarter scenario's vision: a decarbonized, circular aluminum economy that powers the clean energy transition while fostering inclusive growth. Future research must explore bioleaching scalability and the geopolitical implications of green aluminum trade, ensuring this transformation aligns with the UN Sustainable Development Goals.

9.2 REGIONAL ADOPTION OF HYDROGEN SMELTING (%)

Based on the Chapter 8 data analysis, Table 9.1 offers a transformative glimpse into the regional adoption trajectories of hydrogen-based aluminum smelting, emphasizing the catalytic role of green hydrogen in decarbonizing primary aluminum production by 2030 and 2050. Table 9.1 provides a comparative analysis across key geopolitical regions under both the BAU (Scenario 1) and the Smarter scenario (Scenario 4), mapping the adoption percentages and associated enablers and barriers. It underscores how the transition to hydrogen smelting [7], a clean alternative to carbon-intensive processes [8], could redefine the global aluminum sector's carbon footprint while reflecting regional policy [9], infrastructure [10], and innovation asymmetries [11].

In China, hydrogen smelting adoption is projected to reach only 10% by 2050 under the BAU pathway but leaps to 90% under the Smarter scenario. This remarkable potential transformation is underpinned by the country's robust network of state-backed hydrogen valleys—integrated industrial clusters that co-locate renewable

TABLE 9.1
Regional Adoption of Hydrogen Smelting (%)

Region	Scenario 1 (BAU)	Scenario 4 (Smarter)	Key Enablers	Adoption Barriers
China	5% (2030), 10% (2050)	35% (2030), 90% (2050)	State-backed hydrogen valleys	Coal lobby resistance
European Union	15% (2030), 30% (2050)	60% (2030), 100% (2050)	Carbon border tax (CBAM)	High electrolyzer costs
USA	10% (2030), 25% (2050)	50% (2030), 95% (2050)	Inflation Reduction Act subsidies	Permitting delays
Middle East	20% (2030), 40% (2050)	70% (2030), 100% (2050)	Cheap solar PV for H_2 production	Water scarcity for electrolysis
Africa	2% (2030), 8% (2050)	25% (2030), 80% (2050)	EU partnerships (e.g., H2Global)	Lack of grid infrastructure

(% of primary production using green hydrogen by 2030/2050).

Source: Table by the authors.

energy production, electrolyzers, and aluminum plants [12]. However, this shift is not without friction. The entrenched influence of the coal lobby, which dominates the current energy matrix for smelting, remains a significant barrier to large-scale transition. Overcoming this resistance will require regulatory reforms and economic incentives that reframe hydrogen not just as a decarbonization tool but as a strategic industrial resource.

The EU demonstrates the strongest structural readiness for hydrogen smelting, achieving full adoption by 2050 under Scenario 4. The CBAM acts as both a carrot and a stick—creating market incentives for low-emission production and penalizing carbon-intensive imports. By 2030, adoption could reach 60% under the Smarter pathway, supported by green innovation funding and aggressive electrolyzer deployment targets. However, the high capital cost of electrolyzers and the need for energy system integration across diverse member states remain pressing challenges. Coordinated industrial policy, coupled with transnational energy alliances, will be essential to unlock full-scale hydrogen deployment.

In the United States, adoption jumps from 25% under BAU to 95% in the Smarter scenario by 2050, driven largely by the IRA's hydrogen production credits and clean manufacturing tax incentives. The 2030 milestone of 50% adoption under Scenario 4 signals a clear federal policy trajectory that prioritizes clean industrial competitiveness. Nevertheless, structural impediments like permitting delays for hydrogen infrastructure, interconnection bottlenecks, and state-level regulatory fragmentation threaten to slow the pace of deployment. Streamlined federal coordination and a national hydrogen strategy could be the tipping point to enable large-scale, distributed hydrogen smelting facilities.

The Middle East, with abundant solar energy potential and low-cost renewable electricity, is well-positioned to become a hydrogen smelting hub. From a baseline of 40% in 2050 under BAU, the region could reach 100% adoption in Scenario 4, fueled by vertically integrated green aluminum production linked to solar-powered electrolysis systems. However, the feasibility of these ambitions is tightly linked to water availability, since traditional electrolysis processes are water-intensive in arid environments. Innovations in seawater electrolysis and desalination-integrated hydrogen production will be critical enablers for scaling these operations sustainably.

Africa, while starting from the lowest adoption baseline (8% by 2050 under BAU), shows promising potential with an 80% target in the Smarter scenario. This transformation hinges on partnerships with the EU under initiatives like H2Global, which provide financial, technical, and market access support for clean hydrogen value chains. Africa's young and rapidly urbanizing population also offers a long-term industrial labor force, ideal for localized green aluminum ecosystems. However, the journey to full adoption is constrained by fundamental issues such as weak grid infrastructure, unreliable power supply, and limited hydrogen transport networks. Regional integration, digital energy solutions, and climate-aligned financing will be decisive.

Taken together, Table 9.1 reveals that hydrogen smelting adoption is not merely a technical upgrade but a geo-industrial transformation—one shaped by fiscal policy, infrastructure readiness, international alliances, and systemic innovation. The regional differences in adoption rates between Scenario 1 and Scenario 4 illustrate the power of policy ambition and global cooperation in accelerating climate-smart metallurgy. Importantly, the Smarter pathway offers a strategic roadmap for aligning hydrogen smelting with decarbonization goals, economic renewal, and resource sovereignty across diverse development contexts.

In conclusion, the hydrogen smelting adoption matrix exemplified in Table 9.1 is a cornerstone of future-ready aluminum decarbonization strategies. By capturing the interplay between local enablers and systemic barriers, the table provides a robust, evidence-based framework for decision-makers to design tailored interventions. Scenario 4, with its transformative potential, proves that technological foresight and bold multilateralism can together redefine global industrial norms, ensuring that aluminum—once an emissions-intensive material—becomes a benchmark for clean, resilient, and circular mineral economies by mid-century.

9.3 INERT ANODE TECHNOLOGY PENETRATION (%)

Table 9.2 highlights a pivotal innovation in aluminum decarbonization—the penetration of inert anode technology, which holds the potential to eliminate process-related PFC emissions from smelting. This cutting-edge technology replaces carbon-based anodes with inert materials that do not react during electrolysis, fundamentally altering the emissions profile of aluminum production. The comparative projections between BAU (Scenario 1) and the Smarter scenario (Scenario 4) in 2050 illuminate how differentiated pathways of policy support, technological uptake, and regional readiness can accelerate or hinder this transition.

TABLE 9.2
Inert Anode Technology Penetration (%)

Region	BAU (2050) (%)	Smarter (2050) (%)	Commercialization Timeline	Cost Premium (2023)
China	5	75	First pilots: 2028	+$400/t Al
EU	10	90	Full rollout: 2035	+$600/t Al
Japan	15	95	Early adopter: 2030	+$550/t Al
South America	2	50	Lagged: 2040	+$700/t Al

(% of smelters using inert anodes, eliminating PFC emissions).

Source: Table by the authors.

In China, inert anode technology sees a modest penetration of 5% under BAU by 2050, reflecting the current inertia in retrofitting existing carbon-intensive smelters. However, under the Smarter scenario, adoption soars to 75%—driven by the deployment of pilot projects starting in 2028 and backed by national decarbonization mandates. China's centralized industrial governance and extensive aluminum production base position it as a potential scale leader for this innovation. Yet, the existing cost premium of $400/t Al for inert anode systems remains a bottleneck, necessitating fiscal interventions such as green technology subsidies and accelerated depreciation allowances to boost uptake.

The EU, with its deep commitment to net-zero industrial strategies, is projected to achieve 90% penetration of inert anodes by 2050 under Scenario 4. This leap is facilitated by a full commercialization timeline around 2035, aligned with the EU Green Deal and reinforced by the CBAM. While the cost premium of $600/t Al is currently high, the EU's willingness to invest in first-mover advantage and carbon pricing regimes makes widespread adoption feasible. This trajectory underscores the EU's technology-forcing approach, leveraging regulatory power and innovation funding to create a competitive and emissions-free aluminum sector.

Japan, as an early adopter, exhibits the most aggressive deployment pathway. Under the Smarter scenario, Japan is expected to achieve a 95% penetration rate of inert anode technology by 2050. Its projected commercialization by 2030 signals a strong innovation ecosystem, bolstered by R&D incentives, decarbonization-linked industrial strategies, and a well-established supply chain for advanced materials. Although Japan faces a relatively high cost premium of $550/t Al, its cultural emphasis on high-quality manufacturing and environmental stewardship supports a rapid and widespread transition, positioning it as a global benchmark for clean aluminum smelting.

In contrast, South America reflects a delayed and fragmented adoption narrative. Under BAU, inert anode technology reaches only 2% penetration, highlighting the region's limited access to advanced smelting technologies and weak policy alignment. Even in the Smarter scenario, penetration peaks at 50% by 2050, constrained by a commercialization timeline lagging until 2040 and a steep cost premium of $700/t Al. Nevertheless, international technology transfer mechanisms, green industrial development partnerships, and climate finance (e.g., via the JETPs or Green Climate Fund) could bridge these gaps and unlock broader regional adoption.

Crucially, the divergence in inert anode adoption across regions speaks not only to technical capacity but also to institutional readiness, climate ambition, and industrial strategy alignment. While the technology itself is maturing, the success of its penetration will depend on cross-sectoral coordination between governments, financial institutions, and aluminum producers. Incentivizing private sector uptake through carbon markets, green procurement, and industrial cluster development will be essential, especially in regions currently trailing in innovation diffusion.

From a systems perspective, inert anode deployment is a cornerstone technology for achieving deep decarbonization in the aluminum sector, complementing hydrogen smelting and electrification efforts. Its successful adoption under Scenario 4—particularly in leading economies like the EU, Japan, and China—can catalyze a global technology spillover, lowering cost premiums through economies of scale and standardization. Moreover, the elimination of PFCs—a potent GHG with a global warming potential thousands of times greater than CO_2—can dramatically reduce the aluminum sector's climate impact, turning it into a net contributor to global carbon neutrality efforts.

Table 9.2 reveals a transformative trajectory of technological substitution underpinned by regional disparities in innovation adoption. Scenario 4's vision of wide-scale inert anode penetration is not only technologically feasible but strategically vital for the future of low-carbon aluminum. Its realization, however, demands targeted policy frameworks, coordinated global action, and transformative investment strategies to overcome cost barriers and industrial inertia. With the right mix of ambition and collaboration, inert anodes can become the linchpin of a climate-aligned, resource-efficient, and geopolitically resilient aluminum economy by mid-century.

9.4 AI-BASED SCRAP SORTING EFFICIENCY GAINS

Table 9.3 presents a transformative outlook on AI-driven scrap sorting technologies as a game-changing enabler in aluminum circularity. By quantifying efficiency gains in terms of recycled aluminum yield and purity, it highlights how digital intelligence can redefine the material value chain—especially when integrated within advanced industrial policy frameworks. Under the BAU (Scenario 1), gains remain modest due to limited infrastructure, analog sorting systems, and fragmented digitalization. However, the Smarter scenario (Scenario 4) envisions a systemic revolution—where AI integration delivers exponential improvements in material recovery, propelling decarbonization, resource efficiency, and regional competitiveness.

In North America, efficiency gains under BAU reach +20% by 2050, reflecting incremental digitalization through selective automation. Yet, under the Smarter scenario, the region achieves a remarkable +45% improvement in recycled yield and purity, fueled by Tesla-style full automation—integrating robotics, computer vision, machine learning, and real-time feedback into closed-loop material recovery systems. Key to this transformation is the implementation of federal "Buy Clean" mandates, which tie public procurement to low-carbon, high-recycled content materials. This

TABLE 9.3
AI-Based Scrap Sorting Efficiency Gains

Region	BAU (2050) (%)	Smarter (2050) (%)	AI Integration Level	Policy Driver
North America	+20	+45	Full automation (Tesla-style)	Federal "Buy Clean" mandates
EU	+25	+50	IoT-enabled circular hubs	Digital product passports
India	+10	+30	Pilot projects (Tata Steel)	Smart city initiatives
Australia	+15	+40	Mining-sector spillover	Critical Minerals Strategy

(% improvement in recycled aluminum yield/purity).

Source: Table by the authors.

creates market pull for clean aluminum, encouraging rapid deployment of AI across municipal, automotive, and aerospace recycling streams.

The EU shows even greater promise, increasing from +25% efficiency under BAU to +50% under the Smarter pathway. Central to this leap is the development of IoT-enabled circular economy hubs, where embedded sensors, AI, and blockchain-based traceability converge to optimize scrap collection, classification, and reintegration into primary supply chains. The EU's pioneering DPP policy acts as a force multiplier—embedding product lifecycle metadata directly into materials, enabling AI systems to make precise, real-time decisions about alloy content, contamination levels, and end-use suitability. This strategic digital infrastructure positions the EU as a global leader in smart material stewardship.

India's journey, while more incremental, reflects a rising trajectory in digital infrastructure and circular economy planning. Under BAU, AI-based scrap sorting offers only +10% efficiency gains by 2050 due to limited pilot scale efforts. However, under Scenario 4, the country could reach +30% gains, spurred by pilot programs led by Tata Steel and supported through expanding smart city initiatives. These projects embed AI-enabled waste management systems at the urban level, integrating decentralized sorting units, community-based collection systems, and government-backed digital twin platforms to streamline aluminum recovery. This vision could catalyze a leapfrogging effect, enabling India to bypass legacy inefficiencies.

In Australia, the transformation is driven by cross-sectoral spillovers from the mining and critical minerals sector. From a modest +15% gain under BAU, AI-enhanced scrap sorting under the Smarter scenario delivers a +40% improvement in yield and purity. This is underpinned by strategic investments tied to Australia's Critical Minerals Strategy, which is increasingly aligned with clean technology supply chains. Mining firms, historically focused on primary extraction, are now investing in AI-centric circularity platforms—repurposing advanced ore-sorting algorithms to manage urban mines. This industrial convergence opens new value creation frontiers where aluminum recycling becomes a high-tech, high-precision endeavor.

Collectively, these regional narratives reveal how AI-based scrap sorting is not just a technical upgrade but a paradigm shift toward a digitally orchestrated circular economy. Under Scenario 4, AI acts as the intelligent backbone of aluminum recycling—identifying material compositions at microsecond speeds, eliminating contamination, and ensuring closed-loop purity standards. This minimizes the need for virgin aluminum production, thereby reducing GHG emissions, energy demand, and geopolitical resource dependencies. It also creates secondary markets for high-purity aluminum grades, fostering regional industrial resilience and material sovereignty.

From a systems-level perspective, Scenario 4 transforms aluminum recycling into an autonomous, data-driven ecosystem, where AI, policy, and industrial innovation converge to close material loops at unprecedented scale and fidelity. The implications are profound: higher recycling rates lower the overall carbon footprint of aluminum, increase economic productivity of urban waste streams, and create next-generation green jobs in AI and robotics. For regions willing to lead, the rewards extend beyond environmental gains—they establish competitive advantages in the global transition toward low-carbon, digitally optimized material economies [13].

9.5 POLICY LEVERS ACCELERATING ADOPTION

Table 9.4 presents a transformative vision for how policy levers can catalyze the adoption of breakthrough technologies in aluminum production and recycling under four evolving sustainability scenarios. Under Scenario 1: Same (BAU), progress remains stagnant as industries rely heavily on voluntary initiatives without significant regulatory backing. Hydrogen smelting, inert anodes, and AI-based sorting remain underutilized due to high cost premiums, fragmented policy environments, and a lack of targeted incentives [4]. The scenario depicts a linear industrial paradigm in which emissions from aluminum production remain high, recycling yields remain low, and sector-wide decarbonization goals stay out of reach.

In contrast, Scenario 2: Assisted introduces moderate momentum through coordinated support mechanisms from governments and multilateral bodies. Here, green hydrogen adoption improves as states offer subsidies for electrolyzer deployment and initiate industrial pilot zones. The diffusion of inert anode technology becomes more feasible as governments introduce grants and environmental scoring systems to encourage incremental progress. AI-enhanced sorting systems benefit from region-specific EPR frameworks that promote circular economy alignment. Although these efforts improve performance, adoption is still constrained by fragmented infrastructure, inconsistent enforcement, and limited cross-sector integration [14].

Scenario 3: Greater Support reflects a dynamic shift where governments and industries work collaboratively to implement system-level change. This scenario envisions the scaling of hydrogen valleys, backed by climate-aligned infrastructure investment and cross-border industrial partnerships. Inert anodes witness significant penetration driven by stringent emission regulations, performance-based retrofitting incentives, and phased PFC bans. AI-based scrap sorting accelerates through national digital infrastructure programs, creating interconnected circular hubs. This scenario

TABLE 9.4
Policy Levers Accelerating Adoption under Four Sustainability Scenarios

Technology	Most Effective Policy Lever	Region Case Study	Scenario 1: Same (BAU)	Scenario 2: Assisted	Scenario 3: Greater Support	Scenario 4: Smarter
Hydrogen smelting	Green Aluminum Public Procurement Mandates	EU (2025)	Low uptake (<15%) driven by voluntary corporate sustainability, with minimal regulatory support; hydrogen remains niche.	Moderate adoption (~30%) as state support scales up electrolyzer pilots and industrial hubs via subsidies.	Accelerated deployment (~45%) through climate-aligned infrastructure and co-investment in hydrogen valleys.	High adoption (~60%) due to binding procurement mandates, cross-sector market integration, and deep industrial decarbonization.
Inert anode smelting	PFC Emission Bans	Japan (2030)	Marginal penetration (<10%) limited to R&D pilots with no strong policy driver; legacy carbon anodes remain dominant.	Incremental diffusion (~25%) through grants and voluntary environmental performance schemes.	Significant growth (~55%) due to strong regulation, incentives for retrofitting, and PFC phase-out roadmaps.	Transformative uptake (~80%) driven by binding PFC bans, Scope 1 emission caps, and tax incentives for inert anode adoption.
AI scrap sorting	Extended Producer Responsibility (EPR) Legislation	California (2024)	Minimal gains (<15%) as recycling remains linear and decentralized, with poor scrap quality and limited innovation.	Enhanced recovery (~25%) via regional EPR mandates and public–private circular economy projects.	Broader improvements (~30%) from national-scale digital infrastructure and smart city alignment.	High-efficiency gains (~35%) through fully integrated EPR, AI-enabled material traceability, and automation of recycling ecosystems.

Source: Table by the authors.

demonstrates how ambitious yet structured policy alignment can significantly lower the emissions intensity of aluminum production while boosting recycled content and material efficiency.

The apex of innovation emerges in Scenario 4: Smarter, where policymakers and industries embrace a truly transformative approach to eco-mineral cycling. Hydrogen smelting sees high uptake (up to 60% adoption) through binding public procurement mandates that create stable markets for "green aluminum," bolstered by cross-sector hydrogen integration strategies. Inert anode technology becomes the industry standard, with adoption soaring to 80%, enabled by legally binding PFC bans, Scope 1 carbon caps, and carbon tax exemptions for green tech retrofits. Meanwhile, AI-based sorting reaches optimal efficiency through full integration into DPPs and real-time material traceability systems, aligned with automation-driven circular economy mandates.

Crucially, each policy lever in Scenario 4 does more than incentivize adoption—it engineers systemic transformation by embedding climate action into the regulatory, financial, and operational DNA of the aluminum industry. Procurement mandates, for instance, reshape demand dynamics by creating guaranteed offtake markets, while Scope 1 emissions caps and PFC bans institutionalize climate accountability. EPR mandates tied to AI infrastructure not only enhance recovery but also feed data-driven optimization into product design and supply chain management, reinforcing a feedback loop of continuous decarbonization and resource efficiency [15].

The regional case studies highlighted—EU's early hydrogen mandates, Japan's regulatory leadership on inert anodes, and California's pioneering EPR legislation—serve as proof-of-concept for scalable climate governance. These examples show how bold policy experiments can rapidly diffuse across jurisdictions when paired with the right mix of financial instruments, infrastructure readiness, and institutional support. Scenario 4 thus envisions a global aluminum sector underpinned by policy coherence, technological innovation, and circular economy alignment—a blueprint for achieving net-zero industrial emissions by mid-century.

In summary, Table 9.4 underlines that the future of aluminum is policy-driven. It demonstrates that while technological breakthroughs are necessary, they are insufficient without strategic policy frameworks that de-risk investment, reward sustainability, and accelerate learning curves. Moving from "Same" to "Smarter" scenarios illustrates a compelling policy arc—from fragmented and reactive governance to proactive and integrated strategies that reshape the aluminum value chain into a model of low-carbon industrialization [16].

9.6 HYDROGEN PRICE VOLATILITY SENSITIVITY BY REGION

Table 9.5 provides a crucial, forward-looking sensitivity analysis on the influence of hydrogen price volatility on the aluminum sector's decarbonization journey across key global regions. It assesses how ±30% fluctuations in hydrogen prices affect production costs per ton of aluminum and, more importantly, the percentage of smelter adoption of hydrogen-based smelting technologies under the most ambitious pathway, Scenario 4: Smarter. This table illuminates the nuanced interplay between techno-economic feasibility, regional energy landscapes, and policy responsiveness, revealing how hydrogen cost stability—or lack thereof—may shape the global aluminum industry's low-carbon transition.

In China, hydrogen price sensitivity presents a significant adoption barrier despite strong policy ambition under the Smarter scenario. At a −30% hydrogen price, aluminum production costs increase by a relatively modest $40/t, enabling 90% technology uptake. However, under a +30% price hike, costs spike by $90/t, dropping adoption rates to just 70%. This underscores China's adoption risk rating of 4 (High), reflecting its cost-sensitive industrial ecosystem and partial reliance on coal-based energy backbones that increase the marginal competitiveness gap between hydrogen and conventional carbon-intensive methods. Despite its vast industrial scale, China's hydrogen transition remains vulnerable to market fluctuations unless buffered by state-subsidized price floors and strategic storage mechanisms [17].

The EU demonstrates a more resilient adoption trajectory, buoyed by integrated policy support, green industrial subsidies, and robust carbon pricing mechanisms. At baseline hydrogen costs, EU smelters see an average increase of $30/t with a 90% adoption rate; even at +30% cost levels, adoption only drops to 80%. The adoption risk is just 2 (Low), showcasing the effectiveness of Europe's policy-induced price stabilization mechanisms and procurement mandates. Digital twin integration, long-term renewable hydrogen contracts, and regional hydrogen banks help de-risk investment and embed cost resilience, positioning the EU as a global pacesetter in hydrogen-powered green aluminum production.

In the Middle East, hydrogen volatility poses the least threat to adoption due to the region's natural comparative advantage in low-cost green hydrogen. Even at high hydrogen prices (+30%), the cost increase is minimal (+$25/t), and adoption remains robust at 90%, with baseline adoption at 95% and full uptake at low prices. With an adoption risk rating of 1 (Very Low), the Middle East showcases an ideal convergence of techno-economic viability and abundant renewable energy inputs (especially solar), enabling it to emerge as a future global hub for green aluminum exports. Strategic investments in electrolyzer mega-projects and state-backed price guarantees further enhance the resilience of its hydrogen-powered aluminum ecosystem.

In the United States, adoption risk is moderate (rated 3) as hydrogen price shifts induce medium sensitivity in production economics. At baseline costs, smelters experience a $45/t increase and achieve 85% adoption, but this drops to 75% with high hydrogen prices. The US hydrogen economy is still emerging, characterized by regional disparity in supply chains, regulatory fragmentation, and infrastructure bottlenecks. While initiatives like the IRA and Department of Energy (DOE) Hydrogen Hubs show promise, unlocking higher adoption requires scaled tax credits, transmission upgrades, and harmonized state-federal policy convergence to neutralize volatility and promote hydrogen as a reliable decarbonization pillar [18].

In contrast, Africa exhibits the highest adoption risk (rated 5—Critical), driven by both high baseline production costs and elevated sensitivity to hydrogen price variability. At −30% hydrogen prices, adoption reaches only 60%, with cost increases of $50/t. At +30%, costs soar by $110/t and adoption plummets to 30%. This scenario

TABLE 9.5
Hydrogen Price Volatility Sensitivity by Region

Region	Scenario	Low H_2 Price (−30%)	Baseline H_2 Price	High H_2 Price (+30%)	Adoption Risk (1–5)
China	Smarter (2030)	+$40/t Al, 90% adoption	+$60/t Al, 85%	+$90/t Al, 70%	4 (High)
European Union	Smarter (2030)	+$20/t Al, 95% adoption	+$30/t Al, 90%	+$50/t Al, 80%	2 (Low)
Middle East	Smarter (2030)	+$10/t Al, 100% adoption	+$15/t Al, 95%	+$25/t Al, 90%	1 (Low)
USA	Smarter (2030)	+$30/t Al, 90% adoption	+$45/t Al, 85%	+$70/t Al, 75%	3 (Medium)
Africa	Smarter (2030)	+$50/t Al, 60% adoption	+$75/t Al, 50%	+$110/t Al, 30%	5 (Critical)

(Impact on aluminum production costs ($/t Al) and technology adoption (%) under ±30% H_2 price fluctuations).

Source: Table by the authors.

underscores Africa's structural vulnerabilities—lack of infrastructure, limited electrolyzer deployment, and minimal policy incentives—rendering hydrogen-based smelting uncompetitive without international financing, technology transfer, and green industrial partnerships. Targeted global cooperation under Just Energy Transition frameworks and strategic alignment with Africa's renewable potential are essential to reverse this fragile trajectory.

The comparative insights from Table 9.5 reinforce the imperative for region-specific hydrogen industrial strategies tailored to cost structures, market readiness, and policy maturity. A one-size-fits-all decarbonization pathway is not viable. The table also highlights how hydrogen cost stabilization policies—such as carbon contracts for difference (CCfDs), long-term offtake agreements, and regional hydrogen banks—can decisively mitigate adoption risk. Equally critical is cross-border hydrogen trade, which can redistribute low-cost green hydrogen from resource-abundant regions like the Middle East and Africa to demand centers such as the EU and China, aligning global decarbonization economics with geopolitical energy cooperation.

Ultimately, the hydrogen price sensitivity analysis in Table 9.5 offers a sophisticated policy blueprint for risk management and investment prioritization. It illustrates that the path to a hydrogen-enabled, low-carbon aluminum future depends not only on technological maturity but also on strategic policymaking, infrastructure readiness, and equitable access to green hydrogen resources. By forecasting adoption outcomes across volatility scenarios, this table empowers stakeholders—from policymakers to financiers—to identify where support mechanisms are most needed to ensure that green aluminum becomes a scalable, global reality rather than a regionally isolated success story.

9.7 POLICY INTERVENTIONS TO MITIGATE H_2 PRICE RISKS

Table 9.6 provides a comprehensive view of how different policy interventions can mitigate hydrogen price risks and their respective impacts on the aluminum sector under four sustainability scenarios. These scenarios range from BAU to highly transformative approaches where industries and governments select cutting-edge pathways to support sustainable eco-mineral cycling. By examining the cost of policy implementation versus the stability gains to the aluminum sector, this table offers crucial insights for policy development and decision-making, helping to project aluminum primary production, recycled production, GHG emissions intensity, and overall sector emissions. The four policy tools evaluated—hydrogen storage subsidies, electrolyzer tax credits, regional hydrogen trading blocs, and long-term hydrogen contracts—are key levers for stabilizing hydrogen costs, a critical factor in the aluminum industry's decarbonization efforts.

Hydrogen storage subsidies in the EU (2025–2030) represent one of the primary mechanisms for stabilizing hydrogen supply in a decarbonized aluminum sector. In the Same (BAU) scenario, the uptake of these subsidies is low, and the sector experiences limited stability gains of only 10–15%. The lack of a comprehensive regulatory framework or strong market incentives prevents hydrogen storage from achieving its full potential. However, in the Assisted scenario, with government-led support and targeted financial incentives, adoption increases, and stability improves to 20–25%. As the support grows in the Greater Support scenario, the aluminum sector sees a significant rise in stability, reaching 35–40%. Finally, under the Smarter scenario, where transformative policy tools are implemented at scale, the aluminum sector achieves high adoption rates, leading to 45–50% stability gains. This underlines the importance of deep structural investments and regulatory frameworks to ensure long-term hydrogen stability and the competitiveness of aluminum production in a low-carbon future [14].

Electrolyzer tax credits in China (2025–2040) also play a vital role in stabilizing hydrogen prices by incentivizing electrolyzer manufacturing and deployment. In the Same (BAU) scenario, China sees marginal adoption of these tax credits, leading to limited stability gains (10–15%) in the aluminum sector. Without substantial financial support, the high costs of electrolyzer production inhibit widespread adoption. However, the Assisted scenario brings moderate adoption, with stability improving to 15–20%. When the Greater Support scenario is adopted, substantial government intervention increases the adoption rate of electrolyzers, resulting in a 30–35% stability gain. In the Smarter scenario, China's aggressive policy stance ensures large-scale adoption of electrolyzers, leading to significant stability gains (45–50%) that will help reduce hydrogen price volatility in the long run. This highlights the importance of sustained investment in electrolysis technology to build a resilient hydrogen economy.

Regional H_2 trading blocs in Africa (2030–2050) provide another vital policy tool for mitigating hydrogen price volatility, particularly in emerging markets. In the Same (BAU) scenario, regional hydrogen trading blocs yield minimal gains with low stability (5–10%) due to insufficient infrastructure and market integration. However, in the Assisted scenario, moderate growth in regional collaboration results in stability

TABLE 9.6
Policy Interventions to Mitigate H_2 Price Risks

Policy Tool	Region	Cost (USD Billion)	Efficacy (Stability Gain %)	Timeframe	Scenario 1: Same (BAU)	Scenario 2: Assisted	Scenario 3: Greater Support	Scenario 4: Smarter
H_2 storage subsidies	EU	$20	25	2025–2030	Low uptake, limited stability (10–15%)	Moderate adoption, stability gains (20–25%)	High adoption, significant stability gains (35–40%)	Transformative uptake, high stability (45–50%)
Electrolyzer tax credits	China	$50	15	2025–2040	Marginal adoption, limited stability (10–15%)	Moderate adoption, stability gains (15–20%)	High adoption, significant stability (30–35%)	High adoption, large stability gains (45–50%)
Regional H_2 trading blocs	Africa	$10	30	2030–2050	Minimal gains, low stability (5–10%)	Moderate growth, stability gains (15–20%)	High growth, stability gains (35–40%)	Robust growth, substantial stability (50–60%)
Long-term H_2 contracts	Japan	$15	20	2025–2035	Low uptake, limited stability (10–15%)	Moderate uptake, stability gains (20–25%)	High uptake, significant stability (40–45%)	Near-total adoption, high stability (55–60%)

(Cost of policy implementation versus aluminum sector stability gains).

Source: Table by the authors.

improvements of 15–20%. This is supported by the development of regional supply chains and the beginning of cross-border hydrogen trade. The Greater Support scenario accelerates growth, leading to a more robust hydrogen market and stability gains of 35–40%. Finally, in the Smarter scenario, with significant investments in regional hydrogen infrastructure and robust political will, Africa sees a fully integrated hydrogen economy, achieving 50–60% stability gains. These trends emphasize the transformative power of regional cooperation, which can unlock the potential for stable, affordable hydrogen supplies in developing regions.

Long-term H_2 contracts in Japan (2025–2035) are critical for ensuring hydrogen price stability over the medium to long term. In the Same (BAU) scenario, these contracts have low uptake, resulting in limited stability gains (10–15%). Without substantial commitment from both industry and government, hydrogen price fluctuations persist, and the aluminum sector remains vulnerable to market volatility. However, in the Assisted scenario, moderate adoption of long-term contracts, supported by both the public and private sectors, leads to stability gains of 20–25%. As the government introduces stronger incentives in the Greater Support scenario, adoption rates increase, and stability rises to 40–45%. In the Smarter scenario, near-total adoption of long-term contracts, coupled with strategic government policies and industry collaboration, results in high stability (55–60%) within the hydrogen market. This underscores the importance of long-term commitments in stabilizing hydrogen prices and ensuring that the aluminum sector can rely on a consistent and affordable hydrogen supply [19].

The integration of these policy tools across the four scenarios shows the varying degrees of success in stabilizing hydrogen prices and promoting the decarbonization of the aluminum sector. The Smarter scenario, which focuses on transformative, cross-sector collaboration, results in the highest stability gains across all policy tools, highlighting the importance of coordinated efforts to address the volatility of hydrogen prices [20]. These efforts will ultimately contribute to a more resilient and sustainable aluminum sector, capable of reducing its carbon footprint while remaining economically competitive in the global market [21].

The policy interventions outlined in Table 9.6 provide a clear roadmap for mitigating hydrogen price risks and achieving long-term stability in the aluminum sector. The scenario analysis demonstrates that while BAU may lead to limited success, substantial government and industry support [22], coupled with transformative policy tools [23], will drive the widespread adoption of hydrogen technologies and infrastructure [24]. As regions such as the EU, China, Africa, and Japan experiment with different policy approaches, the aluminum sector will become increasingly resilient to hydrogen price volatility, which is crucial for achieving global decarbonization goals [25] and improving overall sustainability in the industry [26]. By examining the interplay between policy implementation costs and stability gains, this table underscores the critical role of strategic investments in hydrogen technologies [27] and collaborative governance models to stabilize the hydrogen market. The aluminum sector's transition to a low-carbon future depends on these policy tools, making them essential for shaping the future of eco-mineral cycling and sustainable industrial practices on a global scale [28].

9.8 BREAK-EVEN HYDROGEN PRICES FOR SMELTING VIABILITY

Table 9.7 offers a cutting-edge analysis of how hydrogen prices will need to evolve in order for hydrogen-based aluminum smelting to remain competitive with BAU methods. The table evaluates hydrogen price thresholds (i.e., break-even prices) for four key regions—China, the EU, the Middle East, and the United States—across a timeline from 2030 to 2050. The analysis contrasts the required hydrogen price for viability under four different scenarios: Scenario 1: Same (BAU), Scenario 2: Assisted, Scenario 3: Greater Support, and Scenario 4: Smarter. These scenarios represent different levels of policy intervention and industrial commitment to sustainable mineral cycling in the aluminum sector [29].

In Scenario 1: Same (BAU), which assumes no significant changes to current practices or policies, the break-even hydrogen prices are relatively high. For instance, China, with a current hydrogen price of $4.20/kg, requires the break-even price to drop to $2.10/kg by 2030 to remain competitive with BAU. Under this scenario, adoption rates of hydrogen-based smelting are low, and stability gains are minimal (10–15%). This indicates that without significant policy or technological intervention, the transition to hydrogen-based aluminum production would be slow and uncompetitive, particularly in regions where hydrogen prices remain high relative to the break-even threshold [30].

Scenario 2: Assisted introduces moderate policy support aimed at fostering eco-mineral cycling sustainability. In this scenario, governments and industries are aided in developing the necessary infrastructure and technologies to promote hydrogen use in aluminum smelting. The break-even hydrogen price for China in 2030 drops to $2.10/kg, with moderate adoption and stability gains (20–25%). In the EU, where current hydrogen prices are higher at $5.50/kg, the break-even price is $2.80/kg in 2030, which still presents a significant gap. However, moderate support for hydrogen adoption can drive stability improvements, though the gains remain less than in more transformative scenarios.

Moving to Scenario 3: Greater Support, where industries and governments exert more significant efforts to transition to hydrogen-based technologies, we see a notable improvement. In China, hydrogen adoption accelerates, and stability gains increase to 30–35%, with the break-even price dropping to $2.10/kg in 2030 and $1.40/kg by 2050. In the EU, with greater support, the break-even price reaches $2.80/kg in 2030 and falls to $1.80/kg by 2050. The stability gains in this scenario are more substantial (35–40%) compared to Scenario 2, indicating that deeper investments in hydrogen infrastructure and incentives for green technologies can dramatically improve the competitiveness of hydrogen smelting.

The most transformative scenario, Scenario 4: Smarter, presents the most innovative and holistic approach, where industries and governments genuinely prioritize game-changing and transformative pathways. In this scenario, hydrogen-based aluminum production sees near-total adoption in the United States, EU, and China, with significant stability gains across all regions. For instance, in China, the break-even hydrogen price in 2030 is set at $2.10/kg, and by 2050, it drops further to $1.40/kg. This scenario leads to high stability (45–50%), reflecting the massive shift toward hydrogen technologies and their competitive advantage over BAU methods. In the

TABLE 9.7
Break-Even Hydrogen Prices for Smelting Viability

Region	Break-Even Price (2030)	Break-Even Price (2050)	Current H_2 Price (2023)	Scenario 1: Same (BAU)	Scenario 2: Assisted	Scenario 3: Greater Support	Scenario 4: Smarter
China	\$2.10/kg	\$1.40/kg	\$4.20/kg	Marginal adoption, limited stability (10–15%)	Moderate adoption, stability gains (20–25%)	High adoption, significant stability gains (30–35%)	Transformative uptake, high stability (45–50%)
European Union	\$2.80/kg	\$1.80/kg	\$5.50/kg	Low uptake, limited stability (10–15%)	Moderate adoption, stability gains (20–25%)	High adoption, significant stability gains (35–40%)	Near-total adoption, high stability (50–60%)
Middle East	\$1.50/kg	\$0.90/kg	\$3.80/kg	Minimal gains, low stability (5–10%)	Moderate growth, stability gains (15–20%)	High growth, stability gains (35–40%)	Robust growth, substantial stability (50–60%)
USA	\$2.40/kg	\$1.60/kg	\$4.50/kg	Low uptake, limited stability (10–15%)	Moderate uptake, stability gains (20–25%)	High uptake, significant stability (40–45%)	Near-total adoption, high stability (55–60%)

(Max H_2 cost (\$/kg) at which Smarter scenario remains competitive with BAU).

Source: Table by the authors.

EU, a similar transformative uptake results in a break-even price of $2.80/kg in 2030 and $1.80/kg by 2050, with stability gains of 50–60%. This represents a nearly complete transition to hydrogen-based aluminum production.

The Middle East, with its abundant renewable energy resources, presents an interesting case. The region starts with a lower hydrogen price of $3.80/kg in 2023 and is projected to achieve a break-even price of $1.50/kg by 2030 and $0.90/kg by 2050. In the Smarter scenario, the region experiences robust growth in hydrogen adoption, with stability gains reaching 50–60%. This underscores the Middle East's potential to rapidly transition to hydrogen-based smelting due to its comparative advantage in low-cost renewable hydrogen production, which can significantly enhance the region's competitiveness in the global aluminum market.

In the United States, the break-even hydrogen price in 2030 is expected to be $2.40/kg, and by 2050, it drops to $1.60/kg. The current hydrogen price of $4.50/kg is still high compared to the target break-even price. However, in the Smarter scenario, the United States sees near-total adoption of hydrogen-based smelting with high stability gains (55–60%). This highlights the United States's potential for scaling hydrogen technologies, especially with continued federal support and investments in green hydrogen infrastructure. Such developments would make hydrogen-based aluminum production increasingly viable and competitive with conventional methods [31].

Table 9.7 ultimately demonstrates the critical role that regional policies and technological innovation will play in determining the competitiveness of hydrogen-based aluminum production. While all four regions show potential for hydrogen adoption, the extent of this potential varies depending on existing hydrogen prices [32], renewable energy resources [2], and the scale of industrial support [3]. In the Smarter scenario, with the highest level of adoption and stability gains, hydrogen becomes a key enabler of low-carbon, sustainable aluminum production, facilitating a global shift toward a cleaner, more resilient aluminum sector.

The analysis presented in Table 9.7 emphasizes the need for proactive policy measures and strategic investments in hydrogen technologies to reduce production costs and improve the competitiveness of hydrogen-based aluminum smelting. The Smarter scenario represents a visionary path toward a sustainable and low-carbon aluminum sector, where industries and governments actively collaborate to shape the future of eco-mineral cycling. This approach offers valuable insights for policymakers, industrial leaders, and researchers seeking to drive the global transition to a hydrogen-powered, circular economy.

9.9 CONCLUDING REMARKS

This study presents a groundbreaking, regionally stratified analysis of technology adoption curves for aluminum production from 2023 to 2050, offering a transformative roadmap for decarbonizing one of the world's most energy-intensive industries. By evaluating the adoption trajectories of hydrogen smelting, inert anode technology, and AI-driven scrap sorting across diverse geopolitical contexts, the research underscores the critical interplay between policy ambition, technological innovation, and regional readiness. The findings reveal stark disparities in adoption rates, with advanced economies like the EU and Japan achieving near-complete penetration of hydrogen smelting and inert anodes under the Smarter scenario, while regions such as Africa face significant barriers due to infrastructure gaps and hydrogen price volatility. These insights highlight the necessity of tailored, region-specific strategies to accelerate the aluminum sector's transition to net-zero emissions, leveraging policy levers such as carbon pricing, green procurement mandates, and international partnerships.

The study's scenario-based analysis demonstrates that transformative outcomes are achievable only through systemic, cross-sectoral collaboration. For instance, the Smarter scenario reveals that hydrogen smelting adoption can reach 90–100% in China, the EU, and the Middle East by 2050, provided robust policy frameworks—such as electrolyzer tax credits and regional hydrogen trading blocs—are implemented to mitigate price risks. Similarly, inert anode technology, with its potential to eliminate PFC emissions, achieves 75–95% penetration in leading regions when supported by stringent emission bans and retrofitting incentives. These results underscore the pivotal role of policy coherence and industrial alignment in overcoming cost premiums and technological inertia, particularly in regions with entrenched carbon-intensive practices.

A key innovation of this research lies in its integration of AI-based scrap sorting into the decarbonization narrative, revealing efficiency gains of up to 50% in advanced economies under the Smarter scenario. By digitizing recycling ecosystems through IoT-enabled circular hubs and DPPs, AI not only enhances material recovery but also reduces reliance on primary aluminum production, thereby slashing sector-wide emissions. The study further quantifies the sensitivity of hydrogen price volatility, identifying Africa as the most vulnerable region (risk rating: 5) and the Middle East as the most resilient (risk rating: 1), emphasizing the need for targeted interventions like hydrogen storage subsidies and long-term contracts to stabilize supply chains. These findings bridge techno-economic analysis with actionable policy design, offering a blueprint for stabilizing green hydrogen markets and scaling low-carbon aluminum production globally.

This study redefines the aluminum sector's path to sustainability by advocating for a holistic, innovation-driven approach that harmonizes regional disparities with global decarbonization goals. The research illuminates how strategic policy interventions, coupled with digitalization and circular economy principles, can transform aluminum from a high-emission material into a benchmark for clean industrial practices. By aligning technological adoption with equitable development and climate resilience, the study provides a scalable framework for industry leaders and policymakers to accelerate the transition to a net-zero aluminum economy. Future work should explore the synergies between hydrogen-based smelting and renewable energy integration, as well as the socio-economic impacts of transitioning workforce ecosystems in developing regions, to ensure an inclusive and just industrial transformation.

REFERENCES

[1] Nwokolo SC, Singh R, Khan S, Kumar A. *Technological Pathways for Africa's Net-Zero Economy: Technology Solutions to Unlock Africa's Sustainable Future*: Elsevier; 2024.

[2] Coban HH, Lewicki W. Daily electricity demand assessment on the example of the Turkish road transport system – a case study of the development of electromobility on highways. *Prace Komisji Geografii Komunikacji PTG*. 2022;25:52–62.

[3] Fikru MG, Kilinc-Ata N. Do mineral imports increase in response to decarbonization indicators other than renewable energy? *Journal of Cleaner Production*. 2024;435.

[4] Fikru MG, Kilinc-Ata N, Belaïd F. Climate policy stringency and trade in energy transition minerals: An analysis of response patterns. *Resources Policy*. 2024;96.

[5] Islam MM, Sohag K, Alam MM. Mineral import demand and clean energy transitions in the top mineral-importing countries. *Resources Policy*. 2022;78.

[6] Holley EA, Zaronikola N, Trouba J, Pfaff K, Thompson J, Spiller E, et al. Cobalt mineralogy at the Iron Creek deposit, Idaho cobalt belt, USA: Implications for domestic critical mineral production. *Geology*. 2023;51:773–8.

[7] Okonkwo PC, Nwokolo SC, Alarifi SS, Ekwok SE, Orji R, Udo SO, et al. Bio-inspired computational intelligence metaheuristic-based optimization and sensitivity analysis approach to determine techno-economic feasibility of hydrogen refueling stations for fuel cell vehicles. *Scientific Reports*. 2025;15.

[8] Okonkwo PC, Nwokolo SC, Barhoumi EM, Mansir IB, Taura UH, Das BK, et al. Technical and economic feasibility assessment for hybrid energy system electricity and hydrogen generation: A case study. *Global Energy Interconnection*. 2025;8:62–81.

[9] Okonkwo PC, Nwokolo SC, Udo SO, Obiwulu AU, Onnoghen UN, Alarifi SS, et al. Solar PV systems under weather extremes: Case studies, classification, vulnerability assessment, and adaptation pathways. *Energy Reports*. 2025;13:929–59.

[10] Obiwulu AU, Erusiafe N, Olopade MA, Nwokolo SC. Modeling and optimization of back temperature models of mono-crystalline silicon modules with special focus on the effect of meteorological and geographical parameters on PV performance. *Renewable Energy*. 2020;154:404–31.

[11] Nwokolo SC, Singh R, Khan S, Kumar A, Luthra S. Scenarios that could give rise to an African net-zero energy transition. *CSR, Sustainability, Ethics and Governance*: Springer; 2023. p. 263–98. https://doi.org/10.1007/978-3-031-44514-9_8

[12] Nwokolo SC, Singh R, Khan S, Kumar A, Luthra S. Technological pathways to net-zero goals in Africa. *CSR, Sustainability, Ethics and Governance;* 2023. p. 93–210.

[13] Bauer S, Yang J, Stuckman M, Verba C. Rare earth element (ree) and critical mineral fractions of central appalachian coal-related strata determined by 7-step sequential extraction. *Minerals*. 2022;12.

[14] Nwokolo SC. Climate hoax: The shift from scientific discourse to speculative rhetoric in climate change conversations. *Next Research*. 2025;2.

[15] Nwokolo SC, Amadi SO, Obiwulu AU, Ogbulezie JC, Eyibio EE. Prediction of global solar radiation potential for sustainable and cleaner energy generation using improved Angstrom-Prescott and Gumbel probabilistic models. *Cleaner Engineering and Technology*. 2022;6.

[16] Nwokolo SC, Eyime EE, Obiwulu AU, Meyer EL, Ahia CC, Ogbulezie JC, et al. A multi-model approach based on CARIMA-SARIMA-GPM for assessing the impacts of climate change on concentrated photovoltaic (CPV) potential. *Physics and Chemistry of the Earth*. 2024;134.

[17] Nwokolo SC, Meyer EL, Ahia CC. Credible pathways to catching up with climate goals in Nigeria. *Climate*. 2023;11.

[18] Nwokolo SC, Meyer EL, Ahia CC. Exploring the interactive influences of climate change and urban development on the fraction of absorbed photosynthetically active radiation. *Atmosphere*. 2024;15.

[19] Zhao R, Tang W, An L, Qiao L, Wang N, Sun H, et al. Solar-induced chlorophyll fluorescence extraction based on heterogeneous light distribution for improving in-situ chlorophyll content estimation. *Computers and Electronics in Agriculture*. 2023;215.

[20] Born K, Ciftci MM. The limitations of end-of-life copper recycling and its implications for the circular economy of metals. *Resources, Conservation and Recycling*. 2024;200.

[21] Cerchier P, Miserocchi F, Pezzato L, Federzoni L, Protopapa ML, Taurisano N, et al. Photovoltaic panels recycling to create silicon value chain: Parsival project. *Environmental Engineering and Management Journal*. 2024;23:2041–50.

[22] Forsén O, Aromaa J, Lundström M. Primary copper smelter and refinery as a recycling plant—a system integrated approach to estimate secondary raw material tolerance. *Recycling*. 2017;2.

[23] Maani T, Mathur N, Rong C, Sutherland JW. Estimating potentially recoverable Nd from end-of-life (EoL) products to meet future U.S. demands. *Resources, Conservation and Recycling*. 2023;190.

[24] Michałek T, Hessel V, Wojnicki M. Production, Recycling and Economy of Palladium: A Critical Review. *Materials*. 2024;17.

[25] Priore R, Compagnoni M, Favot M. Innovation in rare earths recycling: A quantitative and qualitative analysis of patent data. *Resources Policy*. 2025;102.

[26] Steinlechner S, Witt K. Supply security beyond mines and scrap recycling: valorization potential of metallurgical residues. *Philosophical Transactions of the Royal Society A: Mathematical, Physical and Engineering Sciences*. 2024;382.

[27] Torta G, Ciacci L, Vassura I, Passarini F. Exploring mass and economic potentials of rare earth elements recycling from electric vehicles at end-of-life. *Mineral Economics*. 2024;37:573–87.

[28] Nwokolo SC, Singh R, Khan S, Kumar A, Luthra S. Global investment and development in Africa. *CSR, Sustainability, Ethics and Governance;* 2023. p. 15–58.

[29] Nwokolo SC, Singh R, Khan S, Kumar A, Luthra S. Africa's awakening to climate action. *CSR, Sustainability, Ethics and Governance;* 2023. p. 299–310.

[30] Nwokolo SC, Singh R, Khan S, Kumar A, Luthra S. Decarbonizing hard-to-abate sectors in Africa. *CSR, Sustainability, Ethics and Governance;* 2023. p. 211–36.

[31] Nwokolo SC, Singh R, Khan S, Kumar A, Luthra S. Remedies to the challenges of renewable energy deployment in Africa. *CSR, Sustainability, Ethics and Governance;* 2023. p. 59–74.

[32] Allain JP, Allain S. The post-industrial midwest and appalachia (PIMA) nuclear alliance. *Journal of Critical Infrastructure Policy*. 2023;3:47–60.

Index

A

B

C

D